HORMONE RESISTANCE SYNDROMES

CONTEMPORARY ENDOCRINOLOGY

P. Michael Conn, SERIES EDITOR

19. **Human Growth Hormone:** *Basic and Clinical Research,* edited by ROY G. SMITH AND MICHAEL O. THORNER, 1999
18. **Menopause:** *Endocrinology and Management,* edited by DAVID B. SEIFER AND ELIZABETH A. KENNARD, 1999
17. **The IGF System:** *Molecular Biology, Physiology, and Clinical Applications,* edited by RON G. ROSENFELD AND CHARLES ROBERTS, 1999
16. **Neurosteroids:** *A New Regulatory Function in the Nervous System,* edited by ETIENNE-EMILE BAULIEU, MICHAEL SCHUMACHER, AND PAUL ROBEL, 1999
15. **Autoimmune Endocrinopathies,** edited by ROBERT VOLPÉ, 1999
14. **Hormone Resistance Syndromes,** edited by J. LARRY JAMESON, 1999
13. **Hormone Replacement Therapy,** edited by A. WAYNE MEIKLE, 1999
12. **Insulin Resistance:** *Epidemiology, Pathophysiology, and Nondiabetic Clinical Syndromes,* edited by GERALD M. REAVEN AND AMI LAWS, 1999
11. **Endocrinology of Breast Cancer,** edited by ANDREA MANNI, 1999
10. **Molecular and Cellular Pediatric Endocrinology,** edited by STUART HANDWERGER, 1999
9. **The Endocrinology of Pregnancy,** edited by FULLER W. BAZER, 1998
8. **Gastrointestinal Endocrinology,** edited by GEORGE H. GREELEY, JR., 1999
7. **Clinical Management of Diabetic Neuropathy,** edited by ARISTIDIS VEVES, 1998
6. **G Proteins, Receptors, and Disease,** edited by ALLEN M. SPIEGEL, 1998
5. **Natriuretic Peptides in Health and Disease,** edited by WILLIS K. SAMSON AND ELLIS R. LEVIN, 1997
4. **Endocrinology of Critical Diseases,** edited by K. PATRICK OBER, 1997
3. **Diseases of the Pituitary:** *Diagnosis and Treatment,* edited by MARGARET E. WIERMAN, 1997
2. **Diseases of the Thyroid,** edited by LEWIS E. BRAVERMAN, 1997
1. **Endocrinology of the Vasculature,** edited by JAMES R. SOWERS, 1996

HORMONE RESISTANCE SYNDROMES

Edited by

J. LARRY JAMESON, MD, PHD

*Northwestern University Medical School,
Chicago, IL*

HUMANA PRESS
TOTOWA, NEW JERSEY

© 1999 Humana Press Inc.
999 Riverview Drive, Suite 208
Totowa, New Jersey 07512

For additional copies, pricing for bulk purchases, and/or information about other Humana titles, contact Humana at the above address or at any of the following numbers: Tel: 973-256-1699; Fax: 973-256-8341; E-mail: humana@humanapr.com or visit our website at http://humanapress.com

All rights reserved. No part of this book may be reproduced, stored in a retrieval system, or transmitted in any form or by any means, electronic, mechanical, photocopying, microfilming, recording, or otherwise without written permission from the Publisher.

All articles, comments, opinions, conclusions, or recommendations are those of the author(s), and do not necessarily reflect the views of the publisher.

This publication is printed on acid-free paper. ∞
ANSI Z39.48-1984 (American National Standards Institute)
Permanence of Paper for Printed Library Materials.

Cover design by Patricia F. Cleary.

Photocopy Authorization Policy:
Authorization to photocopy items for internal or personal use, or the internal or personal use of specific clients, is granted by Humana Press Inc., provided that the base fee of US $8.00 per copy, plus US $00.25 per page, is paid directly to the Copyright Clearance Center at 222 Rosewood Drive, Danvers, MA 01923. For those organizations that have been granted a photocopy license from the CCC, a separate system of payment has been arranged and is acceptable to Humana Press Inc. The fee code for users of the Transactional Reporting Service is: [0-89603-652-9/99 $8.00 + $00.25].

Printed in the United States of America. 10 9 8 7 6 5 4 3 2 1

Hormone resistance syndromes / edited by J. Larry Jameson.
    p. cm. — (Contemporary endocrinology ; 14)
   Includes index.
   ISBN 0-89603-652-9 (alk. paper)
  1. Hormone resistance. I. Jameson, J. Larry. II. Series: Contemporary endocrinology (Totowa, NJ) ; 14.
   [DNLM: 1. Syndrome. 2. Hormones--physiology. 3. Mutation. 4. Drug Resistance. 5. Receptors, Drug--physiology. QZ 140 H812 1999]
  RC649.H65 1999
  616.4--dc21
  DNLM/DLC
  for Library of Congress   98-41146
                                                                          CIP

PREFACE

In 1942, Fuller Albright proposed that pseudohypoparathyroidism resulted from end-organ resistance to parathyroid hormone (PTH). Since then, the concept of hormone resistance has been expanded to include almost every type of hormone, including peptides like GHRH, large proteins such as insulin, and the glycoprotein hormones, steroid hormones, and even ions such as calcium.

The idea of hormone resistance builds upon thinking of hormone action in terms of an endocrine axis that is controlled by feedback regulation. Most resistance syndromes are characterized by elevated hormone levels because their failure to act normally results in a lack of feedback inhibition. As an example, androgen resistance is typified by elevated gonadotropin and testosterone levels that reflect defective androgen action. Similarly, TSH resistance is characterized by low thyroid hormone levels that lead to reduced feedback inhibition and the elevation of circulating TSH levels. Thus, the diagnosis of most hormone resistance syndromes can be made by examining the hormone levels at various points in a given endocrine axis.

In most cases, hormone resistance syndromes are caused by mutations in receptors. In the case of proteins, the mutant receptors are typically membrane receptors. These include seven transmembrane, G-protein-coupled receptors in the case of resistance to TSH, LH, FSH, ACTH, GHRH, and vasopressin. However, other classes of membrane receptors can also be affected. For example, the insulin receptor is a tyrosine kinase receptor, the GH receptor belongs to a family of cytokine receptors, and the MIS receptor is related to the TGFβ serine kinase group of receptors. G-protein mutations cause Albright's hereditary osteodystrophy, and, as might be expected, there is resistance to multiple different hormones that act through G-protein-coupled receptors. A variety of resistance syndromes involve members of the nuclear receptor superfamily, including resistance to androgens, vitamin D, thyroid hormone, glucocorticoids, and estrogen. Another form of resistance, sometimes referred to as "post-receptor," is not covered in this book, largely because its pathophysiology remains elusive. This type of resistance is typified by acquired forms of insulin resistance that cause type II diabetes mellitus.

A decade ago, studies of hormone resistance syndromes relied almost entirely upon measurements of circulating hormone levels. Although such measurements are still very important for diagnostic and therapeutic purposes, the cloning of hormone receptors has allowed recombinant DNA studies to be used to unequivocally demonstrate the presence of mutations. These molecular studies allow diagnoses to be made in individual patients, and also permit genetic counseling within families. In addition, studies of naturally occurring mutations have provided many insights into the structure and function of receptors. By definition, the mutations that cause disease identify important functional domains in the proteins. In some instances, mutations cause large deletions of the receptor, and not much is learned about structure–function. However, in cases such as the thyroid hormone receptor, the point mutations have identified domains involved in hormone binding, receptor dimerization, and transactivation.

In virtually every resistance syndrome, it is striking that there is tremendous heterogeneity in the locations of mutations in different individuals. Though this finding provides an opportunity to elucidate the structural impact on receptor function, there are also practical implications. First, the heterogeneity suggests that most mutations arise independently, as opposed to being passed down through a founder effect. Thus, simple screening tests cannot be used to detect highly prevalent mutations. Second, it raises the possibility that clinical phenotypes will vary, depending on the location and the type of mutation that is present in the receptor. The correlation of genotype and phenotype is an active area of research and requires sophisticated clinical investigation to tease apart subtle differences in physiology.

In the next decade, we can expect additional resistance syndromes to be described. In addition, there will be a dramatic increase in the number of different mutations described as more patients are characterized at a molecular level. Ultimately, it is likely that databases will exist on the world wide web for each of these syndromes, as is currently the case for thyroid hormone receptor mutations. Although therapy is straightforward in some disorders, it is challenging and poorly studied in others. Whether gene therapy will play a role in some of the resistance syndromes is an interesting topic for speculation, but will require much further study.

J. Larry Jameson

CONTENTS

Preface v
Contributors ix

1 Dwarfism: *GHRH Resistance* 1
  Gerhard Baumann and Hiralal G. Maheshwari

2 Laron Syndrome: *Primary Growth Hormone Resistance* 17
  Zvi Laron

3 Pseudohypoparathyroidism 39
  Ali Al-Zahrani, Michael A. Levine,
     and William F. Schwindinger

4 Hereditary Resistance to Vitamin D 59
  Peter J. Malloy and David Feldman

5 Decreased Responsiveness to Extracellular Ca^{2+} Owing
     to Abnormalities in the Ca^{2+}_o-Sensing Receptor 87
  Edward M. Brown, Olga Kifor, and Mei Bai

6 Resistance to TSH 111
  Peter Kopp

7 Thyroid Hormone Resistance 145
  V. Krishna K. Chatterjee, Roderick J. Clifton-Bligh,
     and Mark Gurnell

8 Diabetes Mellitus: *Insulin Resistance* 165
  Simeon I. Taylor and Elif Arioglu

9 LH Insensitivity Syndrome 185
  Axel P. N. Themmen, John W. M. Martens,
     and Han G. Brunner

10 FSH Resistance 197
   Kristina Aittomäki and Ilpo T. Huhtaniemi

11 Genetic Alterations of Androgen Receptor Function 209
   Ken Brantley, Tianshu Gao, and Michael J. McPhaul

12 Retained Müllerian Ducts: *AMH Resistance Syndrome* 233
   Nathalie Josso, Jean-Yves Picard, and Rodolfo Rey

13 Adrenal Insufficiency: *ACTH Resistance* 245
   Adrian J. L. Clark

14 Glucocorticoid Resistance and Hypersensitivity 259
   Denis P. Franchimont and George P. Chrousos

Index 273

CONTRIBUTORS

KRISTINA AITTOMÄKI, MD, PHD, *Department of Medical Genetics, University of Helsinki, Finland*
ALI AL-ZAHRANI, MD, *Division of Endocrinology and Metabolism, The Johns Hopkins University School of Medicine, Baltimore, MD*
ELIF ARIOGLU, MD, *National Institute of Diabetes and Digestive and Kidney Disease, National Institutes of Health, Bethesda, MD*
MEI BAI, PHD, *Endocrinology–Hypertension Division, Brigham and Women's Hospital, Boston, MA*
GERHARD BAUMANN, MD, *Division of Endocrinology, Metabolism and Molecular Medicine, Department of Medicine, Northwestern University Medical School, Chicago, IL*
KEN BRANTLEY, MD, PHD, *Endocrinology and Metabolism Division, Department of Internal Medicine, University of Texas Southwestern Medical Center, Dallas, TX*
EDWARD M. BROWN, MD, *Endocrinology–Hypertension Division, Brigham and Women's Hospital, Boston, MA*
HAN G. BRUNNER, MD, PHD, *Department of Human Genetics, University Hospital, Nijmegen, The Netherlands*
V. KRISHNA K. CHATTERJEE, MD, BMBCH, FRCP, *Department of Medicine, University of Cambridge, UK*
GEORGE P. CHROUSOS, MD, *National Institute of Child Health and Human Development, National Institutes of Health, Bethesda, MD*
RODERICK J. CLIFTON-BLIGH, BSC (MED), *Department of Medicine, University of Cambridge, UK*
ADRIAN J. L. CLARK, DSC, FRCP, *Molecular Endocrinology Section, Department of Endocrinology, St. Bartholomew's and the Royal London School of Medicine and Dentistry, London, UK*
DAVID FELDMAN, MD, *Department of Medicine, Stanford University, Stanford, CA*
DENIS P. FRANCHIMONT, MD, *National Institute of Child Health and Human Development, National Institutes of Health, Bethesda, MD*
TIANSHU GAO, MD, *Endocrinology and Metabolism Division, Department of Internal Medicine, University of Texas Southwestern Medical Center, Dallas, TX*
MARK GURNELL, BSC, MBBS, MRCP, *Department of Medicine, University of Cambridge, UK*
ILPO T. HUHTANIEMI, MD, PHD, *Department of Physiology, University of Turku, Finland*
NATHALIE JOSSO, MD, *Unité de Recherches sur l'Endocrinologie du Développement (INSERM), Département de Biologie, Montrouge, France*
OLGA KIFOR, MD, *Endocrinology–Hypertension Division, Brigham and Women's Hospital, Boston, MA*
PETER KOPP, MD, *Division of Endocrinology, Metabolism, and Molecular Medicine, Northwestern University Medical School, Chicago, IL*
ZVI LARON, MD, *Endocrine and Diabetes Research Unit, Schneider Children's Medical Center of Israel, Petah Tiqva, Israel*
MICHAEL A. LEVINE, MD, *Division of Endocrinology and Metabolism, The Johns Hopkins University School of Medicine, Baltimore, MD*

HIRALAL G. MAHESHWARI, MBBS, PHD, *Division of Endocrinology, Metabolism and Molecular Medicine, Department of Medicine, Northwestern University Medical School, Chicago, IL*

PETER J. MALLOY, PHD, *Department of Medicine, Stanford University, Stanford, CA*

JOHN W. M. MARTENS, PHD, *Department of Endocrinology and Reproduction, Erasmus University, Rotterdam, The Netherlands*

MICHAEL J. MCPHAUL, MD, *Endocrinology and Metabolism Division, Department of Internal Medicine, University of Texas Southwestern Medical Center, Dallas, TX*

JEAN-YVES PICARD, PHD, *Unité de Recherches sur l'Endocrinologie du Développement (INSERM), Département de Biologie, Montrouge, France*

RODOLFO REY, MD, PHD, *Unité de Recherches sur l'Endocrinologie du Développement (INSERM), Département de Biologie, Montrouge, France*

WILLIAM F. SCHWINDINGER, MD, PHD, *Division of Endocrinology and Metabolism, The Johns Hopkins University School of Medicine, Baltimore, MD*

SIMEON I. TAYLOR, MD, PHD, *National Institute of Diabetes and Digestive and Kidney Diseases, National Institutes of Health, Bethesda, MD*

AXEL P. N. THEMMEN, PHD, *Department of Endocrinology and Reproduction, Erasmus University, Rotterdam, The Netherlands*

1 Dwarfism
GHRH Resistance

Gerhard Baumann, MD
and Hiralal G. Maheshwari, MBBS, PhD

CONTENTS

REGULATION OF GROWTH HORMONE (GH) SECRETION:
  THE GROWTH HORMONE-RELEASING HORMONE
  (GHRH)/SOMATOSTATIN/GROWTH HORMONE-RELEASING
  PEPTIDE (GHRP) SYSTEM
GHRH: STRUCTURE AND ACTIVITY
THE GHRH RECEPTOR: STRUCTURE, EXPRESSION, AND SIGNALING
GHRH AND PITUITARY GROWTH AND DEVELOPMENT
GHRH RESISTANCE
POTENTIAL IMPLICATIONS OF GHRH RESISTANCE SYNDROMES FOR
  SHORT STATURE IN THE GENERAL POPULATION
ACKNOWLEDGMENTS
REFERENCES

REGULATION OF GROWTH HORMONE (GH) SECRETION: THE GROWTH HORMONE-RELEASING HORMONE (GHRH)/SOMATOSTATIN/GROWTH-HORMONE-RELEASING PEPTIDE (GHRP) SYSTEM

Growth hormone (GH) synthesis and secretion are governed by at least two, and probably three, hypothalamic hypophysiotropic factors, delivered to the pituitary gland via the pituitary portal system (Fig. 1). GHRH is a positive regulator stimulating GH production *(1)*. It will be described in detail below. Somatostatin is a negative regulator that inhibits GH secretion. It is widely distributed in the central nervous system and neuroendocrine tissues, where it acts as an inhibitor of a variety of cellular secretory functions *(2)*. Recently, a third factor has been postulated, though not yet identified—the endogenous GHRP analog. GHRP, a small peptide existing as several variants, stimulates GH secretion in conjunction with GHRH *(3)*. Although GHRP is at present merely a pharmacological agent, the recent cloning of a specific receptor for GHRP from pituitary *(4)* lends credence to the view that an endogenous ligand akin to GHRP must exist and probably participates in the regulation of GH secretion.

From: *Contemporary Endocrinology: Hormone Resistance Syndromes*
Edited by: J. L. Jameson © Humana Press Inc., Totowa, NJ

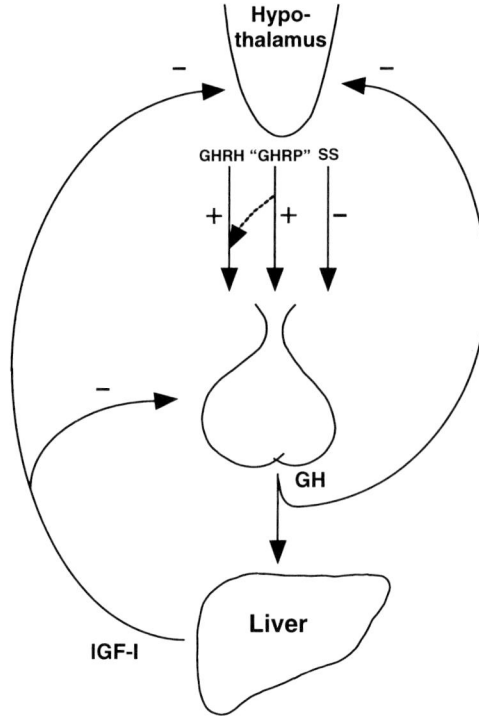

Fig. 1. Regulation of GH secretion at the hypothalamic–pituitary level. "GHRP" denotes the putative endogenous ligand for the GHRP-R, and SS denotes somatostatin. The interaction between GHRP and GHRH signaling is indicated by the dashed arrow. The long (IGF-1) and the short (GH) feedback loops are shown.

All three factors interact with specific receptors of the serpentine (seven-transmembrane domain), G-protein-coupled type. The GHRH receptor (GHRH-R) will be described in detail in this chapter. There are five somatostatin receptors, all of which are expressed in the pituitary *(5)*, but in normal human pituitary, only subtypes 1, 2, and 5 appear to be expressed *(6)*. The GHRP receptor belongs to a class of G-protein-coupled receptors that includes the neurotensin and thyrotropin-releasing hormone (TRH) receptors; it is expressed primarily in the hypothalamus and pituitary, but also in other areas of the brain and peripheral tissues *(4,7,8)*.

The relative contributions of the three hypophysiotropic factors to the elaboration of GH have been difficult to dissect in the intact organism. Administration of GHRH or GHRP results in a GH secretory spike, but the amplitude of that spike depends on the prevailing somatostatin tone. GHRP is a relatively weak stimulus alone, but its effect is amplified severalfold in the presence of GHRH. Thus, it is clear that the two (or three) peptides interact in defining the GH secretory rhythm. In the absence of information about their concentrations in the pituitary portal system—information that is technically difficult or impossible to obtain—their relative roles in generating that rhythm remains speculative. The situation is further complicated by feedback of insulin-like growth factor 1 (IGF-1) and by GH itself on GH secretion, and by the influence of sex steroids, nutrition, and developmental stage on GH production. Therefore, with few exceptions *(9,10)*, most of the concepts about the fine regulation of GH secretion have been derived by inference from indirect evidence.

GHRH:
STRUCTURE AND ACTIVITY

GHRH is a 40–44 amino acid polypeptide first isolated from two pancreatic tumors causing acromegaly by producing GHRH ectopically *(11,12)*. Subsequently, GHRH was rapidly isolated from the hypothalami of a variety of species. Human GHRH exists as $GHRH_{1-40}$ and $GHRH_{1-44}$, with the latter the predominant form. Full biological activity is encompassed in the first 29 amino acids.

The GHRH gene is a member of a gene family that also includes genes for vasoactive intestinal peptide (VIP), pituitary adenyl cyclase activating polypeptide (PACAP), secretin, gastric inhibitory peptide (GIP), and glucagon *(13)*. The gene consists of six exons, with two alternative promoters/exons used in hypothalamus and placenta or testis, respectively *(14,15)*. Exons 2–4 encode a GHRH precursor that is cleaved into an amino-terminal peptide, GHRH, and a putative carboxy-terminal product named GHRH-related peptide (GHRH-RP) *(15)*. The latter has been shown to be expressed in Sertoli cells and may be important for spermatogenesis *(15)*.

GHRH is widely expressed in both central nervous and peripheral tissues, such as pancreas, gut, gonad, and placenta *(16–20)*. However, the most prominent GHRH activity is localized in the median eminence and arcuate nucleus of the hypothalamus *(17)*. The biological significance of GHRH in extrahypothalamic sites is largely unknown.

GHRH positively affects GH production at several levels. GHRH promotes proliferation of somatotrophs *(21)*, enhances GH synthesis at the transcriptional level *(22)*, and stimulates GH secretion *(23–25)*. In vivo, the GH secretory response to a given dose of GHRH is highly variable *(26)*, primarily because of the influence of the prevailing somatostatin tone. GHRH has also been reported to have effects on pancreatic islet hormone and gastrin release as well as feeding behavior, but the physiological importance of these activities is not established.

In plasma, GHRH is rapidly inactivated by a dipeptidylaminopeptidase to $GHRH_{3-44}$, with a half-life of about 7 min *(27)*.

THE GHRH RECEPTOR:
STRUCTURE, EXPRESSION, AND SIGNALING

The GHRH-R is a member of the serpentine, G-protein-coupled receptor class. More specifically, it belongs to the family of receptors for secretin, VIP, glucagon, PACAP, calcitonin, and parathyroid hormone *(28)*. It was first cloned in 1992 by three independent groups in rat, mouse, and man, respectively *(29–31)*. The sequence is about 80% identical in the three species. The mature GHRH-R is a 401 amino acid protein, with an extended extracellular domain containing 7 cysteines and 1 potential N-linked glycosylation site (Fig. 2). In the rat, an alternatively spliced form contains a 41 amino acid insert in the third intracellular loop *(29)*. This form appears fully functional in terms of ligand binding and cAMP response, but is only weakly expressed *(32)*. Similarly, in the mouse, alternative splicing products of unknown biological significance have been demonstrated *(30)*.

The GHRH-R gene has only recently been fully characterized. In humans, its consists of 13 exons and spans 15.5 kb on the short arm of chromosome 7 *(33)*. The rat gene contains an additional exon that codes for the above-mentioned 41 amino acid insert *(29)*.

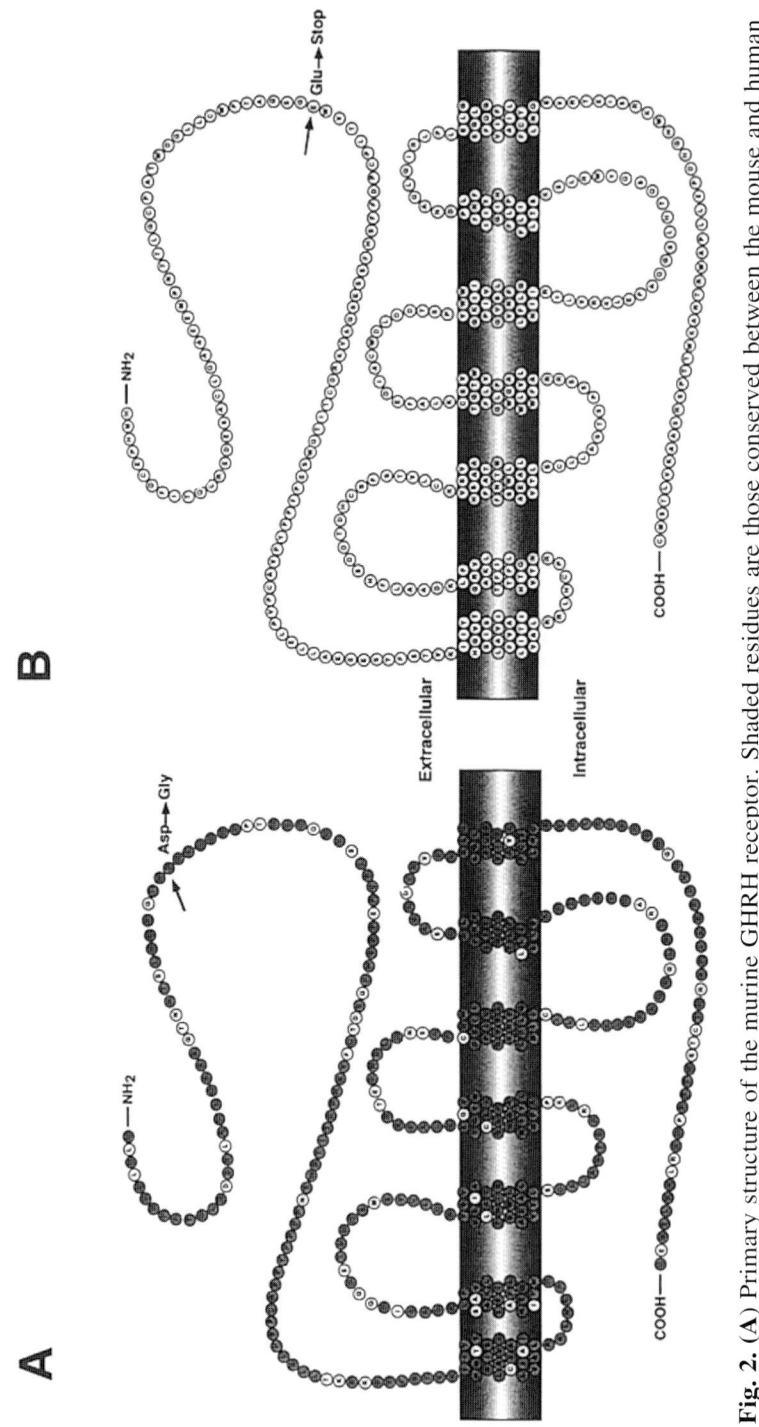

Fig. 2. (**A**) Primary structure of the murine GHRH receptor. Shaded residues are those conserved between the mouse and human GHRH-R. The arrow indicates the mutation in the little mouse. (**B**) Primary structure of the human GHRH-R. The arrow indicates the mutation described in Sindh dwarfism.

Most exons are short (~100 bp), and exon–intron boundaries do not generally correspond to functional domains in the protein.

The GHRH-R is expressed most abundantly in the pituitary, but also in numerous other tissues *(16)*. It is, however, not clear whether the expression level in extrapituitary tissues is sufficiently high to be functional. An interesting exception is placenta, where the GHRH-R has not been found to be expressed in contrast to GHRH, which is highly expressed *(14,16)*. This raises questions regarding the role of GHRH in the placenta, in particular also since an alternative promoter for GHRH transcription is used. The possible existence of another receptor for GHRH-like molecules must be considered. In human (and probably rodent) pituitary, the GHRH-R appears to be expressed only in somatotroph cells *(32,34)*.

The regulation of GHRH-R expression is only incompletely understood. In the pituitary, GHRH-R expression is believed to be under the control of Pit-1, a pituitary-specific transcription factor *(30)*. Lack of Pit-1, such as is seen in Snell dwarf mice, is associated with defective GHRH-R expression. The effect of GHRH itself on GHRH-R expression is controversial, with both upregulation and downregulation being reported *(35,36)*. Glucocorticoids, but not sex steroids, enhance GHRH-R gene transactivation in the rat *(32)*; little is known about GHRH-R regulation in other species.

The GHRH-R signals through the G_s-cAMP pathway and also activates ion channel pathways, which lead to rapid calcium influx and GH release in the somatotroph *(23,37)*. Although activation of protein kinase C and phospholipid turnover can affect the somatotroph response to GHRH, their involvement in GHRH signaling under physiological conditions is not clearly established.

GHRH AND PITUITARY GROWTH AND DEVELOPMENT

As indicated above, GHRH stimulates somatotroph proliferation in vitro *(21)*. This effect is mediated through the GHRH-R, and it would be expected that GHRH-R deficiency results in somatotroph hypoplasia. Indeed, mice who are deficient in the GHRH-R because of Pit-1 deficiency (Snell and Jackson dwarf mice) do have hypoplastic pituitaries, including somatotroph, lactotroph, and thyrotroph aplasia *(38)*. The "little" mouse, which bears a dysfunctional GHRH-R (*see below*), also has significant somatotroph hypoplasia as well sparse GH secretory granules in the remaining somatotrophs *(39)*. Inactivation of the cAMP signaling pathway in somatotrophs by overexpression of a defective, mutant CREB results in severe somatotroph hypoplasia and dwarfism *(40)*. Thus, an intact GHRH/GHRH-R/cAMP/CREB system is critically important for normal somatotroph development. Since somatotrophs are the predominant cell type in the pituitary, the GHRH system is also important for overall pituitary size and anatomy. There is no evidence that the GHRH system is important for the differentiation and development of pituitary cells other than somatotrophs.

The importance of the GHRH system for somatotroph proliferation and function is further corroborated by the observation that overstimulation of the GHRH-R by ectopically produced GHRH, or in transgenic animals by overexpressed GHRH, leads to somatotroph hyperplasia and sometimes adenoma formation, with ensuing acromegaly or gigantism *(41–43)*. Similarly, constitutively active mutations of $G_s\alpha$ can lead to somatotroph adenoma formation and acromegaly *(44)*.

GHRH RESISTANCE
The Little Mouse: A Model for GHRH Resistance

DISCOVERY AND GENETICS

In the mid-1970s, three unusually small mice were observed in a colony of C57BL/6J mice maintained at the Jackson Laboratory in Bar Harbor, ME. This phenotype had arisen spontaneously, and interbreeding and backcross experiments were conducted to determine the mode of inheritance and chromosomal linkage. The small phenotype was propagated in an autosomal-recessive mode with full penetrance and segregated with markers on mouse chromosome 6 *(45)*. The mouse was called "little," and the genetic defect *lit*, to differentiate it from other types of dwarfed mice, such as the Snell *(46)*, Jackson *(47)*, and Ames *(48)* mouse, which had a slightly different phenotype and mapped to different chromosomes (chromosomes 16, 16, and 11, respectively).

MOLECULAR DEFECT AND ITS CONSEQUENCES

The dwarf phenotype of the homozygous (*lit/lit*) little mouse has been the subject of intense investigation since its discovery. Its proportionate dwarfism suggested a lesion in the GH–IGF axis, and studies proceeded in a progressive manner to test several of the critical components of this axis. Initial studies focused on the question of whether little mice produced GH and prolactin. The GH content of the pituitary was found to be substantially decreased *(45)*. Subsequent studies showed that growth could be restored with transplantation of normal pituitary tissue or treatment with GH *(49)*. Somatomedin (IGF) activity was found to be very low in the serum of little mice *(50)*, and hepatic GH receptors were found to be normal *(51)*. Treatment with IGF-1 was effective in promoting linear growth *(52)*. These findings are consistent with genetic GH deficiency. An analysis of the structural GH gene showed the GH locus to be grossly intact, although point mutations could not be excluded *(53)*. However, the later assignment of the mouse GH gene to chromosome 11 *(54)* rendered it an unlikely candidate for the little mouse whose defect mapped to chromosome 6.

The availability of GHRH after its characterization in the early 1980s *(11,12)* permitted testing of pituitary responsiveness to that neuropeptide. Unlike normal mice, the little mouse did not release GH in response to GHRH, given either acutely or chronically *(55)*. Similarly, cultured pituitary cells from little mice failed to secrete GH, or accumulate cAMP, in response to GHRH *(56)*. The fact that GH was secreted by the same cells in response to dibutyryl-cAMP, forskolin, or cholera toxin placed the defect at the GHRH-R or immediately downstream.

The cloning of the GHRH-R *(29–31)* paved the way for the identification of the molecular lesion in the little mouse. In 1993, three groups identified the GHRH-R gene as the site of the defect in the little mouse, and two described the actual mutation *(57–59)*. A single base change (A→G) converts an aspartic acid residue in the extracellular domain to glycine (Fig. 2A) *(57,58)*. The affected Asp^{38} is highly conserved not only among species, but in the entire receptor family to which the GHRH-R belongs.* Therefore, it likely serves a crucial role in receptor function. Indeed, the mutated GHRH-R protein is unable to promote cAMP accumulation in response to GHRH *(57,58)*. Initially, the precise reason for this dysfunction was not readily identified because of technical

---

*Throughout this chapter, the amino acid numbering refers to the mature GHRH-R, starting with the histidine following the 22 amino acid signal peptide as the amino-terminus.

difficulties with GHRH binding assays *(32)*. However, it is now clear that the mutation disables GHRH binding *(60,61)*. Thus, the little mouse represents a classic example of hormone resistance caused by missense mutation that abolishes ligand recognition.

In contrast to the little mouse, the lesion in the Snell or Jackson mouse lies in inactivating mutations in the Pit-1 gene *(38)*, and the Ames mouse harbors a mutation in the Prophet of Pit-1 *(Prop-1)* gene, whose early expression is required for Pit-1 lineage differentiation *(62)*. As indicated above, Pit-1 is required for GHRH-R expression, so that all these dwarf strains ultimately share GHRH-R dysfunction. In addition, pituitary cells in Snell/Jackson and Ames mice fail to differentiate into thyrotrophs and lactotrophs.

PHYSICAL PHENOTYPE

The little mouse is of normal or near-normal size at birth, but exhibits progressive postnatal growth retardation that becomes statistically significant at 2 wk *(45,63)*. As an adult, and its weight is about 60%, its length 70–75% of that of (presumably normal size) heterozygous mice *(45,63)*. This growth pattern is typical of GH deficiency or GH resistance *(64)*. The little mouse is less growth-retarded than the Snell or the Ames mouse, presumably because the latter two are also thyroid-stimulating hormone (TSH)- and prolactin-deficient, with hypothyroidism a likely cause of additional growth impairment. While the little mouse is somewhat less GH-deficient than these other dwarf mice, it appears doubtful that its residual GH secretion *(see* Biochemical Phenotype*)* is biologically meaningful. Dwarfism is proportionate, with long bones and intramembranous bones reduced in size in a commensurate manner, except for the skull, which is relatively spared, but still smaller than normal *(45,49)*. The little mouse has an abnormal body composition, with excess adipose tissue and decreased protein, water, and mineral content *(63)*, similar to what is seen in GH deficiency in humans *(65)*. With the exception of the very small pituitary size, little detailed information is published about organ size. Similarly, information about potential phenotypic differences between heterozygote (*lit/+*) and wild-type (+/+) C57BL/6J mice is scant.

Development of the central nervous system appears abnormal, with significantly smaller brain size. This is somewhat expected in view of the microcephalic skull. Retarded glial proliferation, with hypomyelination and delayed neuronal development, with short and sparsely branched dendrites, has been reported *(66,67)*. The cerebrum and cerebellum appear to be affected primarily. It is not clear whether these abnormalities have any functional consequences, since no neurological or behavioral dysfunction is evident.

The little mouse is fertile, although initial reports suggested that males were subfertile *(45)*. Subsequently, this was attributed to dietary reasons *(57)*. The question about fertility is important because of the expression of GHRH in the gonad and placenta *(14, 16–19)*. It appears that GHRH signaling through the GHRH-R is not critically important for reproduction. Recent evidence has suggested that in the testis, an alternative product of the GHRH gene, the GHRH-RP, participates in spermatogenesis *(15)*. GHRH-RP does not appear to signal through the GHRH-R *(15)*. Sexual maturation is delayed in both sexes *(45)*, as is also seen in human GH deficiency of any etiology. Lactation is reported to be scant in primiparas, but to normalize with subsequent pregnancies *(45)*. This appears to be a consequence of GH deficiency (and possibly secondary prolactin deficiency) as treatment of primiparas with either GH or prolactin normalizes lactation *(68)*.

One particularly interesting feature of rodents in general and the little mouse in particular is the somatic growth that occurs during pregnancy or even pseudopregnancy

(69). This is made possible by the late closure of epiphyseal plates, and in the normal mouse, has been attributed to enhanced GH and perhaps prolactin production during pregnancy *(70)*. The little mouse, like its wild-type counterpart, exhibits marked weight gain during pregnancy or pseudopregnancy, with postpartum weights significantly exceeding those of unmated animals. In the absence of functional somatotrophs, it seems unlikely that pituitary GH is responsible for growth. Whether prolactin can promote weight gain is doubtful *(49)*. Other candidate hormones for this effect are the placental lactogens *(71)* or an as-yet undiscovered murine placental GH *(72)*, though that hypothesis would not explain growth during pseudopregnancy. To date, little information exists about the level of hormone expression during pregnancy in the little mouse, and the reason for this puzzling phenomenon remains elusive.

BIOCHEMICAL PHENOTYPE

The pituitary gland of the (at least 15-d old) little mouse contains 4–9% of the normal complement of immunoreactive GH *(55,56,73)* and 8–18% of GH mRNA *(57,73)*. By contrast, the Snell mouse pituitary contains no detectable GH *(73)*, and the Ames mouse pituitary contains <1% of the normal complement *(62)*. Thus, unlike these two dwarfed strains, there are low, but significant GH stores present in the little mouse, yet that GH is not "releasable." Serum GH levels are very low (0.05–0.6 ng/mL), insufficient to sustain a biological response *(55,73)*. (It should be noted that GH secretion is normally pulsatile, and that the cited GH levels were obtained at random times. However, pulsatility is probably diminished in the little mouse, since pulses are primarily generated by GHRH action.)

The prolactin status of the little mouse is somewhat unclear. The initial report *(45)* suggested that there may be prolactin deficiency, but this has not been further documented. Lactotroph cells are present in normal amounts (*see below*), and mice are able to lactate after their first pregnancy. There may be some limitation in prolactin production secondary to GH deficiency. Other pituitary hormones are considered normal, based on normal fertility and normal serum corticosterone and thyroid hormone levels *(63)*.

Little mice have 5- to 10-fold reduced serum IGF-1 levels and 3- to 10-fold reduced IGF binding protein 3 levels, consistent with GH deficiency *(63)*. Somatomedin bioactivity in serum is also substantially reduced *(50)*. No abnormality in IGF binding proteins 1, 2, and 4 is observed *(63)*. Serum insulin levels in little mice are normal, but blood glucose levels were about 20% below normal in a limited sample of animals *(63)*.

PITUITARY DEVELOPMENT AND MORPHOLOGY

As already indicated, GHRH signaling through the GHRH-R is important not only for GH synthesis and secretion, but also for somatotroph development. Therefore, it would be expected that the little mouse has a pituitary developmental defect. Indeed, in mature animals there is considerable pituitary hypoplasia *(45,58)* that can be attributed to a scarcity of somatotroph cells, which normally occupy 50–60% of the pituitary volume *(39)*. Immunohistochemical and electron microscopic analyses show sparse somatotrophs throughout the pituitary, with an abnormal spatial distribution of remaining cells toward the anterolateral periphery of the gland *(45,58,74)*. In addition, the remaining somatotrophs contain few GH secretory granules *(73,74)*. Quantitative morphometry has yielded estimates of 24% of normal for somatotroph abundance/total tissue, 58% of normal for granules/somatotroph, and 12% of normal for granules/total tissue *(39)*.

The latter corresponds quite well to the GH content as measured by immunoassay. The somatotroph complement of little mice is normal at birth, but begins to fall behind that of normal mice shortly thereafter. The lack of somatotroph development becomes apparent between 8 and 15 d after birth, and persists into adulthood *(74)*. Thus, early pituitary ontogenesis is preserved under the direction of Prop-1 and Pit-1, but later GHRH-mediated somatotroph proliferation and outgrowth is blocked *(58,62)*. (It should be noted in this context that Snell [Pit-1-deficient] and Ames [Prop-1-deficient] mouse pituitaries are virtually devoid of somatotrophs, in addition to lactotroph and thyrotroph hypoplasia.)

In contrast to the scarcity of somatotrophs, lactotroph development and abundance appear normal in the little mouse *(58)*. Similarly, thyrotrophs and corticotrophs are present, though perhaps with altered spatial distribution *(58,75)*.

SUMMARY: THE LITTLE MOUSE

The little mouse has isolated GH deficiency because it harbors a missense mutation in the GHRH-R, with inability of the receptor to bind GHRH and, hence, GHRH resistance. This results in proportionate dwarfism and other sequelae of GH deficiency, such as abnormal body composition. The little mouse is fertile, but may have some degree of prolactin deficiency as illustrated by its inability to lactate at first parturition. Pituitary hypoplasia results from lack of terminal proliferation of somatotrophs; other pituitary cell types are normal. This GHRH-resistant syndrome illustrates the crucial importance of GHRH for GH secretion, with somatostatin and the putative GHRP ligand playing subordinate roles. The phenotype is largely limited to that caused by GH deficiency, calling into question the physiological importance of the GHRH-GHRH receptor system in extrapituitary sites.

GHRH Resistance in Humans: Dwarfism of Sindh

DISCOVERY AND GENETICS

Despite the recognition of the little mouse as a GHRH-resistant model since the mid-1980s, it took more than another decade to identify the equivalent condition in humans. An earlier search for potential mutations in the GHRH-R as a basis for isolated GH deficiency did not yield any abnormalities *(76)*, perhaps in part because the structure of the human GHRH gene was not known and analysis had to be based on cDNA information. In 1996, two teenage siblings with short stature and isolated GH deficiency were reported to harbor a nonsense mutation (Glu→Stop) in the extracellular domain of the GHRH-R *(77)*. They were of Indian descent, living in New York, and had consanguineous parents. Simultaneously, we were investigating a large, highly consanguineous kindred in Pakistan that had recently brought forth 18 dwarfs. We were alerted to this occurrence by a Karachi newspaper article entitled "The Dwarfs of Sindh" *(78)*. The mode of transmission was autosomal-recessive with a high degree of penetrance. Endocrine testing established GH deficiency as the cause of dwarfism, and linkage analysis of several candidate genes implicated the GHRH-R gene on chromosome 7 *(79)*.

MOLECULAR LESION AND ITS CONSEQUENCES

Stepwise amplification and sequencing of the GHRH-R gene in the Pakistani dwarfs showed the gene to be present, but yielded a G to T change in exon 3, which converts a glutamic acid to a stop codon. This predicts a truncation in the extracellular domain at Glu^{50}, resulting in a severely disabled receptor protein (Fig. 2B). Interestingly, this is the

Table 1
**Principal Clinical Features of GHRH-Resistant Patients
with a Nonsense Mutation in the GHRH-R**

Proportionate dwarfism
No dysmorphic features
Progressive postnatal growth retardation
Mean adult height 127 cm for men, 114 cm for women (7.2–8.3 SD below the norm)
Birth size reportedly normal
Small head (head circumference 4.1 SD below the normal mean)
Puberty in males delayed by 2–3 yr (unknown in females)
Delayed bone maturation (mean bone age 4.2 yr behind chronological age)
No microphallus
Fertility possible in both sexes (one instance)
Lactation normal
Intelligence appears normal
No known hypoglycemia
Low blood pressure, asymptomatic

very same mutation that was shown in the two patients in New York *(77)*, although there is no known relationship between the two families. The mutation segregated 100% with the dwarf phenotype. Of interest is the fact that only the latest generation is affected, despite a long-standing practice of consanguineous marriages within the very limited village community where this syndrome arose. Because of the severity of this nonsense mutation, no functional studies have been performed as the mutated protein is unlikely to be expressed.

CLINICAL PHENOTYPE

The clinical phenotype of homozygous individuals with the inactivating GHRH-R mutation largely resembles that of severe, isolated GH deficiency. The salient features are listed in Table 1. There is postnatal growth failure resulting in an adult height of about 4 ft. The affected individuals are normally proportioned and look like miniature versions of normal people. In particular, their head size looks relatively normal in relation to their body size. Puberty is delayed, but fertility and lactation are preserved. Intelligence appears normal, although no formal tests have been performed.

Certain features differ from classical severe GH deficiency resulting from either mutations in the GH gene or organic lesions of the pituitary. In particular, the skull is considerably smaller (~4 SD below the normal mean) than what has been described in classical GH deficiency, where head circumferences are near normal to ~2 SD below normal *(80–82)*. This accounts for the "miniature" aspect of these people and is in agreement with the microcephaly described for the little mouse (*see above*). Another feature that differs from classical GH deficiency is the absence of a microphallus and of childhood hypoglycemia (at least as can be assessed by careful questioning). It is not clear what accounts for these differences. Third, the low blood pressure is not generally seen in GH deficiency, and there is no obvious explanation for this finding. Thus, it appears that the syndrome of GHRH resistance embodies some aspects that go beyond GH deficiency *per se*, although the majority of clinical manifestations are a consequence of GH deficiency.

Table 2
Baseline Serum Endocrine
Parameters in GHRH-Resistant Patients
with a Nonsense Mutation in the GHRH-R[a]

GH	<0.3 ng/mL
IGF-1	5.2 ± 2.0 ng/mL
IGF-2	86 ± 22 ng/mL
IGF-BP3	0.4 ± 0.1 µg/mL
IGF-BP2	440 ± 123 ng/mL
GH binding protein	1.3 ± 0.5 nM
Prolactin	3.3 ± 1.7 ng/mL
Thyroxine	9.0 ± 2.1 µg/dL
TSH	1.4 ± 0.9 µg/dL
Cortisol (9 AM)	9.4 ± 2.3 µg/dL
Testosterone (adult men)	871 ± 219 ng/dL

[a] Data represent mean ± SD.

Of interest is the fact that fertility is preserved, in view of the expression of GHRH in the gonad and placenta *(15,16,18–20)*. We cannot exclude that the affected dwarfs are subfertile, since there is only one union between affected persons. In view of the apparent normal fertility in the little mouse, it is reasonable to postulate that the GHRH/GHRH-R system is probably not crucial for fertility, but rather that the GHRH gene expresses alternative products (e.g., GHRH-RP) that signal through a different receptor *(15)*.

At this writing, no data on pituitary size or morphology in patients with GHRH resistance have been reported.

Subjects who are heterozygous for the mutation are normal or near-normal in height; they display no other abnormalities.

BIOCHEMICAL PHENOTYPE

General blood biochemical parameters are in the normal range. The endocrine profile corresponds to that of isolated GH deficiency (Table 2). Other pituitary hormones are normal. GH release cannot be stimulated by either GHRH, L-dopa, or clonidine (peak values <0.3 ng/mL for all three stimuli). TSH and prolactin responses to TRH are present, but somewhat blunted. The patients respond normally to exogenous GH in terms of IGF-1 and IGF binding protein 3 generation.

Subjects who are heterozygous for the mutation have intermediate levels for IGF-1, IGF-BP3, and IGF-BP2 (103 ± 53 ng/mL, 2.8 ± 1.7 µg/mL, and 389 ± 106 ng/mL, respectively). Responses to GH provocative stimuli are blunted in heterozygotes. Thus, gene dosage does have an effect on GH secretion, but the effect is not sufficient to result in an abnormal physical phenotype.

SUMMARY: DWARFISM OF SINDH

Humans with GHRH resistance owing to an inactivating mutation in the GHRH-R gene resemble the little mouse in many respects. This includes isolated GH deficiency, dwarfism with growth retardation beginning at birth, microcephaly, and fertility. Unlike the mouse, which has difficulty lactating after the first pregnancy, the one existing dwarfed mother nursed normally. Other features of the human syndrome, such as the detailed biochemical phenotype, provide new insights that were not previously available in the little mouse. Future studies should focus on the pituitary size and morphology in humans.

GHRH Resistance: A Synopsis

With respect to GHRH resistance in general, the information in the little mouse and in the dwarfs of Sindh is mutually confirmatory and also complementary. In both species, inactivating mutations in the GHRH-R result in complete GHRH resistance. The resulting syndrome is largely limited to isolated GH deficiency, raising questions about the physiological importance of extrapituitary GHRH-R signaling. On the other hand, these syndromes highlight the crucial role of GHRH for GH secretion. In view of the complex regulation of GH secretion (*see above*), this is an exceptionally clear message. GHRH is involved at several levels of GH production, i.e., somatotroph development, GH synthesis, and GH release. A germline defect in the GHRH-R affects all of these components, and severe GH deficiency results. With the exception of the two genetic causes of GHRH resistance, no other syndromes of GHRH resistance, either congenital or acquired, have been described to date.

POTENTIAL IMPLICATIONS OF GHRH RESISTANCE SYNDROMES FOR SHORT STATURE IN THE GENERAL POPULATION

The discovery of the first human GHRH resistance syndrome should render it possible to identify other, perhaps milder GHRH insensitivity syndromes as a cause of growth retardation and short stature. The elucidation of the structure, genomic organization, and regulatory region of the human GHRH-R gene *(79)* should facilitate the identification of potential mutations with less severe consequences, such as, e.g., splice site mutations or mutations in the promoter region. The characterization of the promoter region should facilitate studies on the regulation of GHRH-R expression. This, in turn, may lead to the recognition of partial GHRH resistance syndromes resulting from abnormal regulation of the GHRH-R expression. Thus, GHRH resistance thus far only described in its most severe, genetic form may well be a more common occurrence than is presently recognized.

The identification of GHRH and its receptor as key regulators of growth and stature also raises the question of their potential role in determining adult height in the general population. Although a strong genetic component in height achievement is well recognized, the biochemical mediators of this genetic determinant remain unknown. The GHRH–GHRH receptor system should be considered one potential candidate pivot where such mediation can occur.

NOTE ADDED IN PROOF

Since submission of this chapter, two reports on the human GHRH receptor mutation described here have appeared (Acta Paediatr 1997;Suppl 423:33–38 and J Clin Endocrinol Metab 1988;83:432–436).

ACKNOWLEDGMENTS

This work was supported in part by grants from the Genentech Foundation for Growth and Development, The Human Growth Foundation, and the Northwestern University Intramural Grant Program.

REFERENCES

1. Beck C, Larkins RG, Martin TJ, Burger HG. Stimulation of growth hormone release from superfused rat pituitary by extracts of hypothalamus and of human lung tumours. J Endocrinol 1973;59:325–333.
2. Reichlin S. Somatostatin. N Engl J Med 1983;309:1556–1563.
3. Bowers CY. Growth hormone releasing peptides—structure and kinetics. J Pediatr 1993;6:21–31.
4. Howard AD, Feighner SD, Cully DF, Arena JP, Liberator PA, Rosenblum CI, et al. A receptor in pituitary and hypothalamus that functions in growth hormone release. Science 1996;273:974–977.
5. Reisine T, Bell GI. Molecular biology of somatostatin receptors. Endocr Rev 1995;16:427–442.
6. Miller GM, Alexander JM, Bikkal HA, Katznelson L, Zervas N, Klibanski A. Somatostatin receptor subtype gene expression in pituitary adenomas. J Clin Endocrinol Metab 1995;80:1386–1392.
7. Guan XM, Yu H, Palyha OC, McKee KK, Feighner SD, Sirinathsinghji DJ, et al. Distribution of mRNA encoding the growth hormone secretagogue receptor in brain and peripheral tissues. Brain Res Mol Brain Res 1977;48:23–29.
8. Muccioli G, Ghe' C, Ghigo MC, Arvat E, Papotti M, Boghen MF, et al. GHRP receptors in pituitary, central nervous system and peripheral human tissues. J Endocrinol Invest 1997;20(Suppl 4):52.
9. Plotzky PM, Vale V. Patterns of growth hormone-releasing factor and somatostatin secretion into the hypophysial-portal circulation in the rat. Science 1985;230:461–463.
10. Frohman LA, Downs TR, Clarke IJ, Thomas GB. Measurement of growth hormone-releasing hormone and somatostatin in hypothalamic-portal plasma of unanesthetized sheep. Spontaneous secretion and response to insulin-induced hypoglycemia. J Clin Invest 1990;86:17–24.
11. Rivier J, Spiess J, Thorner M, Vale W. Characterization of a growth hormone-releasing factor from a human pancreatic islet tumour. Nature 1982;300:276–278.
12. Guillemin R, Brazeau P, Böhlen P, Esch F, Ling N, Wehrenberg WB. Growth hormone-releasing factor from a human pancreatic tumor that caused acromegaly. Science 1982;218:585–587.
13. Mayo KE, Cerelli GM, Lebo RV, Bruce BD, Rosenfeld MG, Evans RM. Gene encoding human growth hormone-releasing factor precursor: structure, sequence, and chromosomal assignment. Proc Natl Acad Sci USA 1985;82:63–67.
14. Gonzalez-Crespo S, Boronat A. Expression of the rat growth hormone-releasing hormone gene in placenta is directed by an alternative promoter. Proc Natl Acad Sci USA 1991;88:8749–8753.
15. Breyer PR, Rothrock JK, Beaudry N, Pescovitz OH. A novel peptide from the growth hormone releasing hormone gene stimulates Sertoli cell activity. Endocrinology 1996;137:2159–2162.
16. Matsubara S, Sato M, Mizobuchi M, Niimi M, Takahara J. Differential gene expression of growth hormone (GH)-releasing hormone (GRH) and GRH receptor in various rat tissues. Endocrinology 1995;136:4147–4150.
17. Bloch B, Brazeau P, Ling N, Böhlen P, Esch F, Wehrenberg WB, et al. Immunohistochemical detection of growth hormone-releasing factor in brain. Nature 1993;301:607–608.
18. Berry SA, Pescovitz OH. Identification of the rat GHRH-like substance and its messenger RNA in rat testis. Endocrinology 1988;123:661–663.
19. Bagnato A, Moretti C, Ohnishi J, Frajese G, Catt KJ. Expression of the growth hormone releasing hormone gene and its peptide product in the rat ovary. Endocrinology 1992;130:1097–1102.
20. Shibasaki T, Kiyosawa Y, Masuda A, Nakahara M, Imaki T, Wakabayashi I, et al. Distribution of growth hormone-releasing hormone-like immunoreactivity in human tissue extracts. J Clin Endocrinol Metab 1984;59:263–268.
21. Billestrup N, Swanson LW, Vale W. Growth hormone-releasing factor stimulates proliferation of somatotrophs in vitro. Proc Natl Acad Sci USA 1986;83:6854–6857.
22. Barinaga M, Yamamoto G, Rivier C, Vale W, Evans R, Rosenfeld MG. Transcriptional regulation of growth hormone gene expression by growth hormone releasing factor. Nature 1983;306:84–85.
23. Cronin MJ, Hewlett EL, Evans WS, Thorner MO, Rogol AD. Human pancreatic tumor growth hormone (GH)-releasing factor and cyclic adenosine 3',5'-monophosphate evoke GH release from anterior pituitary cells: the effects of pertussis toxin, cholera toxin, forskolin, and cycloheximide. Endocrinology 1984;114:904–913.
24. Thorner MO, Rivier J, Spiess J, Borges JL, Vance ML, Bloom SR, et al. Human pancreatic growth-hormone-releasing factor selectively stimulates growth-hormone secretion in man. Lancet 1983;1:24–28.
25. Rosenthal SM, Schriock EA, Kaplan SL, Guillemin R, Grumbach MM. Synthetic human pancreas growth hormone-releasing factor (hpGRF1-44-NH2) stimulates growth hormone secretion in normal men. J Clin Endocrinol Metab 1983;57:677–679.

26. Gelato MC, Pescovitz OH, Cassorla F, Loriaux DL, Merriam GR. Dose–response relationships for the effects of growth hormone-releasing factor-(1-44)-NH2 in young adult men and women. J Clin Endocrinol Metab 1984;59:197–201.
27. Frohman LA, Downs TR, Williams TC, Heimer EP, Pan YC, Felix AM. Rapid enzymatic degradation of growth hormone-releasing hormone by plasma in vitro and in vivo to a biologically inactive product cleaved at the NH2 terminus. J Clin Invest 1986;78:906–913.
28. Segre G, Goldring SR. Receptors for secretin, calcitonin, parathyroid hormone (PTH)/PTH-related peptide, vasoactive intestinal peptide, glucagonlike peptide 1, growth hormone releasing hormone, and glucagon belong to a newly discovered G-protein-linked receptor family. Trends Endocrinol Metab 1993;4:309–314.
29. Mayo KE. Molecular cloning and expression of a pituitary-specific receptor for growth hormone-releasing hormone. Mol Endocrinol 1992;6:1734–1744.
30. Lin C, Lin SC, Chang CP, Rosenfeld MG. Pit-1-dependent expression of the receptor for growth hormone releasing factor mediates pituitary cell growth. Nature 1992;360:765–768.
31. Gaylinn BD, Harrison JK, Zysk JR, Lyons CE, Lynch KR, Thorner MO. Molecular cloning and expression of a human anterior pituitary receptor for growth hormone-releasing hormone. Mol Endocrinol 1993;7:77–84.
32. Mayo KE, Miller TL, DeAlmeida V, Zheng J, Godfrey PA. The growth hormone-releasing hormone receptor: signal transduction, gene expression, and physiological function in growth regulation. Ann NY Acad Sci 1996;805:184–203.
33. Maheshwari HG, Baumann G. Genomic organization and structure of the human growth hormone releasing hormone receptor gene. Program 79th Meeting Endocrine Society 1997, p. 156.
34. Lopes MB, Gaylinn BD, Thorner MO, Stoler MH. Growth hormone-releasing hormone receptor mRNA in acromegalic pituitary tumors. Am J Pathol 1997;150:1885–1891.
35. Horikawa R, Hellmann P, Cella SG, Torsello A, Day RN, Müller EE, Thorner MO. Growth hormone-releasing factor (GRF) regulates expression of its own receptor. Endocrinology 1996;137:2642–2645.
36. Aleppo G, Moskal SF II, De Grandis PA, Kineman RD, Frohman LA. Homologous down-regulation of growth hormone releasing hormone receptor messenger ribonucleic acid levels. Endocrinology 1997;138:1058–1065.
37. Holl RW, Thorner MO, Leong DA. Intracellular calcium concentration and growth hormone secretion in individual somatotropes: effects of growth hormone-releasing factor and somatostatin. Endocrinology 1988;122:2927–2932.
38. Li S, Crenshaw EB, Rawson EJ, Simmons DM, Swanson LW, Rosenfeld MG. Dwarf locus mutants lacking three pituitary cell types result from mutations in the POU-domain pit-1. Nature 1990;347:528–533.
39. Wilson DB, Wyatt DP, Gadler RM, Baker CA. Quantitative aspects of growth hormone cell maturation in the normal and little mutant mouse. Acta Anat 1988;131:150–155.
40. Struthers RS, Vale WW, Arias C, Sawchenko PE, Montminy MR. Somatotroph hypoplasia and dwarfism in transgenic mice expressing a non-phosphorylatable CREB mutant. Nature 1991;350:622–624.
41. Asa SL, Scheithauer BW, Bilbao JM, Horvath E, Ryan N, Kovacs K, et al. A case for hypothalamic acromegaly: a clinicopathological study of six patients with hypothalamic gangliocytomas producing growth hormone-releasing factor. J Clin Endocrinol Metab 1984;58:796–803.
42. Sano T, Asa SL, Kovacs K. Growth hormone-releasing hormone-producing tumors: clinical, biochemical, and morphological manifestations. Endocrine Rev 1988;9:357–373.
43. Mayo KE, Hammer RE, Swanson LW, Brinster RL, Rosenfeld MG, Evans RM. Dramatic pituitary hyperplasia in transgenic mice expressing a human growth hormone-releasing hormone (GHRH) transgene. Mol Endocrinol 1988;9:777–783.
44. Landis CA, Masters SB, Spada A, Pace AM, Bourne HR, Vallar L. GTPase inhibiting mutations activate the a chain of G_s and stimulate adenylyl cyclase in human pituitary tumours. Nature 1989;340:692–696.
45. Eicher EM, Beamer WG. Inherited ateliotic dwarfism in mice: characteristics of the mutation, little, on chromosome 6. J Heredity 1976;67:87–91.
46. Snell GD. "Dwarf," a new Mendelian recessive character of the house mouse. Proc Natl Acad Sci USA 1929;15:733–734.
47. Eicher EM, Beamer WG. New mouse dw allele: genetic location and effects on lifespan and growth hormone levels. J Heredity 1980;71:187–190.
48. Schaible R, Gowen JW. A new dwarf mouse. Genetics 1961;46:896.

49. Beamer WG, Eicher EM. Stimulation of growth in the little mouse. J Endocrinol 1976;71:37–45.
50. Nissley SP, Knazek RA, Wolff GL. Somatomedin activity in sera of genetically small mice. Horm Metab Res 1980;12:158–164.
51. Herington AC, Harrison D, Graystone J. Hepatic binding of human and bovine growth hormones and ovine prolactin in the dwarf "little" mouse. Endocrinology 1983;112:2032–2038.
52. Gillespie C, Read LC, Bagley CJ, Ballard FJ. Enhanced potency of truncated insulin-like growth factor I (des(1-3)IGF-I) relative to IGF-I in lit/lit mice. J Endocrinol 1990;127:401–405.
53. Phillips JA III, Beamer WG, Bartke A. Analysis of growth hormone genes in mice with genetic defects of growth hormone expression. J Endocrinol 1982;92:405–407.
54. Elliott RW, Lee BK, Eicher E. Localization of the growth hormone gene to the distal half of mouse chromosome 11. Genomics 1990;8:591–594.
55. Clark RG, Robinson ICAF. Effects of a fragment of human growth hormone-releasing factor in normal and "Little" mice. J Endocrinol 1985;106:1–5.
56. Jansson J-O, Downs TR, Beamer WG, Frohman LA. Receptor-associated resistance to growth hormone-releasing factor in dwarf "little" mice. Science 1986;232:511–512.
57. Godfrey P, Rahal JO, Beamer WG, Copeland NG, Jenkins NA, Mayo K. GHRH receptor of little mice contains a missense mutation in the extracellular domain that disrupts receptor function. Nature Genet 1993;4:227–232.
58. Lin SC, Lin CR, Gukovsky I, Lusis AJ, Sawchenko PE, Rosenfeld MG. Molecular basis of the *little* mouse phenotype and implications for cell type-specific growth. Nature 1993;364:208–213.
59. Chua SC, Hennessey K, Zeitler P, Leibel RL. The little (*lit*) mutation cosegregates with the growth hormone releasing factor receptor on mouse chromosome 6. Mammalian Genome 1993;4:555–559.
60. Gaylinn BD, Lyons CE, Mayo KE, Thorner MO. The little mouse GHRH receptor mutation (D60 to G) prevents GHRH binding. Program 76th Annual Meeting Endocrine Society 1994; abstract no. 667.
61. Kajkowski EM, Price LA, Pausch MH, Young KH, Ozenberger BA. Investigation of growth hormone releasing hormone receptor structure and activity using yeast expression technologies. J Receptor Signal Transduction Res 1997;17:293–303.
62. Sornson MW, Wu W, Dasen JS, Flynn SE, Norman DJ, O'Connell SM, et al. Pituitary lineage determination by the Prophet of Pit-1 homeodomain factor defective in Ames dwarfism. Nature 1996; 384:327–333.
63. Donahue LR, Beamer WG. Growth hormone deficiency in "little" mice results in aberrant body composition, reduced insulin-like growth factor-I and insulin-like growth factor-binding protein-3 (IGFBP-3), but does not affect IGFBP-2, -1, or -4. J Endocrinol 1993;136:91–104.
64. Zhou Y, Xu BC, Maheshwari HG, He L, Reed M, Lozykowski M, et al. A mammalian model for Laron syndrome produced by targeted disruption of the growth hormone receptor/binding protein gene (The Laron mouse). Proc Natl Acad Sci USA 1997;94:13,215–13,220.
65. Lonn L, Johansson G, Sjostrom L, Kvist H, Oden A, Bengtsson BA. Body composition and tissue distributions in growth hormone deficient adults before and after growth hormone treatment. Obes Res 1996;4:45–54.
66. Noguchi T. Retarded cerebral growth of hormone-deficient mice. Comp Biochem Physiol—C: Comp Pharmacol Toxicol 1991;98:239–248.
67. Morisawa K, Sugisaki T, Kanamatsu T, Aoki T, Noguchi T. Factors contributing to cerebral hypomyelination in the growth hormone-deficient little mouse. Neurochem Res 1989;14:173–177.
68. Keough EM, Wood BG. Mammary gland development during pregnancy in the dwarf mouse mutant, little. Tissue Cell 1979;11:773–780.
69. Hart GH, Cole HH. Studies on the cause of increased growth during pregnancy. Proc Soc Exp Biol Med 1939;41:310–317.
70. Sinha YN, Selby FW, Vanderlaan WP. Relationship of prolactin and growth hormone to mammary function during pregnancy and lactation in the C3H/St mouse. J Endocrinol 1974;61:219–229.
71. Talamantes F. Structure and regulation of secretion of mouse placental lactogens. Prog Clin Biol Res 1990;342:81–85.
72. Jin J, Poole J, Linzer DIH. Two novel members of the prolactin/growth hormone family are expressed in the mouse placenta. Endocrinology 1997;138:5535–5540.
73. Cheng TC, Beamer WG, Phillips JA III, Bartke A, Mallone RL, Dowling C. Etiology of growth hormone deficiency in Little, Ames, and Snell dwarf mice. Endocrinology 1983;113:1669–1678.
74. Wilson D, Wyatt DP. Immunofluorescent analysis of somatotroph distribution in the adenohypophysis of developing *lit/lit* mice. J Anat 1988;156:51–59.

75. Wilson D, Wyatt DP. Adrenocorticotropic cell distribution in adult and embryonic pituitaries of the little (lit) mutant mouse. Anat Embryol 1992;186:347–353.
76. Cao Y, Wagner JK, Hindmarsh PC, Eble A, Mullis PE. Isolated growth hormone deficiency: testing the little mouse hypothesis in man and exclusion of mutations within the extracellular domain of the growth hormone-releasing hormone receptor. Pediatr Res 1995;38:962–966.
77. Wajnraich MP, Gertner JM, Harbison MD, Chua SC Jr, Leibel RL. Nonsense mutation in the human growth hormone-releasing hormone receptor causes growth failure analogous to the little (lit) mouse. Nature Genet 1996;12:88–90.
78. Anonymous. The Dwarfs of Sindh. DAWN (Karachi), February 21, 1994, p. 7.
79. Maheshwari H, Dupuis J, Silverman BL, Baumann G. Dwarfism of Sindh: a novel form of familial isolated GH deficiency linked to the locus for the GH releasing hormone receptor. Program 10th International Congress of Endocrinology, 1996, p. 709.
80. Goodman HG, Grumbach MM, Kaplan SL. Growth and growth hormone. N Engl J Med 1968;278:57–68.
81. Burt L, Kulin HH. Head circumference in children with short stature due to hypothyroidism. Pediatrics 1977;59:628–630.
82. Laron Z, Roitman A, Kauli R. Effect of human growth hormone therapy on head circumference in children with hypopituitarism. Clin Endocrinol 1979;10:393–399.

2 Laron Syndrome
Primary Growth Hormone Resistance

Zvi Laron, MD

CONTENTS

HISTORICAL INTRODUCTION
NOMENCLATURE
THE GH-R
GENETIC ABNORMALITIES OF THE GH-R
THE GHBP
GEOGRAPHICAL DISTRIBUTION OF LS
CLINICAL AND BIOCHEMICAL DIAGNOSTSIC CHARACTERISTICS
TREATMENT
SUMMARY
IGF-1 GENE DELETION
REFERENCES

HISTORICAL INTRODUCTION

In 1996, our group reported three siblings with clinical and biochemical features of growth hormone (GH) deficiency, but who had extremely high serum GH levels *(1)*. Within two years, we were able to collect 22 patients *(2)* of Oriental Jewish descent.

The first hypothesis that the defect is owing to an abnormal GH molecule was disproved by showing that the circulating GH of these patients behaves normally in immunologic *(3,4)* and radioreceptor tests using GH receptors (GH-R) from normal human liver membranes *(5,6)*. Proof that the disease is the result of a GH-R defect leading to GH resistance was shown in 1984 *(7)* by demonstrating that GH does not bind to human GH-R (hGH-R) prepared from liver biopsies performed on two patients.

In the following years, many more patients from around the world were described by others and by us, the number of known patients now being many hundreds. The cloning *(8)* and the characterization *(9)* of the GH-R and the introduction of new techniques in molecular biology enabled the identification of a series of molecular defects of the GH-R or in its downstream pathways.

From: *Contemporary Endocrinology: Hormone Resistance Syndromes*
Edited by: J. L. Jameson © Humana Press Inc., Totowa, NJ

Table 1
Classification of GH Resistance[a]

Primary GH resistance (insensitivity) syndrome = Laron Syndrome (LS) (hereditary)
  GH receptor defects (quantitative and qualitative)
  Abnormalities of GH signal transduction (postreceptor defects)
  Primary defects of synthesis and action of IGF-1
Secondary GH resistance (insensitivity) diseases (acquired conditions; sometimes transitory)
  Circulating antibodies to GH that inhibit GH action
    (hGH gene deletion patients treated with hGH)
  Antibodies to the GH receptor
  GH insensitivity caused by malnutrition
  GH insensitivity caused by liver disease
  GH insensitivity caused by uncontrolled diabetes mellitus
  Other conditions

[a] Modified from Laron et al. (11).

NOMENCLATURE

The name of this syndrome underwent several changes—from pituitary dwarfism with high-serum GH (1) to Laron dwarfism (10). By consensus of several groups (11) GH resistance was divided into primary GH resistance (insensitivity), i.e., Laron syndrome (LS) and secondary GH resistance (Table 1).

A description of the genetic, clinical, pathophysiological, and therapeutic aspects of LS was compiled (12). Having had the opportunity to follow a group of 49 patients, some of them since birth or infancy into childhood and adult age, the following description will be largely based on our own experience.

THE GH-R

The protein and gene structure of the GH-R has been established by purification and cloning studies (13,14). These studies identified the GH-R as a 13-kDa band. A single gene located on the short arm of chromosome 5 (15) encodes the receptor, which contains 620 amino acids, preceded by an 18 amino acid signal sequence (Fig. 1). The GH-R gene proper consists of nine exons and spans over 87 kb (9). Exons 2–7 encode the extracellular domain (EC) of 246 amino acids, exon 8 corresponds to the transmembrane domain (24 amino acids), and exons 9 and 10 form the cytoplasmatic (intracellular) domain (350 amino acids).

The GH-R belongs to the superfamily of cytokine receptors, which includes the receptors for IL-1, IL-2, and erythropoietin (16). All have a single membrane-spanning domain, in contrast to G-protein-related receptors, which have seven membrane-spanning domains (17). Unlike insulin receptors, cytokine receptors do not have intrinsic tyrosine kinase activity. Instead, they associate with protein kinases encoded by other genes (18). In the case of the GH-R, the kinase is Janus kinase 2 (JAK2). Receptor occupancy leads to autophosphorylation of JAK2, association of JAK2 with the GH-R, and subsequent phosphorylation of the receptor itself. The intracellular signaling cascade includes activation of mitogen-activated protein kinase (MAPK) and of latent transcription factors known as signal transducers and activators of transcription (STATs) (19). At the end of the signal transduction cascade is the transcription of specific genes,

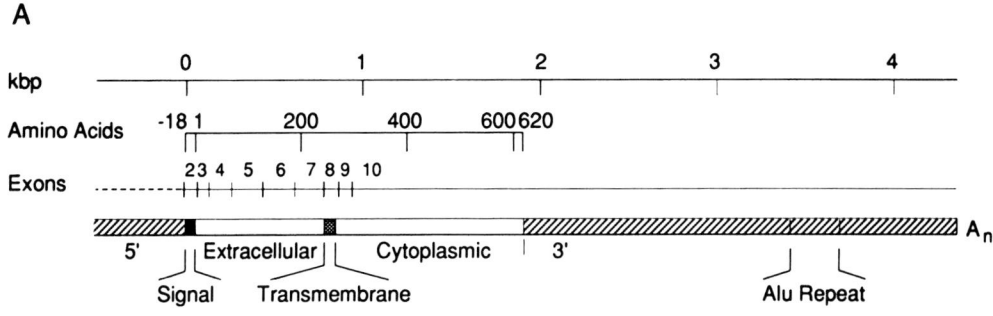

Fig. 1. The GH-R gene.

such as insulin-like growth factor-1 (IGF-1), IGF binding protein-3 (IGFBP-3), and others *(20)*.

GH-R form homodimers in the course of binding a single GH molecule *(21)*. The initial event involves binding of site A of GH to a single receptor molecule, and the second involves binding of site B of GH to a second receptor molecule. Sites A and B on the GH molecule are quite distinct, but the binding sites on the two receptor molecules are nearly identical. Studies are in progress to determine the normal state and abnormalities in the human GH (hGH) binding to its receptor.

GENETIC ABNORMALITIES OF THE GH-R

Deletions and mutations in the GH-R gene are responsible for the classical form of GH resistance, i.e., LS. A variety of gene defects have been observed in patients with LS.

The manifold defects found range from exon deletions to nonsense, frameshift, splice, and missense mutations of exons and introns *(11,22–30)*. The majority are in the extracellular domain of the receptor: exons 3–7 and introns 2–7, resulting in the absence of circulating GH binding protein (GHBP) *(31,32)* *(see later)*.

Only two reports describe mutations affecting the transmembrane domain (exon 8) *(33,34)*, and one relates two mutations in the cytoplasmatic domain (exon 10) *(35)*.

Table 2 compiles most of the published molecular defects of the GH-R in patients with LS. It is of note that despite the great variability in the molecular defects of the gene, they all result in lack of GH signal transmission. Indeed, a single amino acid substitution in the exoplasmic domain of the hGH-R prevents ligand binding to the GH-R *(36)* as does defective membrane expression *(37)*.

Three children from a family with a thus-far unidentified postreceptor defect *(38)* revealed that on administration of exogenous hGH, there was a rise in serum IGFBP-3, but not of IGF-1 (Fig. 2) denoting separate signaling pathways for the transcription of the two genes. In both of the last instances, the circulating GHBP were normal or high.

THE GHBP

Herington et al. *(39)* and Baumann et al. *(40)* independently described and characterized from rabbit and human serum a protein capable of binding GH with high affinity. This protein was subsequently shown to be identical in structure with the extracellular hormone binding domain of the GH-R *(8)*.

Table 2
GH-R Mutations Reported in Patients with LS[a]

Mutations	Molecular defect	Nucleotide change	Exon involved	Domain[b]	GHBP[c]	Reference number
Deletion	Exons 3-5-6		Exons 3-5-6	EC	–	Godowski et al., 1989 (9)
Nonsense	C38X	C→A at 168	4	EC	–	Amselem et al., 1991 (23)
	R43X	C→A at 181	4	EC	–	Amselem et al., 1991 (23)
	Q65X	C→A at 197	4	EC	–	Sobrier et al., 1997 (29)
	W80X	C→A at 293	5	EC	–	Sobrier et al., 1997 (29)
	W157X	C→A at 525	6	EC	?	Sobrier et al., 1997 (29)
	E183X	C→A at 601	6	EC	?	Berg et al., 1994 (26)
	R217X	C→A at 703	7	EC	–	Amselem et al., 1993 (24)
Frameshift	21delTT	delTT at 118	4	EC	–	Counts and Cutler, 1995 (28)
	36delC	delC at 162	4	EC	–	Sobrier et al., 1997 (29)
	46delTT	delTT at 192–193	4	EC	–	Berg et al., 1993 (95)
	230delT	delT at 744	7	EC	–	Sobrier et al., 1997 (unpublished)
	230delAT	delAT at 743–744	7	EC	–	Berg et al., 1993 (95)
Splice	Intron 2	G→A at 70+1		EC	?	Sobrier et al., 1997 (29)
	Intron 4	G→A at 266+1		EC	–	Amselem et al., 1993 (24)
	Intron 5	G→C at 440-1		EC	–	Amselem et al., 1993 (24)
	Intron 6	G→T at 619-1		EC	–	Berg et al., 1993 (95)
	E180Splice	A→G at 594	6	EC	+	Berg et al., 1992 (96)
	Gly236GLY	C→T at 766		EC	–	Baumbach et al., 1997 (30)
	G223G	C→T at 723	7	EC	–	Sobrier et al., 1997 (29)
	Intron 7	G→T at 785-1	7/8	EC/TM	+++	Silbergeld et al., 1997 (34)
	R274T	G→C at 874	8	TM	++	Woods et al., 1996 (33)

Missense	C38S	T→A at 166	4	EC	?	Sobrier et al., 1997 (29)
	S40L	C→T at 173	4	EC	?	Sobrier et al., 1997 (29)
	W50R	T→C at 202	4	EC	−	Sobrier et al., 1997 (29)
	R71K	G→A at 266	4	EC	−	Amselem et al., 1993 (24)
	F96S	T→C at 341	5	EC	−	Amselem et al., 1989 (22)
	V125A	T→C at 428	5	EC	−	Amselem et al., 1993 (24)
	V144D	T→A at 485	6	EC	−	Amselem et al., 1993 (24)
	D152H	G→C at 508	6	EC	+	Duquesnoy et al., 1994 (36)
	R161C	C→T at 535	6	EC	−	Amselem et al., 1993 (24)
	R211G	C→T at 1362	7	EC	−	Amselem et al., 1993 (24)
	C422F[d]	C→T at 1778	10	IC	−	Kou et al., 1993 (35)
	P561T[d]		10	IC	−	Kou et al., 1993 (35)

[a] With additions to the table of Amselem et al. (94).
[b] EC = extracellular; TM = transmembrane; IC = cytoplasmic (intracellular).
[c] GHBP = growth hormone binding protein. ?, not available; +, detectable; ++, high levels; +++, very high levels.
[d] These two mutations were identified on the same GH-R allele.

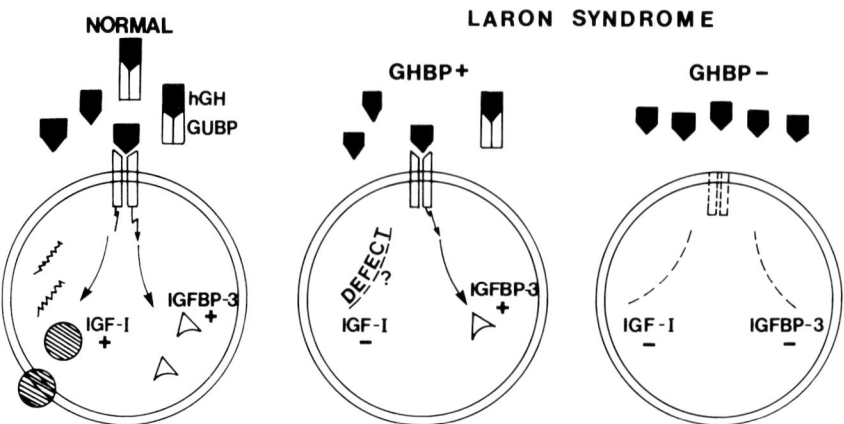

Fig. 2. Schematic illustration of the post-GH-R defect described by Laron et al. *(38)*. These patients have normal serum GHBP, and on administration of exogenous GH, show an increase in serum IGF-BP-3, but not serum IGF-1.

Between 30 and 50% of the circulating GH is bound to this protein. Its quantitative measurements revealed that its serum concentrations changes with age, being low in neonates and reaching maximal values in young adulthood *(41)*.

Determinations of serum GHBP can be used as a simple quantitative determination of the extracellular domain of the GH-R, its absence meaning a defect in this domain of the receptor. A low-serum GHBP concentration in relatives of patients with LS helps identify heterozygous carriers *(42)*.

Typical LS with positive GHBP denotes a defect in the transmembrane, cytoplasmic, or post-GH-R area.

GEOGRAPHICAL DISTRIBUTION OF LS

Since the first description of patients with LS in Israel of Oriental Jewish origin (Yemen, Iran, Afghanistan, Middle East, and North Africa) *(1,2,43)*, many more patients have been diagnosed, the majority originating in the Mediterranean area and Middle East or Arab, Turkish, and Iranian origin, or in descendants of subjects originating in these regions. A large genetic isolate has been reported from Ecuador *(44)* and recently a small one in the Bahamas *(30)*. There are numerous patients in Pakistan and India *(45,46)*. In most of the above populations, consanguinity was or is still in practice. Isolated patients not of Mediterranean or Oriental descent have been reported in Denmark, The Netherlands, Slovakia, Russia, Slovenia, Japan, and the US. A detailed listing appears in recent reviews of this syndrome *(12,43,47,48)*. Analysis of the Israeli cohort led to the conclusion that LS is caused by an autosomal fully penetrant recessive mechanism *(49)*.

CLINICAL AND BIOCHEMICAL DIAGNOSTIC CHARACTERISTICS *(50)*

The patients with this syndrome are indistinguishable from untreated patients with isolated GH deficiency owing to hGH gene deletion *(1,2,51)*. They are very short, i.e., height <4 standard deviation score below the median for sex and age (SDS) (Fig. 3). Progressively more and more obese with small gonads and genitalia, they have hypogly-

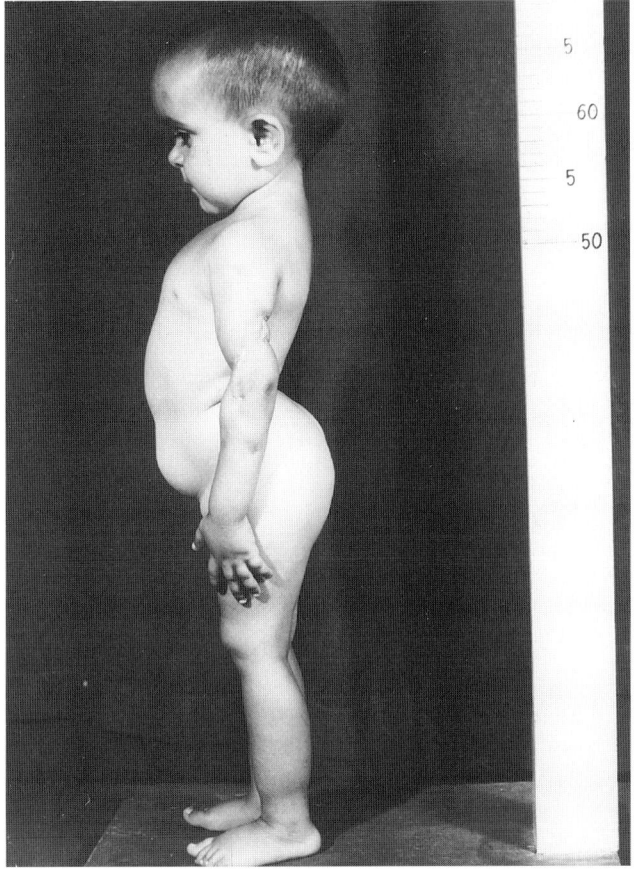

Fig. 3. A 3½-yr-old boy with LS. Note: dwarfism (height 68 cm) and obesity.

cemia and increased serum levels of GH. Their serum IGF-1 is very low to undetectable and does not rise on administration of exogenous GH *(52,53)*.

Neonatal Period

Pregnancy and delivery of our patients were usual. In almost all, for whom information is available, length of gestation was normal. Birth length (not available in all patients) revealed that 12 in our series and some from other reports were short, measuring 42–46 cm. Birth weight is above 2500 g in most newborns with LS.

Congenital Malformations

Few infants with LS have congenital dislocation of the hip joints or Perthes disease. Other malformations reported are aortic stenosis, cataracts, hare lip, and undescended testicles in individual patients.

Features

The head is apparently large (Fig. 4), but actually the head circumference is below normal (Fig. 5) *(43)*. This is owing to the underdevelopment of the facial bones (Fig. 6) *(54,55)*. These changes give the typical appearance of protruding forehead, saddle nose, and small chin (Fig. 4). The hair is sparse, causing deep temporal recessions in young age

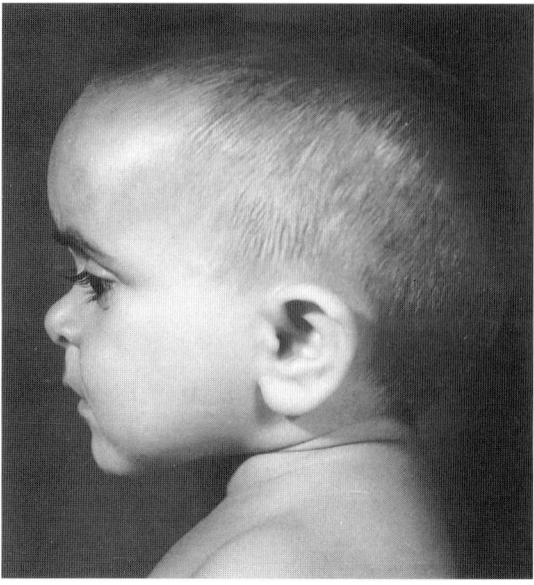

Fig. 4. Characteristic head of patient in Fig. 3. Note: small face, protruding forehead, sparse hair.

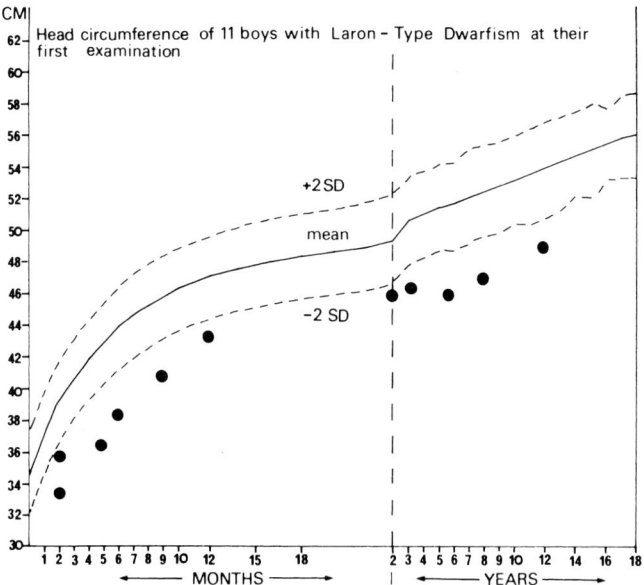

Fig. 5. Small head circumference in 11 boys with untreated LS.

(Fig. 4) *(56)*. The teeth are decayed and subsequently crowded (Fig. 7). The skin has a fine texture and wrinkles at an early age. Histopathological examinations revealed irregular and thickened elastic fibers, which decreased in amount after puberty *(57)*. The voice is typically high pitched because of underdevelopment of the larynx.

Nutritional State

From birth, children with LS are obese; this is not evident by body weight, since their bones are thin and muscular tissue is underdeveloped. They are obese despite eating very

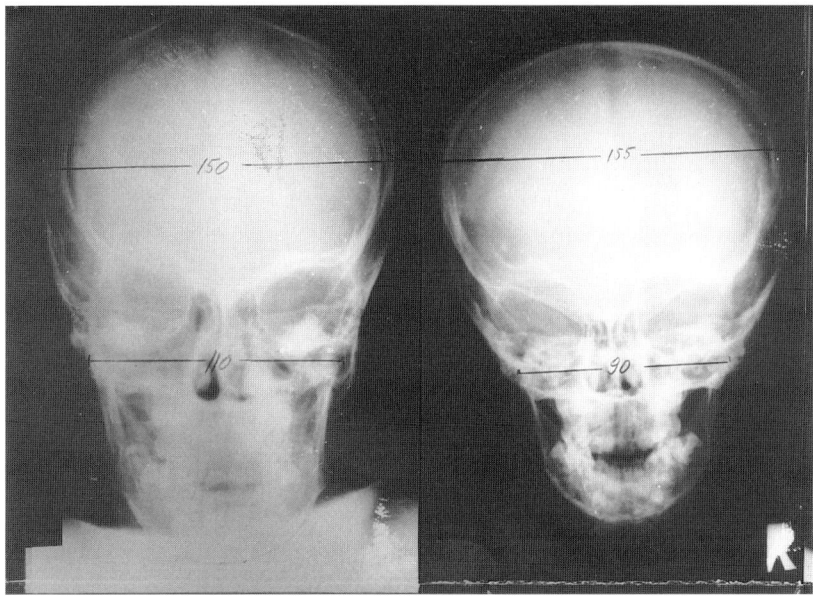

Fig. 6. Anterior–posterior (AP) skull X-ray of a boy with LS (right) compared with a same-aged healthy boy. Note: underdeveloped facial bones in the boy with LS.

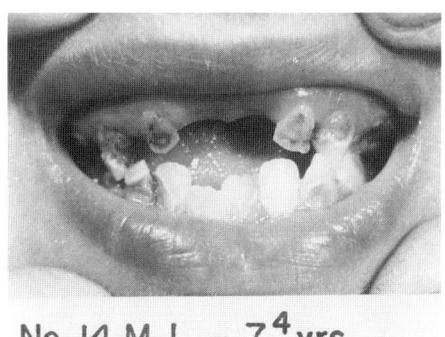

Fig. 7. Defect and crowded teeth in a 7-yr-old boy with LS.

little (Fig. 8). Their obesity is progressive during childhood into adulthood, when the obesity becomes excessive *(58)*.

Body Proportions and Linear Growth

The upper/lower segment ratio is above the norm for sex and age, denoting short limbs for the grown trunk size *(53)* (Fig. 9). The hands and feet are small, i.e., acromicria (in infancy, there is difficulty finding suitable shoe sizes). From infancy on, growth is slow; and if untreated (as most patients with LS are), the final height ranges between 119 and 142 cm in males *(43)* (Fig. 10) and 108 and 136 cm in females. Based on their natural growth, special growth curves for LS have been determined *(59)*. Only a few female first-degree relatives (possibly heterozygotes) are of slight short stature (53). One father is also dwarfed, but he is a patient himself.

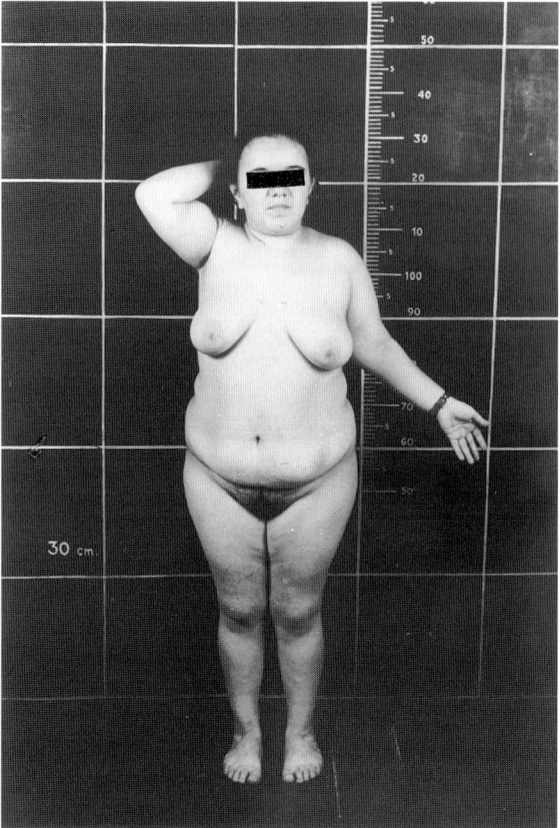

Fig. 8. Obesity in a pubertal girl with LS.

Sexual Maturation

The genitalia and gonads are small from early childhood; this is especially evident in the males *(60)*. Puberty is more delayed in the boys, and there is no pubertal growth spurt *(61,62)*. Nevertheless, both sexes reach full sexual development (Fig. 8). In most instances, there are no difficulties in reproduction. Several of our patients are mothers; one father, himself affected, has two children with LS *(62)*. Similar observations were made in the Ecuadorian cohort *(47)*.

Skeletal Development

Skeletal maturation is retarded and slow starting *in utero*; closure of the epiphyseal cartilage in the long bones occurs after age 16–18 in girls and 20–22 yr in boys. The fontanels and sutures of the skull also close very late. The long bones are thin, and already in young adult age, osteoporosis is evident *(63)*.

Neuromuscular and Psychological Development

Motor development is slow, as is psychological development. Intelligence studies in a large group of patients with LS revealed a lower than normal distribution *(64)*; however,

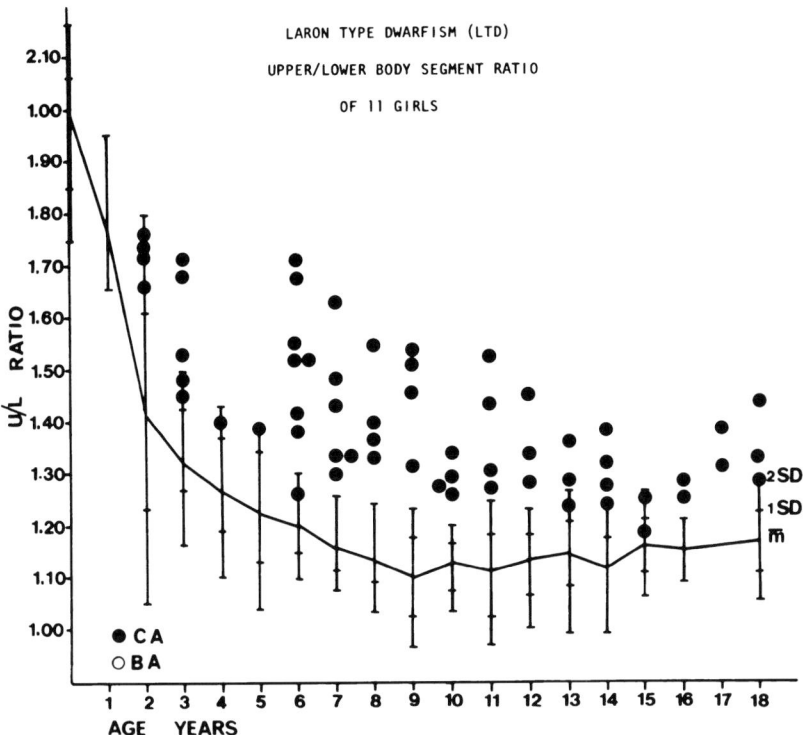

Fig. 9. Increased upper/lower body segment ratio during repeated measurements in 11 girls with LS denoting short limbs.

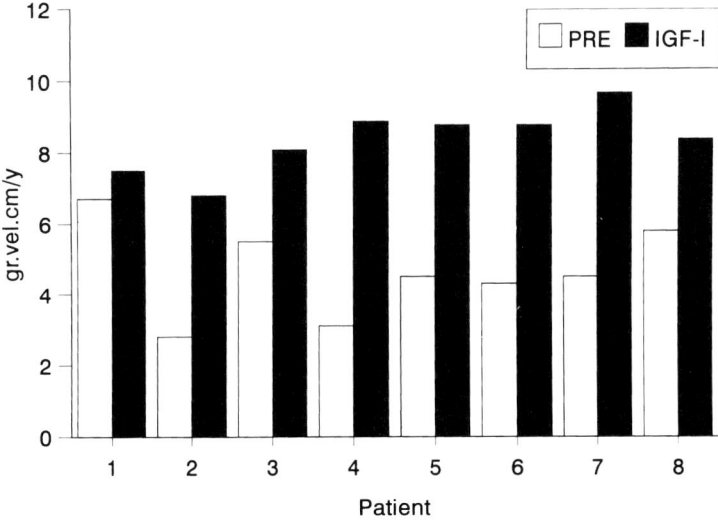

Fig. 10. Growth-stimulating effect of IGF-1 (150–200 µg/kg/d sc) during the first year of treatment in 8 children with LS.

there is great variability; in our group of 49 patients in Israel, there is one PhD and one institutionalized mentally retarded patient. It is obvious that the major cause of the above is the lack of IGF-1 starting *in utero*. Patients with LS also have a reduced muscle strength and endurance *(65)*.

Hormonal and Biochemical Features

GH and IGF-1 secretion are high, reaching levels of 200–300 µg/L during the nocturnal pulses *(66)*. The pattern of secretion, i.e., number of pulses, higher at night than during the day, is normal, and so is the regulation by stimulatory agents (hypoglycemia, arginine, and so forth) *(2)* and suppressive drugs, such as somatostatin (67). Serum IGF-1 is very low, as is IGFBP-3 *(68)*, but IGFBP-1 may be high *(69)*. IGF-1 does not rise on exogenous hGH administration *(52)*, which is one of the major characteristics of this syndrome. Thyroid and adrenal hormone secretions are normal. Gonadotrophin and sex hormone secretions are low, but within normal limits according to the respective biological age. Prolactin when stimulated may be elevated *(70)*. We interpret this as a drift phenomenon to the high GH secretion by the somatomammotrophic cells.

Carbohydrate Metabolism

Patients with LS have hypoglycemia, which is very marked and symptomatic in infancy *(2,43)*, and often asymptomatic during puberty and adult age *(71)*. Concomitantly with the lower than normal glucose levels, these patients have relative hyperinsulinism *(72)*, which may be related to their obesity and low somatostatin. With puberty and adulthood, patients with LS develop glucose intolerance, and the oldest patient has diabetes mellitus.

Lipid Metabolism

With increasing age and obesity, patients with LS develop high total and LDL cholesterol levels *(58)*. In some adult patients, triglycerides are also elevated.

Muscle and bone biochemical parameters vary and are in the lower normal ranges, such as serum phosphorus, alkaline phosphatase, and so forth.

TREATMENT

The only specific treatment of LS is replacement therapy with recombinant IGF-1, which became available only in 1986 *(73)* and even so only in restricted amounts. Untreated patients with LS have an increased number of IGF-1 receptors when compared to healthy subjects *(74)*. One week of IGF-1 administration reduced the number of specific binding sites using red blood cells (RBC) from 5.74 ± 0.86 (SEM) to 2.29 ± 0.64 sites/cell ($p < .004$), i.e., to normal values *(75)*. These studies opened the perspective to IGF-1 clinical trials in patients with LS. Initial investigations showed that an iv bolus injection of IGF-1 after an overnight fasting state in a dose of 75 µg/kg induced marked hypoglycemia in both children and adult patients with LS *(76)* as well as in healthy controls. The concomitant decrease of serum insulin proved that the hypoglycemia was IGF-1 induced and not mediated via insulin *(77)*. The injected IGF-1 was more rapidly eliminated in the patients with LS than in control subjects ($t1/2 = 2.57 \pm 0.67$ vs 4.43 ± 0.52 min) *(78)* owing to the low concentration of serum IGFBP-3 in LS *(68)*. Intravenous IGF-I administration suppresses circulating GHRH, GH, thyroid-stimulating hormone (TSH) *(76,77)*, and glucagon *(79)*. It can thus be concluded that IGF-1 administration stimulates the secretion of somatostatin (Fig. 1) proven also in animal experiments *(80)*.

The effectiveness of the acute administration of IGF-1 by the iv, im, and sc routes led to short- and following long-term clinical trials in patients with LS, first by our group and subsequently by investigators in the US, Europe, and Japan, each group using IGF-1 from a different source, but with an identical structure. There is, however, one difference; whereas our group uses one sc injection/d administered before breakfast, the other groups administer two injections of IGF-1 a day: morning and evening.

Daily sc injections of IGF-1 (120–150 µg/kg) for 7 d to 10 patients with LS (5 children and 5 adults) revealed suppression of serum hGH from 17.5 ± 21.7 to 2.0 ± 1.2 µg/L ($p < .001$), cholesterol from 5.9 ± 1 to 5.7 ± 0.8 mM/L ($p < .04$), LDH from 286 ± 88 to 222 ± 37 U/L ($p < .0005$), and SGOT from 28.9 ± 11.6 to 15.5 ± 7.6 U/L ($p < .01$), and a marked rise in serum procollagen III (PIIINP) from 4.2 ± 0.9 to 7.3 ± 2 µg/L (81).

Long-term administration of IGF-1 over months or years revealed a persistent suppression of serum insulin preventing hypoglycemia, and stabilizing blood glucose levels —provided that meals were regular (82). This was also observed by Walker et al. (83). Sensitive markers of IGF-1 activity are serum alkaline-phosphatase, procollagen I (PICP), and PIIINP (84) serum phosphate and GFR (83,85). Short-term and transitory effects were water and electrolyte retention and calciuria (63,85).

IGF-1 administration was also found to affect its specific circulating binding proteins, thus participating in its own regulation of available free hormone. IGF-1 similar to insulin suppresses IGFBP-1 (69), and after an initial suppression of IGFBP-3 (68), leads to the generation of this largest of the IGFBPs (86,87). Previously it was thought that IGFBP-3 is only under the control of GH. This finding has great practical importance, since it causes a progressively longer biological half-life of the administered IGF-1 during long-term treatment regimens (in preparation) and requires in most patients a progressive reduction of the daily IGF-1 dose in order to prevent overdosage.

One of the major trials of the biological effectiveness of IGF-1 represents the growth-promoting effect. There are four reports on the long-term treatment of children with LS— two 2-yr studies (88,89) and two 3-yr studies (90,91). Table 3 summarizes the published results. It is evident that once daily IGF-1 administration (90) is as effective in promoting growth in these patients as is the IGF-1 twice daily administration (88,89,91,92) even in some with higher doses. On the other hand, twice daily treatment caused more adverse effects (Table 4). These are best explained by either overdosage in some patients and/or as in the case of hypoglycemia, as an inadequate meal after the injection. It is of note that in prepuberty, the growth velocity is similar to that observed when treating GH-deficient children with hGH (Fig. 11), but in the second and third years, the velocity decreases. The possible explanation is controversial. Some authors (89) did not register a rise in IGFBP-3 and blame this; since we found a documented rise in IGFBP-3 in response to IGF-1 (86,87), we consider the reason to be related to prenatal factors similar to those in intrauterine growth retardation.

During IGF-1 treatment, we also registered a fast catch-up of the head circumference (82) (Fig. 12) denoting brain growth even at ages 10–14, and a reduction in adipose tissue as measured by skinfold thickness. The latter was even more accentuated in a group of adult patients with LS, treated for 9 mo (63).

SUMMARY

LS is not a homogenous disorder by etiology. However, the various molecular gene defects lead to dwarfism, and a series of major metabolic abnormalities identical to long-standing GH deficiency. Being a GH-resistant state, the only effective treatment is IGF-1, the hormone these patients cannot generate.

Table 3
Linear Growth Response of Children with LS Treated by IGF-1

Author	Year	Ref.	n	Age (yr) range	At start BA (yr)	Ht SDS m	IGF-1 dose µg/kg/d	0 Before	Growth velocity, cm/yr Year of treatment 1st	2nd	3rd
Ranke et al.	1995	88	31	3.7–19	1.8–13.3	−6.5	40–120 b.i.d.	3.9 ± 1.8	(n = 26) 8.5 ± 2.1	(n = 18) 6.4 ± 2.2	
Backeljauw et al.	1996	89	5	2–11	0.3–6.8	−5.6	80–120 b.i.d.	4.0	(n = 5) 9.3	(n = 5) 6.2	(n = 1) 6.2
Klinger and Laron	1995	90	9	0.5–14	0.2–11	−5.6	150–200 once	4.7 ± 1.3	(n = 9) 8.2 ± 0.8	(n = 6) 6 ± 1.3	(n = 5) 4.8 ± 1.3[a]
Guevarra-Aguirre et al.	1997	91	15	3.1–17	4.5–9.3		120 b.i.d.	3.4 ± 1.4	(n = 15) 8.8 ± 11	(n = 15) 6.4 ± 1.1	(n = 6) 5.7 ± 1.4
Guevarra-Aguirre et al.			8				80 b.i.d.	3.0 ± 1.8	(n = 8) 9.1 ± 2.2	(n = 8) 5.6 ± 2.1	

[a] The younger children had a growth velocity of 5.5 and 6.5 cm/yr.

Table 4
Adverse Effects Reported During IGF-1 Therapy of Children with LS

	Ranke et al. (88) n = 31	*Backeljauw et al. (89)* n = 5	*Klinger & Laron (90)* n = 9	*Guevarra-Aguirre et al. (92)* n = 22
Headaches (early)	21	2	—	?
Nausea, vomiting	?	—	—	3
Hypoglycemia	13	+ early	1 (transitory)	6
Papilledema (transitory)	1	2	—	—
Bell's palsy	1	—	—	—
Lipohypertrophy	7	—	—	—
Enlargement of lymphoid tissue	3	4	1	?
Thickening of nose	5	2	2	?
Local reaction (transitory)	?	?	2	3
Hair loss			—	6
Transient tachycardia	?	?	2	16

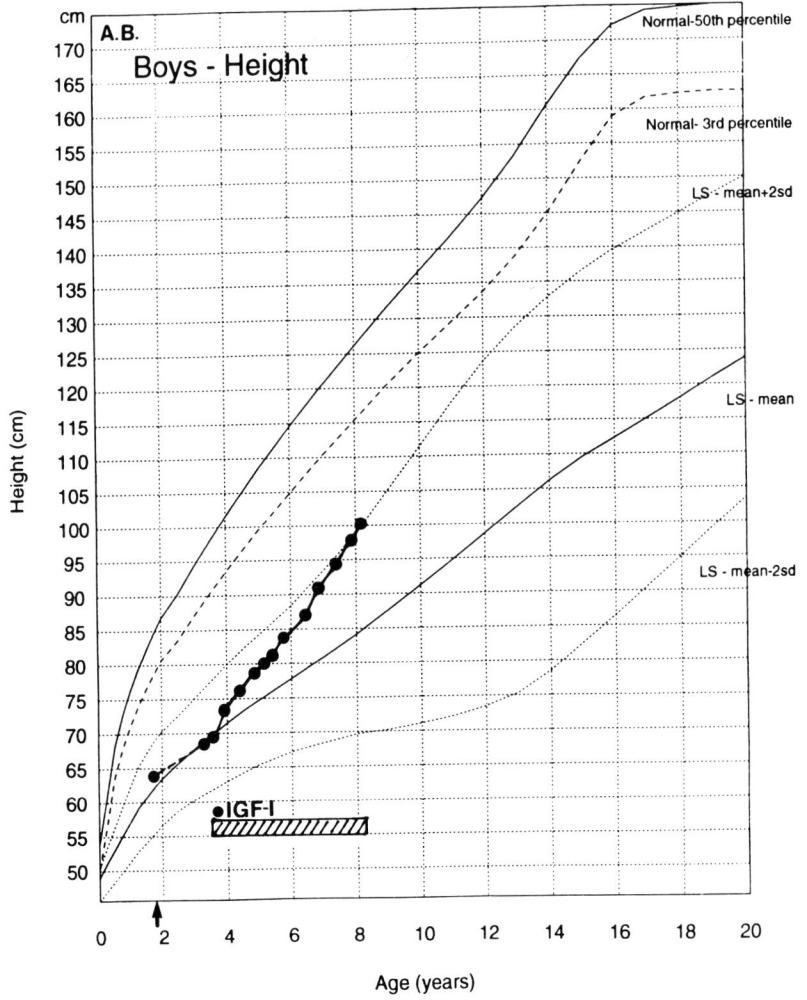

Fig. 11. Growth response to IGF-1 treatment (150–200 µg/kg/d sc) in a boy with LS.

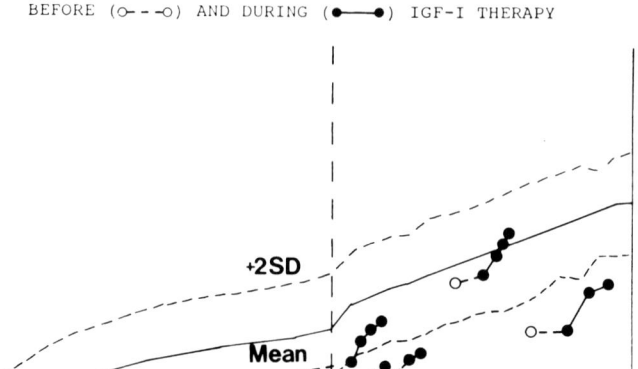

Fig. 12. Head circumference response to IGF-1 treatment (150–200 μg/kg/d sc) in 4 boys with LS.

However, for nonmedical reasons, the availability of IGF-1 for the treatment of patients with LS is more and more restricted, and the more and more frequently diagnosed patients are doomed to suffer irreparable metabolic damage in addition to remaining dwarfed. It is a deplorable situation. Since these patients suffer from osteoporosis *(63)* and muscle weakness *(65)*, it is not certain that they can undergo limb lengthening *(93)* unless they are treated by IGF-1 several years before and during the operation. It is therefore to be hoped that IGF-1 will become available, having been shown to be an effective drug.

The study of these patients who secrete excessive amounts of GH that is prevented from acting, and therefore have very low IGF-1—already *in utero* and postnatally—on one hand, and the treatment with IGF-1 of some of these patients on the other hand, enabled us to learn the identical and differential physiology of GH and IGF-1 along different stages of life.

IGF-1 GENE DELETION

Recently, Woods et al. *(97)* described a boy, the product of a consanguineous marriage, with marked growth retardation who they found to have a deletion of exons 4 and 5 of the IGF-1 gene. He has the extreme characteristic of primary IGF-1 (LS) *(2)*, i.e., marked growth retardation, small head circumference, acromicria, delayed motor development, hypoglycemia in infancy, high serum GH (94 μg/mL), and low IGF-1, which did not rise on hGH administration. Although a low IQ is a frequent finding in LS, mental retardation is rare. The lack of obesity may be explained by existing GH action and by the fact that the proband was delivered prematurely by cesarean section at 37 wk of gestation: birthweight was 1400 g, length was 37.8 cm, and head circumference was 27 cm. One year of treatment with hGH showed that he was GH-resistant.

REFERENCES

1. Laron Z, Pertzelan A, Mannheimer S. Genetic pituitary dwarfism with high serum concentration of growth hormone. A new inborn error of metabolism? Isr J Med Sci 1966;2:152–155.
2. Laron Z, Pertzelan A, Karp M. Pituitary dwarfism with high serum levels of growth hormone. Isr J Med Sci 1968;4:883–894.
3. Eshet R, Laron Z, Brown M, Arnon R. Immunoreactive properties of the plasma hGH from patients with the syndrome of familial dwarfism and high plasma IR-hGH. J Clin Endocrinol Metab 1973;37:819–821.
4. Eshet R, Laron Z, Brown M, Arnon R. Immunological behaviour of hGH from plasma of patients with familial dwarfism and high IR-hGH in a radioimmunoassay system using the cross-reaction between hGH and HCS. Horm Metab Res 1974;6:79–81.
5. Jacobs LS, Sneid DS, Garland JT, Laron Z, Daughaday WH. Receptor-active growth hormone in Laron dwarfism. J Clin Endocrinol Metab 1976;43:403–407.
6. Eshet R, Peleg S, Josefsberg Z, Fuchs C, Arnon R. and Laron Z. Some properties of the plasma hGH activity in patients with Laron-type dwarfism determined by a radioreceptor assay using human liver tissue. Horm Res 1985;22:276–283.
7. Eshet R, Laron Z, Pertzelan A, Dintzman M. Defect of human growth hormone in the liver of two patients with Laron type dwarfism. Isr J Med Sci 1984;20:8–11.
8. Leung DW, Spencer SA, Cachianes G, Hammonds RG, Collins C, Henzel WJ, et al. Growth hormone receptor and serum binding protein: purification, cloning and expression. Nature 1987;330:537–543.
9. Godowski PJ, Leung DW, Meacham LR, Galgani JP, Hellmiss R, Keret R, et al. Characterization of the human growth hormone receptor gene and demonstration of a partial gene deletion in two patients with Laron type dwarfism. Proc Natl Acad Sci USA 1989;86:8083–8087.
10. Elders MJ, Garland JT, Daughaday WH, Fisher DA, Whitney JE, Hughes ER. Laron's dwarfism: Studies on the nature of the defect. J Pediatr 1973;83:253–263.
11. Laron Z, Blum W, Chatelain P, Ranke M, Rosenfeld R, Savage M, et al. Classification of growth hormone insensitivity syndrome. J Pediatr 1993;122:241.
12. Laron Z, Parks, JS, eds. Lessons from Laron Syndrome (LS) 1966–1992. Pediatric and Adolescent Endocrinology, vol. 24. Basel, New York, 1993, pp. 367.
13. Bass SH, Mulkerrin MG, Wells JA. A systematic mutational analysis of hormone-binding determinants in the human growth hormone receptor. Proc Natl Acad Sci USA 1991;88:4498–4502.
14. Spencer SA, Hammonds RG, Henzel WJ, Rodriquez H, Waters MJ, Wood WI. Rabbit liver growth hormone receptor and serum binding protein: purification, characterization and sequence. J Biol Chem 1988;263:7862–7867.
15. Barton DE, Foellmer BE, Woods WI, Francke U. Chromosome mapping of the growth hormone receptor gene in man and mouse. Cytogenet Cell Genet 1989;50:137–141.
16. Kelly PA, Goujon L, Sotiropoulos A, Dinerstein H, Esposito N, Edery M, et al. The GH receptor and signal transduction. Horm Res 1994;42:133–139.
17. Parks JS, Faase E. GH and GH-receptor genes. In: Merimee TJ, Laron Z, eds. Growth Hormone, IGF-I and Growth. New Views of Old Concepts. Modern Endocrinology and Diabetes, vol. 4. Freund Publishing House, London, Tel Aviv, 1996, pp. 23–43.
18. Argetsinger LS, Campbell GS, Yang X, Witthuhn BA, Silvennoinen O, Ihle JN, et al. Identification of JAK2 as a growth hormone receptor-associated tyrosine kinase. Cell 1993;74:237–244.
19. Horseman NS. Editorial: Famine to feast—growth hormone and prolactin signal transducers. Endocrinology 1994;135:1289–1291.
20. Daughaday WH. Are there direct, non-IGF-I-mediated effects of hGH? In: Laron Z, Parks JS, eds. Lessons from Laron Syndrome (LS) 1966–1992. Pediatric and Adolescent Endocrinology, vol. 24. Karger, Basel, 1993, pp. 338–345.
21. de Vos AM, Ultsch M, Kossiakoff AA. Human growth hormone and extracellular domain of its receptor: crystal structure of the complex. Science 1992;255:306–312.
22. Amselem S, Duquesnoy P. Attree O, Novelli G, Bousnina S, Postel-Vinay MC, et al. Laron dwarfism and mutations of the growth hormone-receptor gene. N Engl J Med 1989;321:989–995.

23. Amselem S, Sobrier ML, Duquesnoy P, Rappaport R, Postel-Vinay MS, Gourmelen M, et al. Recurrent nonsense mutations in the growth hormone receptor from patients with Laron dwarfism. J Clin Invest 1991;87:1098–1102.
24. Amselem S, Duquesnoy P, Duriez B, Dastot F, Sobrier ML, Valleix S. Spectrum of growth hormone receptor mutations and associated haplotypes in Laron syndrome. Hum Mol Genet 1993;2:355–359.
25. Meacham WR, Brown MR, Murphy TL, Keret R, Silbergeld A, Laron Z, et al. Characterization of a noncontiguous gene deletion of the growth hormone receptor in Laron's syndrome. J Clin Endocrinol Metab 1993;77:1379–1383.
26. Berg MA, Peoples R, Perez-Jurado L, Guevara-Aguirre J, Rosenbloom AL, Laron Z, et al. Receptor mutations and haplotypes in growth hormone receptor deficiency: a global survey and identification of the Ecuadorean E180splice mutation in an oriental Jewish patient. Acta Paediatr Suppl 1994; 399:112–114.
27. Rosenbloom AL, Berg MA, Kasatkina EP. Volvoka TN, Skorobogatova VF, Sokolovskaya VN, et al. Severe growth hormone insensitivity (Laron syndrome) due to nonsense mutation of the GH receptor in brothers from Russia. J Pediatr Endocrinol Metab 1995;8:159–166.
28. Counts DR, Cutler GB. Growth hormone insensitivity syndrome due to point deletion and frame shift in the growth hormone receptor. J Clin Endocrinol Metab 1995;80:1978–1981.
29. Sobrier ML, Dastot F, Duquesnoy P, Kandemir N, Yordam N, Goossens M, et al. Nine novel growth hormone receptor gene mutations in patients with Laron syndrome. J Clin Endocrinol Metab 1997; 82:435–437.
30. Baumbach L, Schiavi A, Bartlerr R, Perera E, Day J, Brown MR, et al. Clinical, biochemical and molecular investigations of a genetic isolate of growth hormone insensitivity (Laron's syndrome). J Clin Endocrinol Metab 1997;82:444–451.
31. Daughaday WH, Trivedi B. Absence of serum growth hormone binding protein in patients with growth hormone receptor deficiency (Laron dwarfism). Proc Natl Acad Sci USA 1987;84:4636–4640.
32. Baumann G, Shaw MA, Winter RJ. Absence of plasma growth hormone-binding protein in Laron-type dwarfism. J Clin Endocrinol Metab 1987;65:814–816
33. Woods KA, Fraser NC, Postel-Vinay MC, Dusquenoy P, Savage MO, Clark AJL. A homozygous splice site mutation affecting the intracellular domain of the growth hormone (GH) receptor resulting in Laron syndrome with elevated GH-binding protein. J Clin Endocrinol Metab 1996;81:1686–1690.
34. Silbergeld A, Dastot F, Klinger B, Kanety H, Eshet R, Amselem S, et al. Intronic mutation in the growth hormone (GH) receptor gene from a girl with Laron Syndrome and extremely high serum GH binding protein: extended phenotypic study in a very large pedigree. J Pediatr Endocrinol Metab 1997;10:265–274.
35. Kou K, Lajara R, Rotwein P. Amino acid substitutions in the intracellular part of the growth hormone receptor in a patient with the Laron syndrome. J Clin Endocrinol Metab 1993;76:54–59.
36. Duquesnoy P, Sobrier ML, Duriez B, Dastot F, Buchanan CR, Savage MO, et al. A single amino acid substitution in the exoplasmic domain of the human growth hormone (GH) receptor confers familial GH resistance (Laron syndrome) with positive GH-binding activity by abolishing receptor homodimerization. Embo J 1994;13:1386–1395.
37. Duquesnoy P, Sobrier ML, Amselem S, Goossens M. Defective membrane expression of human growth hormone (GH) receptor causes Laron-type GH insensitivity syndrome. Proc Natl Acad Sci USA 1991; 88:10,272–10,276.
38. Laron Z, Klinger B, Eshet R, Kaneti H, Karasik A, Silbergeld A. Laron syndrome due to a post-receptor defect: response to IGF–I treatment. Isr J Med Sci 1993;29:757–763.
39. Herington AC, Ymer S, Stevenson J. Identification and characterization of specific binding proteins for growth hormone in normal human sera. J Clin Invest 1986;77:1817–1823.
40. Baumann G, Stolar MN, Amburn K, Barsano CP, DeVries BC. A specific GH-binding protein in human plasma: Initial characterization. J Clin Endocrinol Metab 1986;62:134–141.
41. Silbergeld A, Lazar L, Erster B, Keret R, Tepper R, Laron Z. Serum growth hormone binding protein activity in healthy neonates, children and young adults—correlation with age, height and weight. Clin Endocrinol 1989;31:295–303.
42. Laron Z Klinger B, Erster B, Silbergeld A. Serum GH binding protein activity identifies the heterozygous carriers for Laron type dwarfism. Acta Endocrinol 1989;121:603–608.
43. Laron Z. Laron type dwarfism (hereditary somatomedin deficiency): a review. In: Frick P, Von Harnack GA, Kochsiek GA, Prader A, eds. Advances in Internal Medicine and Pediatrics, Springer-Verlag, Berlin, 1984, pp. 117–150.

44. Rosenbloom AL, Guevara-Aguirre J, Rosenfeld RG, Fielder PJ. The little women of Loja: growth hormone receptor deficiency in an inbred population of Southern Ecuador. N Engl J Med 1990;323: 1367–1374.
45. Maheshwari HG, Clayton PE, Mughal Z, Price DA, Norman M. Laron type dwarfism: the GH-BP positive phenotype. In: Laron Z, Parks JS, eds. Lessons from Laron Syndrome (LS) 1966–1992. Pediatric and Adolescent Endocrinology, vol. 24. Karger, Basel, 1993, pp. 160–166.
46. Desai M, Choksi C, Colaco P, Ambadkar M. Familial growth hormone deficiency (GHD) and growth hormone insensitivity (GHI) in Indian children. [Abstract #442]. Horm Res 1997;48(Suppl. 2):90.
47. Rosenfeld RG, Rosenbloom AL, Guevara-Aguirre J. Growth hormone (GH) insensitivity due to primary GH receptor deficiency. Endocr Rev 1994;15:369–390.
48. Krzisnik C, Battelino T. Five year treatment with IGF-I of a patient with Laron Syndrome in Slovenia (a follow-up report). J Pediatr Endocrinol Metab 1997;10:443–447.
49. Pertzelan A, Adam A, Laron Z. Genetic aspects of pituitary dwarfism due to absence of biological activity of growth hormone. Isr J Med Sci 1968;4:895–900.
50. Laron Z, Pertzelan A, Karp M, Keret R, Eshet R, Silbergeld A. Laron syndrome—A unique model of IGF-I deficiency. In: Laron Z, Parks JS, eds. Lessons from Laron Syndrome (LS) 1966–1992. Pediatric and Adolescent Endocrinology, vol. 24. Karger, Basel, 1993, pp. 3–23.
51. Laron Z. Prismatic cases: Laron Syndrome (primary growth hormone resistance). From patient to laboratory to patient. J Clin Endocrinol Metab 1995;80:1526–1531.
52. Daughaday WH, Laron Z, Pertzelan A, Heins JN. Defective sulfation factor generation: a possible etiological link in dwarfism. Trans Assoc Am Phys 1969;82:129–138.
53. Laron Z, Pertzelan A, Karp M, Kowadlo-Silbergeld A, Daughaday WH. Administration of growth hormone to patients with familial dwarfism with high plasma immunoreactive growth hormone. Measurement of sulfation factor, metabolic, and linear growth responses. J Clin Endocrinol Metab 1971; 33:332–342.
54. Scharf A, Laron Z. Skull changes in pituitary dwarfism and the syndrome of familial dwarfism with high plasma immunoreactive growth hormone. A roentgenologic study. Horm Metab Res 1972;4:93–97.
55. Konfino R, Pertzelan A, Laron Z. Cephalometric measurements of familial dwarfism and high plasma immunoreactive growth hormone. Am J Orthod 1975;68:196–201.
56. Laron Z, Klinger B, Grunebaum M. Laron type dwarfism. Special feature—Picture of the month. Am J Dis Child 1991;145:473–474.
57. Abramovici A, Josefsberg Z, Mimouni M, Liban E, Laron Z. Histopathological features of the skin in hypopituitarism and Laron-type dwarfism. Isr J Med Sci 1983;19:515–519.
58. Laron Z, Klinger B. Body fat in Laron syndrome patients: Effect of insulin-like growth factor treatment. Horm Res 1993;40:16–22.
59. Laron Z, Lilos P, Klinger B. Growth curves for Laron syndrome. Arch Dis Child 1993;68:768–770.
60. Laron Z, Sarel R. Penis and testicular size in patients with growth hormone insufficiency. Acta Endocrinol 1970;63:625–633.
61. Laron Z, Sarel R, Pertzelan A. Puberty in Laron type dwarfism. Eur J Pediatr 1980;134:79–83.
62. Pertzelan A, Lazar L, Klinger B, Laron Z. Puberty in 15 patients with Laron Syndrome: a longitudinal study. In: Laron Z, Parks JS, eds. Lessons from Laron Syndrome (LS) 1966–1992. Pediatric and Adolescent Endocrinology, vol. 24. Karger, Basel, 1993, pp. 27–33.
63. Laron Z, Klinger B. IGF-I treatment of adult patients with Laron syndrome: preliminary results. Clin Endocrinol 1994;41:631–638.
64. Galatzer A, Aran O, Nagelberg N, Rubitzek J, Laron Z. Cognitive and psychosocial functioning of young adults with Laron syndrome. In: Laron Z, Parks JS, eds. Lessons from Laron Syndrome (LS) 1966–1992. Pediatric and Adolescent Endocrinology, vol. 24. Karger, Basel, 1993, pp. 53–60.
65. Brat O, Ziv I, Klinger B, Avraham M, Laron Z. Muscle force and endurance in untreated and human growth hormone or insulin-like growth factor-I-treated patients with growth hormone deficiency or Laron Syndrome. Horm Res 1997;47:45–48.
66. Keret R, Pertzelan A, Zeharia A, Zadik Z, Laron Z. Growth hormone (hGH) secretion and turnover in three patients with Laron type dwarfism. Isr J Med Sci 1988;24:75–79.
67. Laron Z, Pertzelan A, Doron M, Assa S, Keret R. The effect of dihydrosomatostatin in dwarfism with high plasma immunoreactive growth hormone. Horm Metab Res 1977;9:338–339.
68. Laron Z, Klinger B, Blum WF, Silbergeld A, Ranke MB. IGF binding protein 3 in patients with LTD: effect of exogenous rIGF-I. Clin Endocrinol 1992;36:301–304.

69. Laron Z, Suikkari AM, Klinger B, Silbergeld A, Pertzelan A, Seppala M, et al. Growth hormone and insulin-like growth factor regulate insulin-like growth factor binding protein in Laron type dwarfism, growth hormone deficiency and constitutional growth retardation. Acta Endocrinol 1992;127:351–358.
70. Silbergeld A, Klinger B, Schwartz H, Laron Z. Serum prolactin in patients with Laron-type dwarfism: effect of insulin-like growth factor I. Horm Res 1992;37:160–164.
71. Laron Z. (1997) Hypoglycemia due to hormone deficiencies. J Pediatr Endocrinol Metab 1998;11: 117–120.
72. Laron Z, Avitzur Y, Klinger B. Insulin resistance in Laron Syndrome (primary insulin-like growth factor-I [IGF-I] deficiency) and effect of IGF-I replacement therapy. J Pediatr Endocrinol Metab 1997;10(Suppl 1):105–115.
73. Niwa M, Sato Y, Saito Y, Uchiyama F, Ono H, Yamashita M, et al. Chemical synthesis, cloning and expression of genes for human somatomedin C (insulin like growth factor I) and 59Val somatomedin C. Ann NY Acad Sci 1986;469:31–52.
74. Eshet R, Dux Z, Silbergeld A, Koren R, Klinger B, Laron Z. Erythrocytes from patients with low serum concentrations of IGF-I have an increase in receptor sites for IGF-I. Acta Endocrinol (Copenh.) 1991; 125:354–358.
75. Eshet R, Klinger B, Silbergeld A, Laron Z. Modulation of insulin like growth factor I (IGF-I) binding sites on erythrocytes by IGHF-I treatment in patients with Laron syndrome (LS). Regul Pep 1993;48: 233–239.
76. Laron Z, Klinger B, Erster B, Anin S. Effects of acute administration of insulin like growth factor I in patients with Laron-type dwarfism. Lancet 1988;ii:1170–1172.
77. Laron Z, Klinger B, Silbergeld A, Lewin B, Erster B, Gil-Ad I. Intravenous administration of recombinant IGF-I lowers serum GHRH and TSH. Acta Endocrinol 1990;123:378–382.
78. Klinger B, Garty M, Silbergeld A, Laron Z. Elimination characteristics of intravenously administered rIGF-I in Laron type dwarfs (LTD). Dev Pharmacol Ther 1990;15:196–199.
79. Takano H, Hizuka N, Asakawa K, Sukegawa J, Shizume K. Effects of s.c. administration of recombinant human insulin like growth factor-I (IGF-I) on normal human subjects. Endocrinol Jpn 1990;37: 309–317.
80. Gil-Ad I. Koch, Y. Silbergeld A, Dickerman Z, Kaplan B, Weizman A. et al. Differential effect of insulin-like growth factor I (IGF-I) and growth hormone (GH) on hypothalamic regulation of GH secretion in the rat. J Endocrinol Invest 1996;19:542–547.
81. Laron Z, Klinger B, Jensen LT, Erster B. Biochemical and hormonal changes induced by one week of administration of rIGF-I to patients with Laron type dwarfism. Clin Endocrinol 1991;35:145–150.
82. Laron Z, Anin S, Klipper-Aubach Y, Klinger B. Effects of insulin-like growth factor on linear growth, head circumference and body fat in patients with Laron-type dwarfism. The Lancet 1992;339:1258–1261.
83. Walker JL, Van Wyk JJ, Underwood LE. Stimulation of statural growth by recombinant insulin-like growth factor I in a child with growth hormone insensitivity syndrome (Laron type). J Pediatr 1992;121: 641–646.
84. Klinger B, Jensen LT, Silbergeld A, Laron Z. Insulin-like growth factor-I raises serum procollagen levels in children and adults with Laron syndrome. Clin Endocrinol 1996;45:423–429.
85. Klinger B, Laron Z. Renal function in Laron syndrome patients treated by insulin-like growth factor-I. Pediatr Nephrol 1994;8:684–688.
86. Kaneti H, Karasik A, Klinger B, Silbergeld A, Laron Z. Long-term treatment of Laron type dwarfs with insulin-like growth factor I increases serum insulin-like growth factor—binding protein 3 in the absence of growth hormone activity. Acta Endocrinol 1993;128:144–149.
87. Kaneti H, Silbergeld A, Klinger B, Karasik A, Baxter RC, Laron Z. Long-term effects of insulin-like growth factor (IGF)-I on serum IGF-I, IGF-binding protein-3 and acid labile subunit in Laron syndrome patients with normal growth hormone binding protein. Eur J Endocrinol 1997;137:626–630.
88. Ranke MB, Savage MO, Chatelain PG, Preece MA, Rosenfeld RG, Blum WF, et al. Insulin-like growth factor I improves height in growth hormone insensitivity: Two years' results. Horm Res 1995; 44:253–264.
89. Backeljauw PF, Underwood LE, the GHIS Collaborative Group. Prolonged treatment with recombinant insulin-like growth factor-I in children with growth hormone insensitivity syndrome—a clinical Research Center study. J Clin Endocrinol Metab 1996;81:3312–3317.
90. Klinger B, Laron Z. Three year IGF-I treatment of children with Laron Syndrome. J Pediatr Endocrinol Metab 1995;8:149–158.

91. Guevara-Aguirre J, Rosenbloom AL, Vasconez O, Martinez V, Gargosky SE, Allen L, et al. Two year treatment of growth hormone (GH) receptor deficiency with recombinant insulin-like growth factor-I in 22 children: comparison of two dosage levels and to GH treated GH deficiency. J Clin Endocrinol Metab 1997;82:629–633.
92. Guevara-Aguirre J, Vasconez O, Martinez V, Martinez AL, Rosenbloom AL, Diamond FB Jr. et al. A randomized, double blind, placebo-controlled trial on safety and efficacy of recombinant human insulin-like growth factor-I in children with growth hormone receptor deficiency. J Clin Endocrinol Metab 1995;80:1393–1398.
93. Laron Z, Klinger B. Are patients with Laron syndrome candidates for limb lengthening? In: Laron Z, Mastragostino S, Romano C, Boero S, Cohen A. eds. Limb Lengthening: For Whom, When and How?, Freund Publishing House, London, 1995, pp. 79–91.
94. Amselem S, Sobrier M-L, Dastot F, Duquesnoy P, Duriez B, Goossens M. Molecular basis of inherited growth hormone resistance in childhood. In: Ross RJM, Savage MO, Guest eds. Growth Hormone Resistance. Balliere's Clin Endocrinol Metab Int Pract Res 1996;10:353–369.
95. Berg MA, Argente J, Chernausek S, Gracia R, Guevara-Aguirre J, Hopp M, et al. Diverse growth hormone receptor gene mutations in Laron syndrome. Am J Hum Genet 1993;52:998–1005.
96. Berg MA, Guevara-Aguirre J, Rosenbloom AL, Rosenfeld RG, Francke U. Mutation creating a new splice site in the growth hormone receptor genes of 37 Ecuadorean patients with Laron syndrome. Hum Mutat 1992;1:124–134.
97. Woods KA, Camach-Hubner C, Savage MO, Clark AJL. Intrauterine growth retardation and postnatal deletion of the insulin-like growth factor I gene. N Engl J Med 1996;335:1363–1367.

3 Pseudohypoparathyroidism

Ali Al-Zahrani, MD, Michael A. Levine, MD, and William F. Schwindinger, MD, PhD

CONTENTS

HISTORICAL OVERVIEW
CLINICAL FEATURES
GENETICS
MOLECULAR PATHOPHYSIOLOGY
DIAGNOSIS
TREATMENT
CONCLUSION
REFERENCES

HISTORICAL OVERVIEW

In 1942, Fuller Albright and his associates described three patients who presented with tetany and seizures, and were found to have hypocalcemia *(1)*. These clinical and biochemical features were consistent with the diagnosis of hypoparathyroidism. However, unlike patients with parathyroid hormone (PTH) deficiency, these patients failed to show a rise in serum calcium or urinary phosphate in response to injected parathyroid extract *(1)*. Albright speculated that this condition was owing to target organ (bone and kidney) resistance to PTH action, and dubbed this syndrome pseudohypoparathyroidism (PHP). This was the first description of hormone resistance in humans.

In addition to the clinical and biochemical features directly related to resistance to PTH action, the patients described by Albright had unusual skeletal and developmental manifestations. These included obesity, short stature, brachydactyly, heterotopic ossifications, and dental hypoplasia. In later studies, other subjects were found who had similar developmental abnormalities, but who lacked the hormone resistance present in their relatives with PHP *(2)*. Albright termed this condition pseudo-pseudohypoparathyroidism (pseudoPHP). This constellation of skeletal and developmental features has subsequently been referred to as Albright hereditary osteodystrophy (AHO). Although much has been learned in the ensuing years about the basis for hormone resistance in PHP, the pathophysiologic basis for AHO remains unclear.

Albright's hypothesis that the defect in PHP is in the target organ was strongly supported by the finding that serum PTH concentrations are elevated in patients with PHP

From: *Contemporary Endocrinology: Hormone Resistance Syndromes*
Edited by: J. L. Jameson © Humana Press Inc., Totowa, NJ

(3). In the late 1960s, Chase et al. showed that the infusion of PTH in normal subjects or patients with hypoparathyroidism increases urinary excretion of nephrogenous cAMP and phosphate. In contrast, infusion of PTH in patients with PHP produces markedly blunted or absent urinary cAMP and phosphate responses *(4)*. Later investigations showed that administration of dibutyryl cAMP, a synthetic analog of cAMP, produced a normal phosphaturic response in most patients with PHP *(5)*, indicating that failure to generate cAMP in the kidney was the defect in these patients. PHP in which the urinary cAMP response to PTH infusion is blunted is classified as PHP type I. Insight into the molecular basis for reduced renal cAMP response to PTH came with the observation that addition of GTP to a preparation of kidney membranes from a subject with PHP type I corrected the biochemical defect *(6)*. Shortly thereafter, it was demonstrated that biological activity of the stimulatory G-protein of adenylyl cyclase (G_s) was reduced by 50% in patients with PHP type I who also had AHO *(7,8)*. This form of PHP is classified as PHP type Ia, or AHO with hormone resistance. PHP type Ib describes a distinct group of patients who have hypocalcemia and a blunted increase in urinary cAMP in response to PTH, but who lack features of AHO and have normal levels of G_s. PHP type Ic applies to other patients who have AHO, but in whom biochemical evidence is consistent with a defect in the adenylyl cyclase enzyme itself *(9)*. Still other patients with PHP show a normal urinary cAMP response to PTH, but have little or no phosphaturic response *(10)*. These patients are classified as PHP type II, and likely have a biochemical defect distal to the PTH-directed production of nephrogenous cAMP (Fig. 1).

CLINICAL FEATURES

AHO

Subjects with AHO have skeletal defects which include short stature, a round face, and brachydactyly, characterized by shortening of the metacarpals, metatarsals and the distal phalanx of the thumb (Fig. 2). One or more subcutaneous ossifications are present in a subset of subjects with AHO. Most subjects with AHO are obese. Many subjects with AHO have cognitive deficits, and a variety of neuro-sensory deficits have been described in subjects with AHO.

Skeletal Deformities

Short stature is prominent feature of AHO, and 2/3 of children and 80% of adults are below the 10th percentile for height. This results from disproportionate shortening of the limbs, such that arm span is less than height in the majority of patients with AHO. Other deformities of the long bones include short ulna, bowed radius, deformed elbow, and varus or valgus deformities of the elbows, hips, or knees *(11)*. Deformities of the long bones are not present at birth and may not be evident before 5 yr of age *(12)*. The bony abnormalities become evident only as the skeleton grows. Bone age is usually advanced 2–3 yr. Skeletal deformities of the head and neck lead to a round face, a flattened bridge of the nose, and a short neck. X-rays of the skull commonly reveal hyperostosis frontalis interna, thick calvarium, or widened spongy diploe. Head circumference is above the 90th percentile in nearly half of children with AHO *(11)*. Dental abnormalities are common and include dentin and enamel hypoplasia, short roots, and delayed or absent tooth eruption *(13)*.

Brachydactyly is a common, but not universal feature of AHO. A careful study of the pattern of shortening of the bones has been made *(14)*. In this survey of 79 subjects with

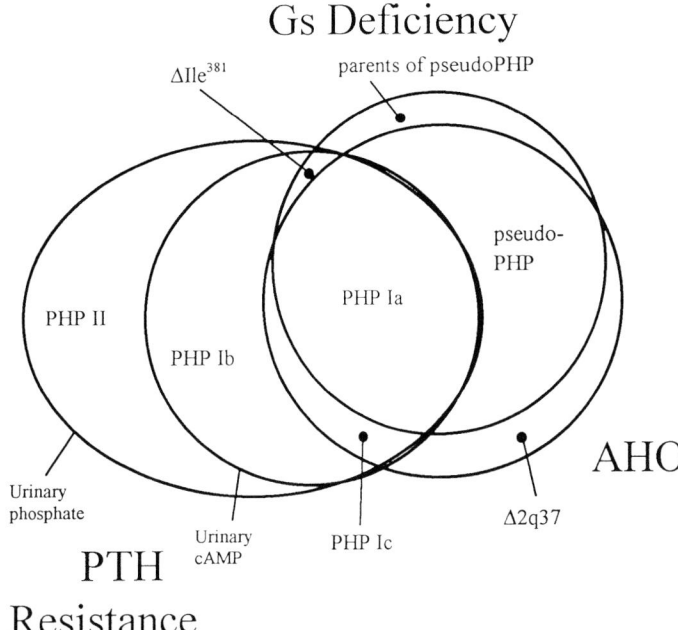

Fig. 1. Parathyroid hormone resistance syndromes. Venn diagram illustrates the various forms of PHP. Syndromes are compared based on four criteria: response of urinary phosphate to PTH infusion, response of urinary cAMP to PTH infusion, presence of features of AHO, and presence of mutations in the *GNAS1* gene that lead to deficiency of $G_s\alpha$. Subjects with PHP type Ia exhibit all of these features. Subjects with PHP type Ic lack only mutations in $G_s\alpha$. Subjects with PHP type Ib have renal resistance to PTH without features of AHO or mutations in $G_s\alpha$. Subjects with pseudoPHP have *GNAS1* gene mutations and AHO without resistance to hormones. A phenocopy of the AHO syndrome has been described with a deletion in chromosome 2q37. Some members of AHO kindreds have mutations in *GNAS1* with no apparent features of AHO or hormone resistance. One kindred has been described with features of PHP type Ib, testotoxicosis, and a 3-bp deletion in $G_s\alpha$.

PHP or pseudoPHP, approx 75% of patients had shortening of one or more metacarpals. The fourth metacarpal was the most commonly and most severely shortened, followed by the fifth. Shortening of the distal phalanx of the thumb occurred in approx 70% of patients. Clinically, metacarpal shortening can be recognized as Archibald's sign, a dimpling of the skin rather than the expected knuckle above an affected metacarpal. Radiographically, shortening of the 4th metacarpal leads to the metacarpal sign, which occurs when a tangent drawn to the heads of the 4th and 5th metacarpals intercepts the 3rd metacarpal. Shortening of the distal phalanx of the thumb can be recognized when the nail plate is broader than it is long. Shortening of metatarsals also occurs in AHO. Brachydactyly may be asymmetric and may affect only the feet.

Heterotopic ossification occurs in one-quarter to one-half of patients with AHO. The most common site of involvement is the skin and subcutaneous tissue. Osteosis cutis presents as multiple, bluish, hard papules or nodules (pinpoint to 5 cm), and often is the first sign of AHO *(15)*. Biopsy reveals spicules of mature bone lined by osteoblasts, partially mineralized osteoid, and amorphous calcium deposits. Multiple osteomas are strongly suggestive of AHO. However, when solitary or localized osteomas occur, other causes of primary osteosis cutis must be considered, including trauma and idiopathic syndromes: primary osteomas of the distal extremities, multiple facial osteomas, and

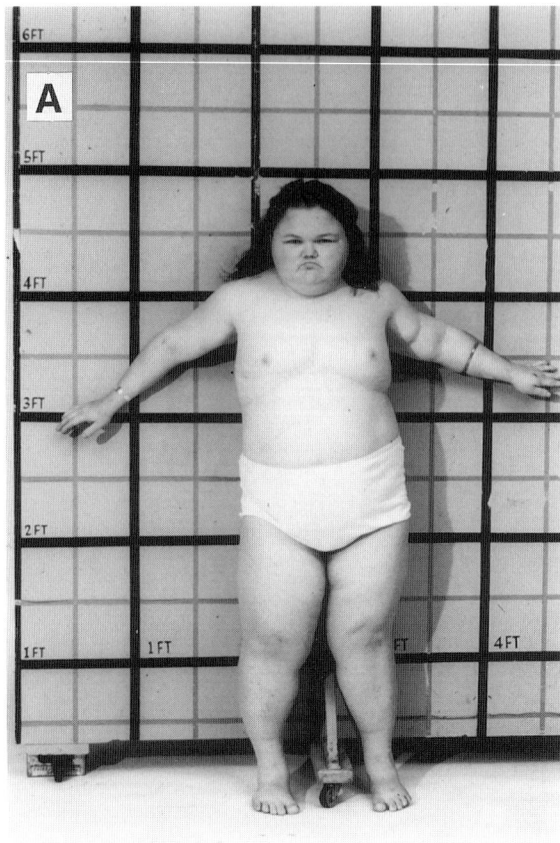

Fig. 2. Albright hereditary osteodystrophy. (**A**) Typical appearance of a patient with AHO; note short stature, disproportionate shortening of the limbs, round face, and obesity.

diffuse multiple osteomas *(16)*. Secondary causes of osteosis cutis include tumors, inflammatory processes, and myositis ossificans progressiva. True ossification must also be distinguished from other causes of cutaneous calcification, including metastatic calcification, which occurs when the serum calcium and phosphate are elevated, and dystrophic calcifications, which occur in the presence of normal levels of calcium and phosphate *(16)*.

OBESITY

Obesity is a common feature of AHO. About half of children and one-quarter of adults with PHP type Ia or pseudoPHP are above the 90th percentile for weight *(11)*. One study has demonstrated reduced stimulation of adenylyl cyclase by the β-adrenergic agonist isoproterenol in membranes prepared from abdominal wall fat of four subjects with PHP type Ia in comparison to fat from 12 unaffected subjects *(17)*.

ASSOCIATED NEUROLOGIC MANIFESTATIONS

Neurosensory abnormalities have been described in patients with PHP type Ia. These include olfactory *(18)*, gustatory *(19)*, and auditory defects *(20)*. The relationship between these disturbances and $G_s\alpha$ deficiency is unclear, since signal transduction for olfaction and taste is mediated by unique G-protein α-subunits that are present in neurolfactory epithelium, $G_{olf}\alpha$ *(21)*, and in taste buds, gustducin *(22)*. More recent

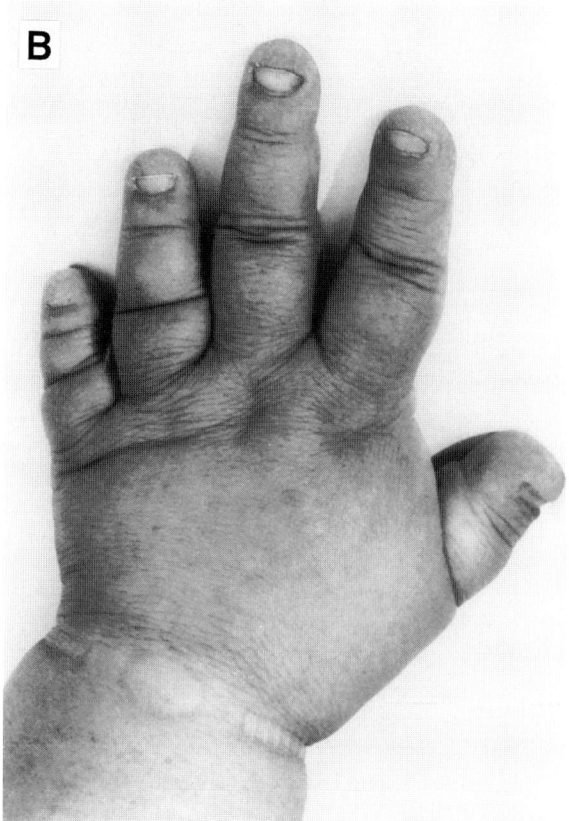

Fig. 2. *(Continued)* (**B**) Brachydactyly of the hand; note thumb sign and shortening of 4th and 5th digits.

studies have demonstrated that similar hearing *(23)* and olfactory *(24)* deficits occur in subjects with other causes of chronic hypocalcemia that are unrelated to deficiency of $G_s\alpha$.

Seizures as well as pseudoseizures related to generalized tetany can occur in subjects with PHP *(25–28)*. However, it is unclear if seizures are more common in PHP than in other disorders associated with hypocalcemia. Hypocalcemia lowers the seizure threshold, and can exacerbate an underlying seizure disorder or induce seizures in susceptible individuals. Pseudoseizures, which represent generalized hypocalcemic tetany, are commonly mistaken for true seizures.

Mild to moderate mental retardation is present in one-half to three-quarters of patients with PHP type Ia. Cognitive deficits are associated with $G_s\alpha$ deficiency in PHP type Ia, but do not occur in patients with PHP Ib *(29)*. Hypothyroidism may also contribute to abnormal cognitive development; unfortunately, treatment of hypothyroidism from birth does not prevent cognitive deficits.

PHP Type Ia

PTH resistance is the hallmark of PHP type Ia. Resistance to other hormones also occurs in these patients, however. Hypothyroidism *(30)* and hypogonadism *(31)* are the most common additional endocrine disturbances in patients with PHP type Ia. It seems likely that the generalized deficiency of $G_s\alpha$ and consequent impairment of receptor-activated generation of the intracellular second messenger cAMP account for resistance to multiple hormones. However, clinically significant resistance does not occur to all hormones that act by stimulating cAMP. For example, diabetes insipidus as a consequence of vasopressin resistance has not been described in subjects with PHP type Ia, and the urinary cAMP response to vasopressin infusion is normal in subjects with PHP type Ia *(32)*. Patients with PHP type Ia have normal morning serum cortisol levels and a normal response to infusion of ACTH *(33)*, although a few cases of adrenal insufficiency have been described in PHP type Ia *(34,35)*. Hepatic glucose response to glucagon is normal, and hypoglycemia is not a feature of PHP type Ia, despite reduced generation of cAMP in response to glucagon infusion *(33,36)*.

Primary hypothyroidism owing to thyroid-stimulating hormone (TSH) resistance occurs in most patients with PHP type Ia *(33)*. Patients usually present with a low serum thyroxine level, an elevated serum TSH level, and an exaggerated TSH response to thyrotrophin-releasing hormone (TRH). Patients with PHP type Ia do not have goiter or antithyroid antibodies. Hypothyroidism may affect the newborn, and may even precede the development of hypocalcemia. Early treatment with thyroid hormone does not prevent the development of mental retardation.

Hypogonadism is also common in patients with PHP type Ia. Women may have delayed puberty, oligomenorrhea, and infertility *(33)*. Plasma gonadotropin levels may be normal or elevated, and the response to GnRH infusion is usually normal *(37)*. In men, serum testosterone may be normal or low, testes may not descend normally, and fertility is usually decreased.

PHP Type Ib

PHP type Ib is characterized by resistance to the renal actions of PTH. Subjects with PHP type Ib do not have resistance to other hormones that act by stimulating cAMP, nor do they have AHO. Features of hyperparathyroid bone disease are frequently associated with PHP type Ib, including osteopenia, subperiosteal bone resorption, and occasionally osteitis fibrosa cystica *(38)*. The molecular basis for PHP type Ib is unknown. Subjects with PHP type Ib have normal levels of $G_s\alpha$, and mutations in the GNAS1 gene have not been detected in subjects with this disorder. Mutations have also not been found in either of the two known PTH receptors. No significant mutations that are likely to be disease-causing are present in the coding sequences *(39,40)*, intron–exon boundaries *(39)*, or promoter region *(41)* of the PTH/PTHrP receptor.

Other Forms of PHP

Subjects with PHP type Ic have PTH resistance and AHO, but do not have deficiency of $G_s\alpha$. Biochemical evidence suggests a defect in the adenylyl cyclase enzyme itself in PHP type Ic *(9)*. Subjects with PHP type II show a normal urinary cAMP response to PTH, but have little or no phosphaturic response *(10)*. Subjects with PHP type II likely have a biochemical defect in the PTH-responsive signal transduction pathway that is distal to the production of cAMP.

GENETICS

Because they typically occur as sporadic conditions, little is known about the genetics of PHP type II and PHP type Ic. Genetic studies in several large kindreds indicates that PHP type Ib is inherited in an autosomal-dominant disorder, although the gene defect remains unknown *(42)*. AHO is inherited in an autosomal-dominant manner, consistent with its linkage to the GNAS1 gene located on chromosome 20q13. However, two non-Mendelian features have been noted in the inheritance of AHO that may provide clues to the basis for the distinctive phenotypes of PHP type Ia and pseudoPHP. In general, AHO patients who inherit a *GNAS1* mutation from their mother have hormone resistance, but those who receive the mutation from their father do not. This has led to the suggestion that parental imprinting of the *GNAS1* gene is responsible for the different phenotypes of PHP type Ia and pseudoPHP *(43)*. However, examples of AHO patients with hormone resistance who inherited the *GNAS1* mutation from their father have been reported *(44)*, and both alleles of *GNAS1* are expressed in a wide variety of human fetal tissues *(45)*. Studies in mice have demonstrated expression of the *Gnas* gene is modified by parental imprinting in renal glomerular tufts, but not in PTH- responsive renal tubules *(46)*. The second non-Mendelian feature noted in the transmission of AHO is reminiscent of genetic anticipation, i.e., the phenotype gets worse in succeeding generations. Identification of the *GNAS1* gene defect in affected members of several large multigenerational families has permitted characterization of transmission of the mutant allele. These studies have shown that the index case has the most severe phenotype (i.e., PHP type Ia), whereas the index case's affected mother has a less severe phenotype (i.e., pseudoPHP). Although molecular studies typically show that the index case's maternal grandfather has the same mutation, features of AHO will be subtle or absent in this subject.

MOLECULAR PATHOPHYSIOLOGY

PTH resistance is the biological hallmark of PHP. Typically, PTH resistance reflects an inability of the hormone to stimulate cAMP synthesis in the renal tubule, although in some patients, PTH resistance is manifest as a deficient phosphaturic response to nephrogenous cAMP. A discussion of the molecular pathophysiology of PHP thus must include a brief description of the receptors, G-proteins, and signal effector molecules that constitute the PTH-responsive signal transduction pathway.

PTH Receptors

PTH regulates calcium homeostasis in the adult. The related polyhormone PTH-related peptide (PTHrP) was initially recognized as the basis for humoral hypercalcemia of malignancy, but subsequent studies have shown that its paracrine actions in skeletal development *(47–49)* and endocrine actions in calcium homeostasis in the fetus are even more important. The region of PTHrP that is responsible for maternal–fetal calcium transport is distinct from the region that modulates humoral hypercalcemia of malignancy *(50)*, suggesting that PTHrP may be a precursor for more than one peptide hormone. PTH and PTHrP share sequence homology only at the amino-terminus, in a region important for binding to a common PTH/PTHrP receptor. The PTH/PTHrP receptor is believed to be the principal PTH receptor in kidney and bone *(51)*. A second receptor, the PTH2 receptor, also binds PTH, but does not bind PTHrP and is not expressed in the kidney *(52)*. Both receptors are members of a large family of G-protein-coupled receptors that share a common predicted topology. Each receptor consists of an extracellular amino-terminus, an intracellular carboxy-terminus, and seven α-helical, transmembrane-spanning domains (7TMS). The PTH/PTHrP receptor is a member of a smaller subfamily that includes receptors for glucagon, GLP-I, GIP, GHRH, VIP, secretin, PACAP, calcitonin, and CRF *(52)*. The 7TMS receptors can interact with more than one G-protein to

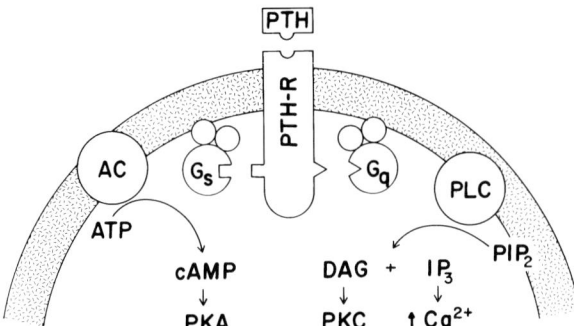

Fig. 3. Intracellular signaling pathways for PTH. Binding of PTH to the PTH/PTHrP receptor (PTH-R activates two classes of heterotrimeric G-proteins: G_s and G_q. Signaling via G_s activates adenylyl cyclase (AC) to produce the intracellular second messenger cAMP, which in turn activates protein kinase (PKA). Signaling via members of the G_q family activates β isoforms of phospholipase C (PLC) and produces two additional intracellular second messengers: inositol triphosphate (IP_3), which releases calcium from intracellular stores, and diacylglycerol (DAG), which activates protein kinase C.

regulate different signal-generating effectors. Activation of the PTH/PTHrP receptor leads to the stimulation of both adenylyl cyclase by G_s and phospholipase C β by members of the G_q family *(53)* (Fig. 3).

Heterotrimeric G-proteins

Heterotrimeric guanine nucleotide binding proteins (G-proteins) couple 7TMS cell-surface receptors for diverse signals to effector molecules that produce intracellular second messengers. G-proteins share a common structure consisting of an α-subunit and a tightly coupled βγ-dimer. The α-subunit interacts with receptor and effector molecules, binds GTP, and possesses intrinsic GTPase activity. At least 16 genes encode mammalian α-subunits; additional protein diversity results from the generation of alternatively spliced mRNAs. The α-subunits are classified into four families: the G_s family ($G_s\alpha$, $G_{olf}\alpha$) associated with the stimulation of adenylyl cyclase, the G_i family ($G_{i1}\alpha$, $G_{i2}\alpha$, $G_{i3}\alpha$, $G_o\alpha$, $G_z\alpha$, rod transducin, cone transducin, gustducin) associated with the inhibition of adenylyl cyclase and with regulation of other effectors, the G_q family ($G_q\alpha$, $G_{11}\alpha$, $G_{14}\alpha$, $G_{15/16}\alpha$) associated with stimulation of phospholipase C β, and the G_{12} family ($G_{12}\alpha$, $G_{13}\alpha$) for which second messengers are unknown. The α-subunits associate with smaller groups of at least 5 β-subunits and 10 γ-subunits. Although this leads to a staggering number of possible heterotrimeric combinations, the situation is somewhat simplified by tissue-specific expression of some subunits and by preferential interactions between subunits. The β- and γ-subunits combine preferentially with one another *(54)*, and the resulting βγ-dimers demonstrate preferred associations with specific α-subunits *(55)*. This combinatorial specificity in the associations between G-protein subunits contributes to the specificity of the interactions of G-protein heterotrimers with receptors *(56)*.

The G-protein α-subunit functions as a molecular switch. In the basal (nonstimulated) state, a G-protein exists in a heterotrimeric form with GDP bound to the α-subunit. On receptor activation, the α-subunit is "turned on" by a conformational change that facilitates the exchange of bound GDP for GTP, with subsequent dissociation of the α-subunit from the receptor and βγ-dimer. The free α- and βγ-subunits are now able to interact with effector enzymes and ion channels to regulate the production of intracellular second messengers. The interaction with effector molecules is "turned off" by hydrolysis of GTP to GDP by the endogenous GTPase activity of the α-subunit. This GTPase activity is slow for an enzymatic process, allowing for prolonged activation of the effector. For some α-subunits, the GTPase activity is accelerated by interaction with effector molecules, e.g., phospholipase C β and cGMP phosphodiesterase. For other α-subunits, GTPase activity is accelerated by interaction with a regulator of G-protein signaling protein (RGS protein) *(57)*. However, no native RGS protein has been identified that accelerates the rate of GTPase activity for $G_s\alpha$.

G-Protein Structure

New insights into G-protein function have come with the recent solution of the X-ray crystal structures of the $G_{i1}\alpha$ subunit bound to GDP (the resting conformation) (58), bound to GDP and AlF_4^- (which approximates the transition state) *(59)*, and bound to the nonhydrolyzable GTPγS (the active conformation) *(59)*. The crystal structure of the $\alpha_{i1}\beta_1\gamma_2$ heterotrimer has also been solved *(60)*. A comparison of these four structures of $G_{i1}\alpha$ provides a picture of a G-protein α-subunit in action *(61)*. The crystal structure of the heterotrimer (Fig. 4) shows the amino-terminus of the α-subunit is in extended contact with the βγ dimer, and positioned for insertion of lipid modification into the plasma membrane. The switch regions (so named because they change conformation with GTP hydrolysis), which interact with effector molecules, are buried by the interaction with βγ-subunit. In the GTP-bound form, the amino- and carboxy-termini are disordered, a structural parallel to their functional inability to interact with βγ or receptor in this conformation. In the AlF_4^--bound form, an arginine (corresponding to Arg^{201} of $G_s\alpha$) has swung into the catalytic site and stabilizes the transition state during GTP hydrolysis. In the GDP-bound form, the amino- and carboxy-termini form an ordered structure around a sulfate molecule in the crystal (perhaps a surrogate for a released γ phosphate that might stabilize the amino- and carboxy-termini before they interact with the βγ-dimer and receptor).

The Stimulatory G-proteins and Adenylyl Cyclase

The α-subunit of the stimulatory G-protein of adenylyl cyclase, $G_s\alpha$, is encoded by the *GNAS1* gene located on human chromosome 20q13.1-13.2 *(62)*. In addition to its role in stimulating adenylyl cyclase, $G_s\alpha$ may directly stimulate the opening of calcium channels *(63)*. *GNAS1* contains 13 coding exons *(64)* and can be alternatively spliced to yield four different protein isoforms with similar biological activities *(65)*. Moreover, there are at least two alternative first exons that produce transcripts of unrecognized function. One encodes a truncated, presumably nonfunctional $G_s\alpha$ protein that uses an initiator ATG sequence in exon 2 *(66)*. Another encodes a larger (92-kDa) protein, termed $XL_s\alpha$, that is produced by the fusion of a novel 51-kDa domain to the protein domain encoded by exons 2–13 of $G_s\alpha$. $XL_s\alpha$ is associated with the *trans*-Golgi apparatus *(67)*.

The stimulatory G-proteins, G_s and G_{olf}, can activate all forms of adenylyl cyclase. There are at least nine distinct membrane-bound isoforms of the adenylyl cyclase enzyme (types I–IX), each with a unique pattern of tissue distribution and biochemical characteristics. Activity of adenylyl cyclase is regulated by both G-protein-dependent and G-protein-independent mechanisms *(68,69)*. Activation of receptors coupled to $G_s\alpha$ (e.g., PTH/PTHrP, TSH, luteinizing hormone [LH], or β-adrenergic receptors) or $G_{olf}\alpha$ (e.g., odorant receptors) leads to stimulation of adenylyl cyclase with increased synthesis

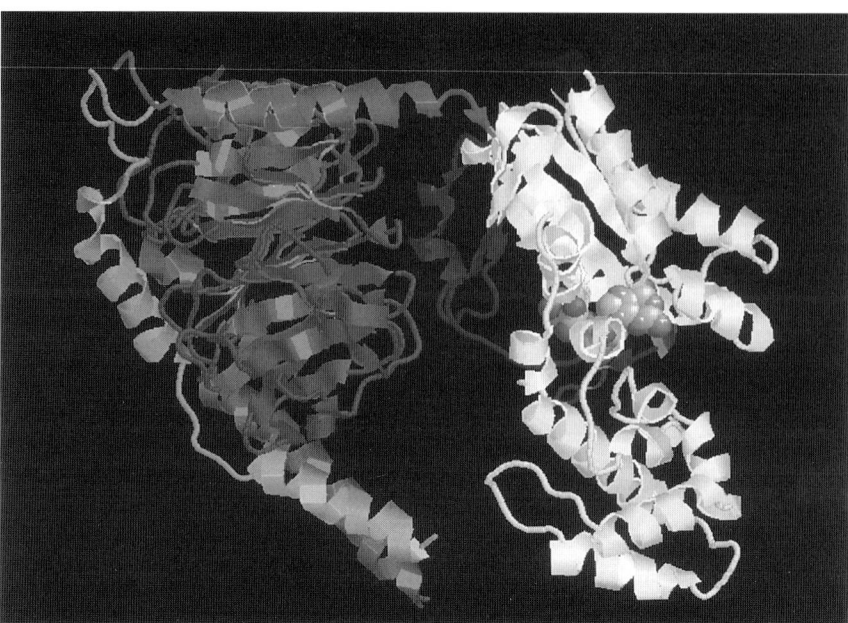

Fig. 4. Ribbon diagram of a heterotrimeric G-protein. The α-subunit is shown in white, with the bound GDP depicted as a space-filling model. The β-subunit is shown in gray, and the γ-subunit in light gray. The amino-terminus of the α-subunit is shaded and seen to be in contact with the β-subunit. The switch regions of the α are shaded and are buried by contact with the β-subunit. The illustration was produced with RasMol (a public domain program for molecular visualization by R. Sayle) based on the crystal structure of the Gαi1β1γ2 heterotrimer (1GP2) deposited in the Protein Data Base (www.pdb.bnl.gov) by M. A. Wall and S. R. Sprang.

of cAMP. Activation of receptors that are coupled to members of the $G_i\alpha$ family (e.g., muscarinic receptors) leads to inhibition of most forms of adenylyl cyclase activity, with the exception of type II. βγ-dimers released by activation members of the G_i family inhibit the activity of some types of adenylyl cyclase (type I) but potentiate the stimulation of other types of adenylyl cyclase (type II and type IV) by $G_s\alpha$. Additional complexity derives from the observation that activity of several forms of adenylyl cyclase (type VII, type II, and possibly type V) are stimulated by protein kinase C, which is activated via the $G_q\alpha$-coupled phospholipase C signaling pathway. Some forms of adenylyl cyclase (types I, III, and VIII) are stimulated by calcium calmodulin, but others (types II, IV, V, and VI) are not. Thus, adenylyl cyclase activity is determined by a complex and coordinate interplay between multiple G-protein subunits and other regulators.

Role of $G_s\alpha$ Mutations in Human Disease

INACTIVATING MUTATIONS OF $G_s\alpha$

A number of different mutations have been identified in the *GNAS1* gene in patients with AHO. These include missense and nonsense mutations, mutations in sequences necessary for correct splicing, and small deletions or insertions (Table 1). Most of these mutations lead to decreased levels of $G_s\alpha$ protein. The mechanism by which one such mutation, $Ser^{250} \rightarrow Arg$, destabilizes the $G_s\alpha$ protein has recently been investigated *(70)*.

Table 1
Mutations Identified in the Gene Encoding G$_s$α (*GNAS*) in AHO or PHP Type Ia and in the Novel Overlap Syndrome (PTH Resistance Associated with Testotoxicosis)

Mutation	Location	Type	Effect	Reference
Met^1(ATG)→Val(GTG)	Exon 1	Missense	Initiator codon	(90)
Gln^{31}(CAG)→Ter(TAG)	Exon 1	Nonsense		(91)
Tyr^{37}(TAG)→Ter(TAG)	Exon 1	Nonsense		(91)
GA/GTAAGT→GA/GAGT	Intron 2	Splicing donor site		(91)
Leu^{99}(ATC)→Pro(ACC)	Exon 4	Missense		(92)
	Exon 4	Deletion 43bp		(93)
AA/GTACGT→AA/GTACAT	Intron 4	Splicing donor site		(91)
Pro^{115}(CCC)→(_CC)	Exon 5	Deletion 1bp		(94)
Leu^{153}(CTG T)→(CT_ _)	Exon 6	Deletion 2bp		(91)
Arg^{165}(CGC)→Cys(TGC)	Exon 6	Missense		(92)
ATT GAC TGT→ATT GT	Exon 6	Deletion 4bp		(91)
CAG GCT GAC→CAG GC	Exon 7	Deletion 4bp		(95)
				(96)
				(91)
Tyr^{190}(TAT)→Asp(GAT)	Exon 7	Missense		(91)
AG/GTGT→AG/GCGT	Intron 7	Splicing donor site		(91)
CAG GTG GAC→CAG AC	Exon 8	Deletion 4 bp		(92)
Arg^{231}(CGC)→His(CAC)	Exon 9	Missense	βγ interaction	(72)
Ser^{250}(AGC)→Arg()	Exon 10	Missense	Stability	(70)
Arg^{258}(CGG)→Trp(TGG)	Exon 10	Missense	Stability	(70)
Glu^{259}(GAG)→Val(GTG)	Exon 10	Missense	Stability	(97)
Gln^{267}(CAG)→CCAG	Exon 10	Insertion 1bp		(94)
Leu^{271}(CTC)→(_TC)	Exon 10	Deletion 1 bp		(98)
AG/GT→AG/CT	Intron 10	Splicing donor site		(98)
Arg^{385}(CGC)→His(CAC)	Exon 13	Missense	Receptor coupling	(71)
Ala^{366}(GCT)→Ser(TCT)	Exon 13	Missense	GDP release stability	(83)
ΔIle^{382}	Exon 13	Deletion 3 bp		(99)

Other missense mutations result in the production of a dysfunctional G$_s$α protein. For example, the Arg^{385}→His mutation results in uncoupling of the G$_s$α from hormone receptors *(71)*, although the Arg^{281}→His mutation inhibits receptor-mediated activation of G$_s$ by interfering with the interaction between G$_s$α and the βγ-dimer *(72)*.

Although the molecular basis of AHO is a mutation of the *GNAS1* gene that leads to decreased expression or function of G$_s$α, we still do not understand the relationship of this defect to all the clinical and biochemical features of this syndrome. Some patients with AHO have resistance to multiple hormones that act by stimulating adenylyl cyclase (PHP type Ia), but other affected members in the same family will lack any evidence of hormone resistance (pseudoPHP). Moreover, subjects with PHP type Ia have resistance to some, but not all hormones that act by stimulating the intracellular production of cAMP. Even less well understood is the role of G$_s$α deficiency in the development of the AHO phenotype.

Given that the number of G$_s$α molecules in a cell is estimated to be hundreds of times greater than the number of receptors, it is not obvious why a 50% deficiency of G$_s$α should cause hormone resistance. Studies of hormone responsiveness of tissues from human subjects with G$_s$α deficiency have produced conflicting results (Table 2). By

Table 2
Studies Examining Adenylyl Cyclase Activity or cAMP Accumulation in Response to Appropriate Agonists in Various Tissues from Subjects with PHP Type Ia

Tissue (agonist)	End-organ resistance (number of subjects)	Normal responsiveness (number of subjects)
Cultured fibroblasts (PGE$_1$)	Levine et al. *(100)* (5)	Bourne et al. *(101)* (5)
Kidney (PTH)	Drezner and Burch *(6)* (1)	Marcus et al. *(102)* (1)
Bone (PTH)		Ish-Shalom et al. *(103)* (1)
Thyroid (TSH)	Mallet et al. *(104)* (1)	
Adipose tissue (isoproterenol)	Kaartinen et al. *(17)* (4)	
Transformed lymphoblasts (PGE$_1$)		Farfel et al. *(105)* (3)
Platelets (PGI$_2$)	Motulsky et al. *(106)* (3)	

contrast, similar studies in murine embryonic stem cells in which one Gnas allele has been disrupted to reduce G$_s\alpha$ expression by 50% clearly show decreased cAMP accumulation in response to the β-adrenergic agonist isoproterenol *(73)*.

Activating Mutations of G$_s\alpha$

Somatic mutations in *GNAS1* that lead to constitutive (i.e., ligand-independent) activation have been identified in a variety of endocrine tumors and in the McCune-Albright syndrome (MAS). A subset of human growth hormone-secreting pituitary tumors exhibit increased adenylyl cyclase activity in vitro, in the absence of GHRH stimulation. These tumors have mutations in either Arg^{201} or Gln^{227} of *GNAS1* *(74,75)* that enable the G$_s\alpha$ subunit to remain in the active, GTP-bound state. Such activating mutations, termed *gsp* oncogenes, also occur in subsets of functioning and nonfunctioning thyroid neoplasms, but are rare in other endocrine tumors.

MAS is a sporadic syndrome characterized by the clinical triad of polyostotic fibrous dysplasia, *café-au-lait* pigmented skin lesions, and endocrine dysfunction *(76,77)*. Diverse endocrinopathies are present in patients with MAS, including gonadotropin-independent precocious puberty, autonomous thyroid nodules, growth hormone excess and hyperprolactinemia, hypercortisolism, and hyperphosphaturic hypophosphatemic rickets or osteomalacia. In patients with MAS, two specific mutations have been identified in Arg^{201} of G$_s\alpha$ Arg^{201} (CGT)→His(CAT) and Arg^{201} (CGT)→Cys(TGT). These mutations have been found in affected endocrine tissues *(78)*, skin *(79)*, and bone *(80)*, as well as in tissues not classically involved in MAS, including blood, liver, heart, and others *(81)*. These mutations are not present in all cells of an individual patient, and even in an affected organ, they are more highly represented in the histologically abnormal tissue. These observations indicate that subjects with MAS are mosaic and are consistent with the hypothesis that this disorder is caused by a somatic mutation that occurs during early embryogenesis *(82)*.

A Novel Overlap Syndrome

An unusual syndrome with features of both PHP type Ia and MAS has been described in two unrelated boys. In both patients, the same mutation, Ala^{366}→Ser, was identified

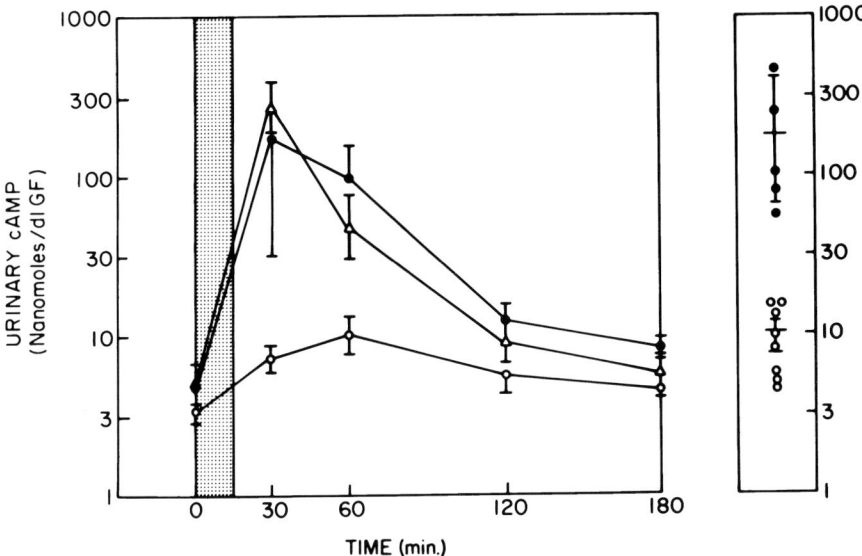

Fig. 5. cAMP excretion in the urine in response to bovine parathyroid extract (300 USP units). The peak response in normal subjects (triangles) as well as those with pseudoPHP (filled circles) is 50- to 100-fold times basal. Subjects with PHP type Ia (open circles) or others types of PHP type I (not shown) show only a two- to fivefold increase. Urinary cAMP is expressed per dL of glomerular filtrate (GF). A time-course is shown in the left panel, and peak response in the right panel.

in the *GNAS1* gene *(83)*. This mutation constitutively activates adenylyl cyclase in vitro, causing hormone-independent cAMP accumulation when expressed in cultured cells. The Ala^{366}→Ser G$_s$α protein is stable at 32°C, the temperature of testes, but is rapidly degraded at 37°C. This temperature-sensitive mutation accounts for the PHP type Ia phenotype in most tissues and also for male-limited precocious puberty (testotoxicosis) in the cooler testes *(83)*.

DIAGNOSIS

The diagnosis of PHP should be considered in every patient who has hypocalcemia or biochemical hypoparathyroidism. Hypocalcemia accompanied by hyperphosphatemia and an elevated PTH level strongly suggests the diagnosis of PHP. PTH levels in PHP may be normal or low owing to hypomagnesemia or other factors. The diagnosis of PHP is confirmed by demonstration of normal renal function and normal levels of magnesium and vitamin D metabolites. A positive family history further supports the diagnosis. Features of AHO or evidence of resistance to multiple hormones suggest PHP type Ia.

The definitive test for PHP involves demonstration of end-organ resistance to PTH (Fig. 5). Several protocols have been developed that are based on the infusion of bovine parathyroid extract or synthetic human PTH (1-34) *(84–86)*. However, at the present time, pharmacological preparations of PTH that are suitable for diagnostic studies in humans are not commercially available in the US.

Our standard protocol for the analysis of renal resistance to PTH involves infusion of synthetic hPTH(1-34) at 9AM in a subject who has not eaten since midnight. Oral hydration is maintained by encouraging the subject to drink 8 oz of water every 30 min. Two baseline urine collections are made, from 60 min to 31 min and from 30 min to 1 min, preceding the infusion. hPTH(1-34), 200 U (35–50 mcg) in adults or 3 U/kg (200 U maximum) in children over the age of 3, is infused over 10 min. Urine collections are made from 0–30, 31–60, 61–90, and 91–120 min. Urine is analyzed for cAMP, phosphate, and creatinine. In addition, blood samples are collected at 0 and 90 min for measurement of serum creatinine and phosphorus. Urinary cAMP should be expressed as nmols of cAMP/dL of glomerular filtrate:

$$U_{cAMP} \text{ (nmols/dL GF)} = U_{cAMP} \text{ (noml/dL)} \times S_{Cre} \text{ (mg/dL)} / U_{Cre} \text{ (mg/dL)} \quad (1)$$

Urinary cAMP in the first 30 min provides the best discrimination between subjects with PHP and normal subjects or subjects with hypoparathyroidism. The tubular reabsorption of phosphate is calculated from the concentrations of phosphate (Phos) and creatinine (Cre) in urine and serum:

$$TRP = 1 - [S_{Cre} \text{ (mg/dL)} \times U_{Phos} \text{ (mg/dL)}] / [S_{Phos} \text{ (mg/dL)} \times U_{Cre} \text{ (mg/dL)}] \quad (2)$$

The phosphaturic response to PTH is then calculated as the percent decline in the tubular maximum for reabsorption of phosphate (T_mP/GF) using a standard nomogram (87). In the first hour after infusion of PTH, the percent decline in T_mP/GF is less than normal in subjects with PHP, but greater than normal in subjects with hypoparathyroidism.

The diagnosis of PHP type Ia can be confirmed by research tests that identify a mutation in the GNAS1 gene or that demonstrate reduced expression or activity of $G_s\alpha$ in erythrocyte membranes. No similar biochemical or molecular tests exist for other forms of PHP.

TREATMENT

The treatment of PHP follows the treatment of the individual endocrinopathies that comprise it, i.e., hypocalcemia, hypothyroidism, and hypogonadism. In addition, the physician must be aware of the possible existence or development of additional endocrine manifestations, and should take care to examine other family members for the presence of this disorder. Our insight into the pathogenesis of PHP has not led to any specific therapies that will improve signal transduction in the affected tissues; therefore the treatment of PHP is still to bypass the defect.

Severe or symptomatic hypocalcemia should be treated by the administration of calcium intravenously. In the acute setting, the goal of therapy is to administer sufficient calcium to relieve symptoms of tetany. This can usually be accomplished by infusing 1 g of calcium gluconate (10 mL of 10% calcium gluconate) over a 15-min period, followed by continuous infusion of a 1% calcium gluconate solution, prepared by adding 5 ampules (100 mL) of 10% calcium gluconate to 1 L of 5% dextrose in water. The rate of infusion is subsequently adjusted to achieve low normal serum calcium levels.

Treatment of chronic hypocalcemia requires the oral administration of calcium supplements and vitamin. In this setting, the goal of therapy is to maintain serum calcium in the low-normal range, to avoid hypercalciuria, and to reduce elevated PTH levels into the normal range. Calcium supplements are given as 1–3 g of elemental calcium/d in divided doses. Absorption is best when taken with food in the stomach. Calcium carbonate is the least expensive supplement and has the highest content of elemental calcium (40%). Calcium citrate (21% elemental calcium) is well absorbed even in the absence of stomach acid, and causes fewer gastrointestinal side effects. All patients with PHP who are hypocalcemic require vitamin D in addition to calcium supplements. Calcitriol is the most physiological treatment. Most patients are maintained on 0.25–1 mcg twice a day. Ergocalciferol is a less-expensive form of vitamin D therapy, and is effective in patients with PHP at doses (50,000–100,000 U/d) that are often less than those required to treat other forms of hypoparathyroidism.

There is no specific therapy for AHO. Heterotopic ossifications occur in about one-third of subjects with AHO. Although osteomas do not usually require treatment, painful osteomas may require surgical removal. Skeletal abnormalities, such as deformed elbows, varus and valgus deformity of the knees, and bowed tibia, may require orthopedic evaluation and treatment. Referral to a podiatrist should be recommended when bony abnormalities in the feet cause painful hyperkeratotic lesions, bursitis of metatarsal joints, or dislocated toes, or when ulcerative lesions occur from pressure between ill-fitting shoes and heterotopic ossifications.

CONCLUSION

It has been more than 50 years since Fuller Albright first described patients with end organ resistance PTH action. Although the first quarter-century was one of important clinical observations (88), the second quarter-century has been one of remarkable advance in our understanding of fundamental pathophysiology. During the past 25 years, we have come to appreciate the role of cAMP as a second messenger, to understand the requirement of a G-protein for coupling of a 7TMS cell-surface receptor to an intracellular signal generator, and to identify mutations in the gene encoding $G_s\alpha$ as the molecular basis of PHP type Ia. Despite this progress, many questions about the pathogenesis of PHP remain unanswered. Why do all subjects with $G_s\alpha$ deficiency not have hormone resistance; is it related to genomic imprinting? What is the molecular basis of AHO; are skeletal deformities and heterotopic ossifications caused by abnormal osteoblast differentiation as a result of resistance to PTHrP? What is the pathogenesis of PTH type Ib; are there undiscovered PTH receptors, or is abnormal signal transduction through $G_q\alpha$-coupled pathways involved? The continued interest in the study of PHP by groups around the world has led to the development of an animal model of AHO (89) and a genome-wide search for the PHP type Ib gene (42). We predict that the next quarter century of study of PHP will continue to provide valuable insights into the physiology of hormone action.

REFERENCES

1. Albright F, Burnett CH, Smith PH, Parson W. Pseudo-hypoparathyroidism—an example of "Seabright-Bantam syndrome." Endocrinology 1942;30:922–932.
2. Albright F, Forbes AP, Henneman PH. Pseudo-pseudohypoparathyroidism. Trans Assoc Am Physicians 1952;65:337.
3. Tashjian AH Jr, Frantz AG, Lee JB. Pseudohypoparathyroidism: assays of parathyroid hormone and thyrocalcitonin. Proc Natl Acad Sci USA 1966;56:1138–1142.
4. Chase LR, Melson GL, Aurbach GD. Pseudohypoparathyroidism: defective excretion of 3',5'-AMP in response to parathyroid hormone. J Clin Invest 1969;48:1832–1844.
5. Bell NH, Avery S, Sinha T, Clark CM Jr, Allen DO, Johnston C Jr. Effects of dibutyryl cyclic adenosine 3',5'-monophosphate and parathyroid extract on calcium and phosphorus metabolism in hypoparathyroidism and pseudohypoparathyroidism. J Clin Invest 1972;51:816–823.
6. Drezner MK, Burch WM Jr. Altered activity of the nucleotide regulatory site in the parathyroid hormone-sensitive adenylate cyclase from the renal cortex of a patient with pseudohypoparathyroidism. J Clin Invest 1978;62:1222–1227.
7. Farfel Z, Brickman AS, Kaslow HR, Brothers VM, Bourne HR. Defect of receptor-cyclase coupling protein in pseudohypoparathyroidism. N Engl J Med 1980;303:237–242.

8. Levine MA, Downs RW Jr, Singer M, Marx SJ, Aurbach GD, Spiegel AM. Deficient activity of guanine nucleotide regulatory protein in erythrocytes from patients with pseudohypoparathyroidism. Biochem Biophys Res Commun 1980;94:1319–1324.
9. Barrett D, Breslau NA, Wax MB, Molinoff PB, Downs RW Jr. New form of pseudohypoparathyroidism with abnormal catalytic adenylate cyclase. Am J Physiol 1989;257:E277–283.
10. Drezner M, Neelon FA, Lebovitz HE. Pseudohypoparathyroidism type II: a possible defect in the reception of the cyclic AMP signal. N Engl J Med 1973;289:1056–1060.
11. Fitch N. Albright's hereditary osteodystrophy: a review. Am J Hum Genet 1982;11:11–29.
12. Steinbach HL, Rudhe U, Jonsson M, Young DA. Evolution of skeletal lesions in pseudohypoparathyroidism. Radiology 1965;85:670–676.
13. Croft LK, Witkop CJ Jr, Glas JE. Pseudohypoparathyroidism. Oral Surg Oral Med Oral Pathol 1965;20:758–770.
14. Poznanski AK, Werder EA, Giedion A. The pattern of shortening of the bones of the hand in PHP and PPHP-a comparison with brachydactyly E, Turner syndrome, and acrodysostosis. Radiology 1977;123:707–718.
15. Prendiville JS, Lucky AW, Mallory SB, Mughal Z, Mimouni F, Langman CB. Osteoma cutis as a presenting sign of pseudohypoparathyroidism. Pediatr Dermatol 1992;9:11–18.
16. Trüeb RM, Panizzon RG, Burg G. Cutaneous ossification in Albright's hereditary osteodystrophy. Dermatology 1993;186:205–209.
17. Kaartinen JM, Kaar ML, Ohisalo JJ. Defective stimulation of adipocyte adenylate cyclase, blunted lipolysis, and obesity in pseudohypoparathyroidism 1a. Pediatr Res 1994;35:594–597.
18. Weinstock RS, Wright HN, Spiegel AM, Levine MA, Moses AM. Olfactory dysfunction in humans with deficient guanine nucleotide-binding protein. Nature 1986;322:635–636.
19. Henkin RI. Impairment of olfaction and of the tastes of sour and bitter in pseudohypoparathyroidism. J Clin Endocrinol Metab 1968;28:624–628.
20. Koch T, Lehnhardt E, Bottinger H, Pfeuffer T, Palm D, Fischer B, et al. Sensorineural hearing loss owing to deficient G proteins in patients with pseudohypoparathyroidism: results of a multicentre study. Eur J Clin Invest 1990;20:416–421.
21. Jones DT, Reed RR. Golf: an olfactory neuron specific-G protein involved in odorant signal transduction. Science 1989;244:790–795.
22. McLaughlin SK, McKinnon PJ, Margolskee RF. Gustducin is a taste-cell-specific G protein closely related to the transducins. Nature 1992;357:563–569.
23. Garty BZ, Daliot D, Kauli R, Arie R, Grosman J, Nitzan M, et al. Hearing impairment in idiopathic hypoparathyroidism and pseudohypoparathyroidism. Isr J Med Sci 1994;30:587–591.
24. Doty RL, Fernandez AD, Levine MA, Moses A, McKeown DA. Olfactory dysfunction in type I pseudohypoparathyroidism: dissociation from Gs alpha protein deficiency. J Clin Endocrinol Metab 1997;82:247–250.
25. Faig JC, Kalinyak J, Marcus R, Feldman D. Chronic atypical seizure disorder and cataracts due to delayed diagnosis of pseudohypoparathyroidism. West J Med 1992;157:64–65.
26. Glynne A, Hunter IP, Thomson JA. Pseudohypoparathyroidism with paradoxical increase in hypocalcaemic seizures due to long-term anticonvulsant therapy. Postgrad Med J 1972;48:632–636.
27. Guberman A, Jaworski ZF. Pseudohypoparathyroidism and epilepsy: diagnostic value of computerized cranial tomography. Epilepsia 1979;20:541–553.
28. Pollak L, Klein C, Tieder M, Arlazoroff A. Therapy-resistant seizures in pseudohypoparathyroidism. A case report. J Pediatr Endocrinol Metab 1995;8:209–211.
29. Farfel Z, Friedman E. Mental deficiency in pseudohypoparathyroidism type I is associated with Ns-protein deficiency. Ann Intern Med 1986;105:197–199.
30. Marx SJ, Hershman JM, Aurbach GD. Thyroid dysfunction in pseudohypoparathyroidism. J Clin Endocrinol Metab 1971;33:822–828.
31. Wolfsdorf JI, Rosenfield RL, Fang VS, Kobayashi R, Razdan AK, Kim MH. Partial gonadotrophin-resistance in pseudohypoparathyroidism. Acta Endocrinol (Copenh) 1978;88:321–328.
32. Moses AM, Weinstock RS, Levine MA, Breslau NA. Evidence for normal antidiuretic responses to endogenous and exogenous arginine vasopressin in patients with guanine nucleotide-binding stimulatory protein-deficient pseudohypoparathyroidism. J Clin Endocrinol Metab 1986;62:221–224.
33. Levine MA, Downs RW Jr. Moses AM, Breslau NA, Marx SJ, Lasker RD, et al. Resistance to multiple hormones in patients with pseudohypoparathyroidism. Association with deficient activity of guanine nucleotide regulatory protein. Am J Med 1983;74:545–556.

34. Tsai KS, Chang CC, Wu DJ, Huang TS, Tsai IH, Chen FW. Deficient erythrocyte membrane Gs alpha activity and resistance to trophic hormones of multiple endocrine organs in two cases of pseudohypoparathyroidism. Taiwan I Hsueh Hui Tsa Chih 1989;88:450–455.
35. Ridderskamp P, Schlaghecke R. Pseudohypoparathyroidism and adrenal cortex insufficiency. A case of multiple endocrinopathy due to peripheral hormone resistance. Klin Wochenschr 1990;68: 927–931.
36. Brickman AS, Carlson HE, Levin SR. Responses to glucagon infusion in pseudohypoparathyroidism. J Clin Endocrinol Metab 1986;63:1354–1360.
37. Namnoum AB, Merriam GR, Moses AM, Levine MA. Reproductive dysfunction in women with Albright's hereditary osteodystrophy. J Clin Endocrinol Metab 1998; 83:824–829.
38. Murray TM, Rao LG, Wong MM, Waddell JP, McBroom R, Tam CS, et al. Pseudohypoparathyroidism with osteitis fibrosa cystica: direct demonstration of skeletal responsiveness to parathyroid hormone in cells cultured from bone. J Bone Miner Res 1993;8:83–91.
39. Schipani E, Weinstein LS, Bergwitz C, Iida-Klein A, Kong XF, Stuhrmann M, et al. Pseudohypoparathyroidism type Ib is not caused by mutations in the coding exons of the human parathyroid hormone (PTH)/PTH-related peptide receptor gene. J Clin Endocrinol Metab 1995;80:1611–1621.
40. Fukumoto S, Suzawa M, Takeuchi Y, Kodama Y, Nakayama K, Ogata E, et al. Absence of mutations in parathyroid hormone (PTH)/PTH-related protein receptor complementary deoxyribonucleic acid in patients with pseudohypoparathyroidism type Ib. J Clin Endocrinol Metab 1996;81:2554–2558.
41. Bettoun JD, Minagawa M, Kwan MY, Lee HS, Yasuda T, Hendy GN, et al. Cloning and characterization of the promoter regions of the human parathyroid hormone (PTH)/PTH-related peptide receptor gene: analysis of deoxyribonucleic acid from normal subjects and patients with pseudohypoparathyroidism type 1b. J Clin Endocrinol Metab 1997;82:1031–1040.
42. Jan de Beur SM, LaBuda MC, Timberlake RJ, Levine MA. Genome-wide linkage analysis of pseudohypoparathyroidism 1b. In: Program and Abstracts 79th Annual Meeting. The Endocrine Society Press, Bethesda, MD, 1997, p. 242 [Abstract].
43. Davies SJ, Hughes HE. Imprinting in Albright's hereditary osteodystrophy. J Med Genet 1993;30: 101–103.
44. Schuster V, Kress W, Kruse K. Paternal and maternal transmission of pseudohypoparathyroidism type Ia in a family with Albright hereditary osteodystrophy: no evidence for genomic imprinting. J Med Genet 1994;31:84 [Letter].
45. Campbell R, Gosden CM, Bonthron DT. Parental origin of transcription from the human GNAS1 gene. J Med Genet 1994;31:607–614.
46. Williamson CM, Schofield J, Dutton ER, Seymour A, Beechy CV, Edwards YH, et al. Glomerular-specific imprinting of the mouse Gsα gene: How does this relate to hormone resistance in Albright hereditary osteodystrophy. Genomics 1996;36:280–287.
47. Karaplis AC, Luz A, Glowacki J, Bronson RT, Tybulewicz VLJ, Kronenberg HM, et al. Lethal skeletal dysplasia from targeted disruption of the parathyroid hormone-related peptide gene. Genes Dev 1994;8:277–289.
48. Vortkamp A, Lee K, Lanske B, Segre GV, Kronenberg HM, Tabin CJ. Regulation of rate of cartilage differentiation by Indian hedgehog and PTH-related protein. Science 1996;273:613–622.
49. Amizuka N, Karaplis AC, Henderson JE, Warshawsky H, Lipman ML, Matsuki Y, et al. Haploinsufficiency of parathyroid hormone-related peptide (PTHrP) results in abnormal postnatal bone development. Dev Biol 1996;175:166–176.
50. Kovacs CS, Lanske B, Hunzelman JL, Karaplis AC, Kronenberg HM. Parathyroid hormone-related peptide (PTHrP) regulates fetal-placental calcium transport through a receptor distinct from PTH/PTHrP receptor. Proc Natl Acad Sci USA 1996;93:15,233–15,238.
51. Abou Samra A-B, Jüppner H, Force T, Freeman MW, Kong X-F, Schipani E, et al. Expression cloning of a common receptor for parathyroid hormone and parathyroid hormone-related peptide from rat osteoblast-like cells: A single receptor stimulates intracellular accumulation of both cAMP and inositol triphosphates and increases intracellular free calcium. Proc Natl Acad Sci USA 1992;89: 2732–2736.
52. Usdin TB, Gruber C, Bonner TI. Identification and functional expression of a receptor selectively recognizing parathyroid hormone, the PTH2 receptor. J Biol Chem 1995;270:15,455–15,458.
53. Offermanns S, Iida-Klein A, Segre GV, Simon MI. Gαq family members couple parathyroid hormone (PTH)/PTH-related peptide and calcitonin receptors to phospholipase C in COS-7 cells. Mol Endocrinol 1996;10:566–574.

54. Yan K, Kalyanaraman V, Gautam N. Differential ability to form the G protein β–γ complex among members of the β and γ subunit families. J Biol Chem 1996;271:7141–7146.
55. Rahmatullah M, Ginnan R, Robishaw JD. Specificity of G protein α–γ subunit interactions. J Biol Chem 1995;270:2946–2951.
56. Kleuss C, Scherubl H, Hescheler G. Schultz G, Wittig B. Different β-subunits determine G-protein interactions with transmembrane receptors. Nature 1992;358:424–426.
57. Dohlman HG, Thorner J. RGS proteins and signaling by heterotrimeric G proteins. J Biol Chem 1997;272:3871–3874.
58. Mixon MB, Lee E, Coleman DE, Berghuis AM, Gilman AG, Sprang SR. Tertiary and quaternary structural changes in Giα1 induced by GTP hydrolysis. Science 1995;270:954–959.
59. Coleman DE, Berghuis AM, Lee E, Linder ME, Gilman AG, Sprang SR. Structure of active confromations of Giα1and the mechanism of GTP hydrolysis. Science 1994;265:1405.
60. Wall MA, Coleman DE, Lee E, Iniguez-Lluhi JA, Posner BA, Gilman AG, et al. The structure of the G protein heterotrimer Giα1β1γ2. Cell 1995;83:1047–1058.
61. Sprang S. G protein mechanisms: insights from structural analysis. Ann Rev Biochem 1997;66: 639–678.
62. Levine MA, Modi WS, O'Brien SJ. Mapping of the gene encoding the α subunit of the stimulatory G protein of adenylyl cyclase (GNAS1) to 20q13.2→13.3 in human by in situ hybridization. Genomics 1991;11:478–479.
63. Yatani A, Brown AM. Rapid β-adrenergic modulation of cardiac calcium channel currents by a fast G protein pathway. Science 1989;245:71–74.
64. Kozasa T, Itoh H, Tsukamoto T, Kaziro Y. Isolation and characterization of the human Gsα gene. Proc Natl Acad Sci USA 1988;85:2081–2085.
65. Bray P, Carter A, Simons C, Guo V, Puckett C, Kamholz J, et al. Human cDNA clones for four species of Gsa signal transduction protein. Proc Natl Acad Sci USA 1986;83:8893–8897.
66. Ishikawa Y, Bianchi C, Nadal-Ginard B, Homcy CJ. Alternative promoter and 5' exon generate a novel $G_s\alpha$ mRNA. J Biol Chem 1990;265:8458–8462.
67. Kehlenbach RH, Matthey J, Huttner WB. XLα$_s$ is a new type of G protein. Nature 1994;372:804–808.
68. Cooper DMF, Mons N, Karpen JW. Adenylyl cyclases and the interaction between calcium and cAMP signaling. Nature 1995;374:421–424.
69. Taussig R, Gilman AG. Mammalian membrane-bound adenylyl cyclases. J Biol Chem 1995;270:1–4.
70. Warner DR, Gejman PV, Collins RM, Weinstein LS. A novel mutation adjacent to the switch III domain of G(S alpha) in a patient with pseudohypoparathyroidism [In Process Citation]. Mol Endocrinol. 1997;11:1718–1727.
71. Schwindinger WF, Miric A, Zimmerman D, Levine MA. A novel Gs alpha mutant in a patient with Albright hereditary osteodystrophy uncouples cell surface receptors from adenylyl cyclase. J Biol Chem. 1994;269:25,387–25,391.
72. Farfel Z, Iiri T, Shapira H, Roitman A, Mouallem M, Bourne HR. Pseudohypoparathyroidism, a novel mutation in the βγ-contact region of Gsα impairs receptor stimulation. J Biol Chem 1996;271: 19,653–19,655.
73. Schwindinger WF, Reese KJ, Lawler AM, Gearhart JD, Levine MA. Targeted disruption of Gnas in embryonic stem cells. Endocrinology 1997;138:4058–4063.
74. Landis CA, Masters SB, Spada A, Pace AM, Bourne HR, Vallar L. GTPase inhibiting mutations activate the α chain of Gs and stimulate adenylyl cyclase in human pituitary tumours. Nature 1989; 340:692–696.
75. Vallar L, Spada A, Giannattasio G. Altered Gs and adenylate cyclase activity in human GH-secreting pituitary adenomas. Nature 1987;330:566–568.
76. McCune DJ, Bruch H. Osteodystrophia fibrosa. Am J Dis Child 1937;54:806–848.
77. Albright F, Butler AM, Hampton AO, Smith P. Syndrome characterized by osteitis fibrosa disseminata, areas of pigmentation, and endocrine dysfunction, with precocious puberty in females. N Engl J Med 1937;216:727–741.
78. Weinstein LS, Shenker A, Gejman PV, Merino MJ, Friedman E, Spiegel AM. Activating mutations of the stimulatory G protein in the McCune-Albright syndrome. N Engl J Med 1991;325:1688–1695.
79. Schwindinger WF, Yang SQ, Miskovsky EP, Diehl AM, Levine MA. An activating $G_s\alpha$ mutation is present in McCune-Albright syndrome and increases hepatic adenylyl cyclase activity. In: Program and Abstracts 75th Annual Meeting. The Endocrine Society Press, Bethesda, MD, 1994, p. 517 [Abstract].

80. Shenker A, Sweet DE, Spiegel AM, Weinstein LS. An activating $G_s\alpha$ mutation is present in fibrous dysplasia of bone in the McCune-Albright syndrome. J Clin Endocrinol Metab 1994;79: 750–775.
81. Shenker A, Weinstein LS, Moran A, Pescovitz OH, Charest NJ, Boney CM, et al. Severe endocrine and nonendocrine manifestations of the McCune-Albright syndrome associated with activating mutations of the stimulatory G protein G_s. J Pediatr 1993;123:509–518.
82. Happle R. The McCune-Albright syndrome: a lethal gene surviving by mosaicism. Clin Genet 1986;29:321–324.
83. Iiri T, Herzmark P, Nakamoto JM, van Dop C, Bourne HR. Rapid GDP release from Gs alpha in patients with gain and loss of endocrine function. Nature 1994;371:164–168.
84. Furlong TJ, Seshadri MS, Wilkinson MR, Cornish CJ, Luttrell B, Posen S. Clinical experiences with human parathyroid hormone 1-34. Aust NZ J Med 1986;16:794–798.
85. McElduff A, Lissner D, Wilkinson M, Cornish C, Posen S. A 6-hour human parathyroid hormone (1-34) infusion protocol: studies in normal and hypoparathyroid subjects. Calcif Tissue Int 1987; 41:267–273.
86. Mallette LE, Kirkland JL, Gagel RF, Law WM Jr, Heath HD. Synthetic human parathyroid hormone-(1-34) for the study of pseudohypoparathyroidism. J Clin Endocrinol Metab 1988;67:964–972.
87. Walton RJ, Bijvoet OLM. Nomogram for derivation of renal threshold phosphate concentration. Lancet 1975;2:309–310.
88. Mautalen CA, Dymling J-F, Harwith M. Pseudohypoparathyroidism 1942–1966. A negative progress report. Am J Med 1967;42:977–985.
89. Yu S, Yu D, Lee E, Eckhaus M, Accili D, Westphal H, et al. Pleiotropic and parent-of-origin specific abnormalities in $Gs\alpha$ knockout mice. In: 79th Annual Meeting Program & Abstracts. The Endocrine Society Press, Bethesda, MD, 1997, p. 110.
90. Patten JL, Johns DR, Valle D, Eil C, Gruppuso PA, Steele G, et al. Mutation in the gene encoding the stimulatory G protein of adenylate cyclase in Albright's hereditary osteodystrophy. N Engl J Med 1990;322:1412–1419.
91. Jan de Beur SM, Deng Z, Levine MA. 1997, unpublished results.
92. Miric A, Vechio JD, Levine MA. Heterogeneous mutations in the gene encoding the alpha-subunit of the stimulatory G protein of adenylyl cyclase in Albright hereditary osteodystrophy. J Clin Endocrinol Metab 1993;76:1560–1568.
93. Luttikhuis ME, Wilson LC, Leonard, JV, Trembath RC. Characterization of a de novo 43-bp deletion of the Gs alpha gene (GNAS1) in Albright hereditary osteodystrophy. Genomics 1994;21:455–457.
94. Shapira H, Mouallem M, Shapiro MS, Weisman Y, Farfel Z. Pseudohypoparathyroidism type Ia: two new heterozygous frameshift mutations in exons 5 and 10 of the Gs alpha gene. Hum Genet 1996;97: 73–75.
95. Weinstein LS, Gejman PV, de Mazancourt P, American N, Spiegel AM. A heterozygous 4-bp deletion mutation in the Gs alpha gene (GNAS1) in a patient with Albright hereditary osteodystrophy. Genomics 1992;13:1319–1321.
96. Yokoyama M, Takeda K, Iyota K, Okabayashi T, Hashimoto K. A 4-base pair deletion mutation of Gs alpha gene in a Japanese patient with pseudohypoparathyroidism. J Endocrinol Invest 1996;19: 236–241.
97. Dixon PH, Ahmed SF, Bonthron DT, Barr DGD, Kelnar CJH, Thakker RV. Mutational analysis of the GNAS1 gene in pseudohypoparathyroidism. J Bone Min Res 1996;11(Suppl 1), S494.
98. Weinstein LS, Gejman PV, Friedman E, Kadowaki T, Collins RM, Gershon ES, et al. Mutations of the Gs alpha-subunit gene in Albright hereditary osteodystrophy detected by denaturing gradient gel electrophoresis. Proc Natl Acad Sci USA 1990;87:8287–90.
99. Smallwood PM, Aparicio LF, Schwindinger WF, Levine MA. Pseudohypoparathyroidism type IB caused by a novel missense mutation in the gene encoding the alpha subunit of the stimulatory G protein of adenylyl cyclase. In: Program and Abstracts 76th Annual Meeting. The Endocrine Society Press, Bethesda, MD, 1994, p. 651 [Abstract].
100. Levine MA, Eil C, Downs RW Jr, Spiegel AM. Deficient guanine nucleotide regulatory unit activity in cultured fibroblast membranes from patients with pseudohypoparathyroidism type I. a cause of impaired synthesis of 3',5'-cyclic AMP by intact and broken cells. J Clin Invest 1983;72:316–324.
101. Bourne HR, Kaslow HR, Brickman AS, Farfel Z. Fibroblast defect in pseudohypoparathyroidism, type I: reduced activity of receptor-cyclase coupling protein. J Clin Endocrinol Metab 1981;53:636–640.
102. Marcus R, Wilber JF, Aurbach GD. Parathyroid hormone-sensitive adenylyl cyclase from the renal cortex of a patient with pseudohypoparathyroidism. J Clin Endocrinol Metab 1971;33:537–541.

103. Ish-Shalom S, Rao LG, Levine MA, Fraser D, Kooh SW, Josse RG, et al. Normal parathyroid hormone responsiveness of bone-derived cells from a patient with pseudohypoparathyroidism. J Bone Miner Res 1996;11:8–14.
104. Mallet E, Carayon P, Amr S, Brunelle P, Ducastelle T, Basuyau JP, et al. Coupling defect of thyrotropin receptor and adenylate cyclase in a pseudohypoparathyroid patient. J Clin Endocrinol Metab 1982;54:1028–1032.
105. Farfel Z, Abood ME, Brickman AS, Bourne HR. Deficient activity of receptor-cyclase coupling protein is transformed lymphoblasts of patients with pseudohypoparathyroidism, type I. J Clin Endocrinol Metab 1982;55:113–117.
106. Motulsky HJ, Hughes RJ, Brickman AS, Farfel Z, Bourne HR, Insel PA. Platelets of pseudohypoparathyroid patients: evidence that distinct receptor-cyclase coupling proteins mediate stimulation and inhibition of adenylate cyclase. Proc Natl Acad Sci USA 1982;79:4193–4197.

4 Hereditary Resistance to Vitamin D

Peter J. Malloy, PhD and David Feldman, MD

CONTENTS

    INTRODUCTION
    VITAMIN D PHYSIOLOGY
    THE VDR
    MECHANISM OF 1,25(OH)$_2$D ACTION
    CLINICAL FEATURES OF HVDRR
    CELLULAR EVIDENCE OF 1,25(OH)$_2$D RESISTANCE
    MUTATIONS AFFECTING THE DBD
    MUTATIONS AFFECTING THE LBD
    OTHER MUTATIONS
    APPROACHES TO TREATMENT OF HVDRR
    CONCLUDING COMMENTS
    REFERENCES

INTRODUCTION

Vitamin D, a primary regulator of calcium homeostasis in the body, is particularly important in skeletal development and in bone mineralization. The active form of vitamin D, 1,25-dihydroxyvitamin D [1,25(OH)$_2$D], binds with high affinity to specific vitamin D receptors (VDR) located in the nucleus of target cells. The VDR is a member of the steroid-thyroid-retinoid receptor gene superfamily of nuclear transcription factors that mediate the biological effects of the hormone through regulation of target genes. Hereditary vitamin D-resistant rickets (HVDRR) is a rare genetic disorder caused by a generalized resistance to 1,25(OH)$_2$D action *(1–3)*. Heterogeneous mutations in the VDR that alter the function of the receptor have been shown to be the molecular basis of HVDRR. A variety of mutations have been identified, some of which cause the VDR to be nonfuctional and therefore result in a total hormone-resistant state, whereas other mutations reduce VDR activity and result in a hyporesponsive state. This chapter will describe the clinical manifestations of HVDRR and the genetic defects in the VDR that result in this hormone-resistant syndrome.

In 1937, Albright et al. *(4)* described a patient with rickets who was hypophosphatemic but had normal serum calcium levels. The patient was resistant to vitamin D therapy. Albright et al. *(4)* suggested that the rickets was caused by end-organ resistance to vitamin D, and thus, the concept of hormone resistance was conceived. In 1978, Brooks

From: *Contemporary Endocrinology: Hormone Resistance Syndromes*
Edited by: J. L. Jameson © Humana Press Inc., Totowa, NJ

et al. *(5)* described a patient with osteomalacia who was hypocalcemic, hypophosphatemic, and had secondary hyperparathyroidism. Interestingly, the patient had markedly increased serum levels of $1,25(OH)_2D_3$. Brooks et al. *(5)* suggested that the rickets was the result of impaired responsiveness of target organs to $1,25(OH)_2D_3$. They termed this disease vitamin D-dependent rickets, type II (VDDR II) to distinguish it from vitamin D-dependent rickets, type I (VDDR I), which is due to a defect in $1,25(OH)_2D_3$ production (*see* 1α-Hydroxylase and PDDR). Later the same year, Marx et al. *(6)* reported similar findings in two children, and again the authors suggested that the disease was due to end-organ resistance to $1,25(OH)_2D_3$. Since these initial studies, there have been many reports of patients with apparent target tissue resistance to $1,25(OH)_2D_3$. Over the years, a number of different terms have been used to describe this syndrome in the literature. As noted, the original reports referred to this disease as vitamin D-dependent rickets, type II. The disease also has been called pseudo-vitamin D deficiency type II (PDDR II), calcitriol-resistant rickets (CRR), and hereditary hypocalcemic vitamin D-resistant rickets (HHVDRR). We prefer HVDRR, since the syndrome is caused by genetic resistance to vitamin D.

VITAMIN D PHYSIOLOGY

Metabolism

Vitamin D is a fat-soluble seco-steroid that exists in two forms, vitamin D_3 (cholecalciferol) from animal sources and vitamin D_2 (ergocalciferol) from plant sources. Vitamin D can be synthesized in the skin so it is not a true vitamin. In animals, the precursor (provitamin) molecule, 7-dehydrocholesterol, is cleaved between carbon-9 and 10 in the B ring using the energy derived from UV B-rays of sunlight, which opens the ring and creates the seco-steroid structure (Fig. 1) *(7)*. Vitamins D_2 and D_3 are essentially biologically inactive and must be converted to hydroxylated metabolites to gain hormonal activity. On entering the circulation, vitamin D (D_2 or D_3) binds to the vitamin D binding protein (DBP), a 58-kDa plasma α globulin *(8,9)*. In the liver, vitamin D is hydroxylated at the carbon-25 position to form 25-hydroxyvitamin D or 25(OH)D by the enzyme 25-hydroxylase *(10)*. In the kidney, the enzyme 25-hydroxy D-1α-hydroxylase (1α-hydroxylase) adds a second hydroxyl group to 25(OH)D to form 1α,25-dihydroxyvitamin D or $1,25(OH)_2D$ (this notation will be used to signify either D_2 or D_3), the hormonally active form of vitamin D.

1α-Hydroxylase and PDDR

The level of 1α-hydroxylase activity is tightly regulated by parathyroid hormone (PTH) *(11)*. It is interesting to note that in patients with hyperparathroidism, the circulating levels of $1,25(OH)_2D$ are elevated, whereas patients with hypoparathyroidism show reduced levels of the hormone *(11,12)*. Other factors, including phosphate and $1,25(OH)_2D$ itself, also regulate the level of 1α-hydroxylase activity. High phosphate levels suppress enzyme activity, whereas low phosphate levels increase enzyme activity. $1,25(OH)_2D$ also regulates its own production with low $1,25(OH)_2D$ levels, leading to increased 1α-hydroxylase activity and high $1,25(OH)_2D$ levels inhibiting the enzyme, probably through feedback inhibition of the 1α-hydroxylase activity, by direct suppression of 1α-hydroxylase gene expression, or a combination of the two *(11)*. Changes in serum calcium concentrations may also have a direct effect on the 1α-hydroxylase activity, but clearly act indirectly via regulation of PTH levels.

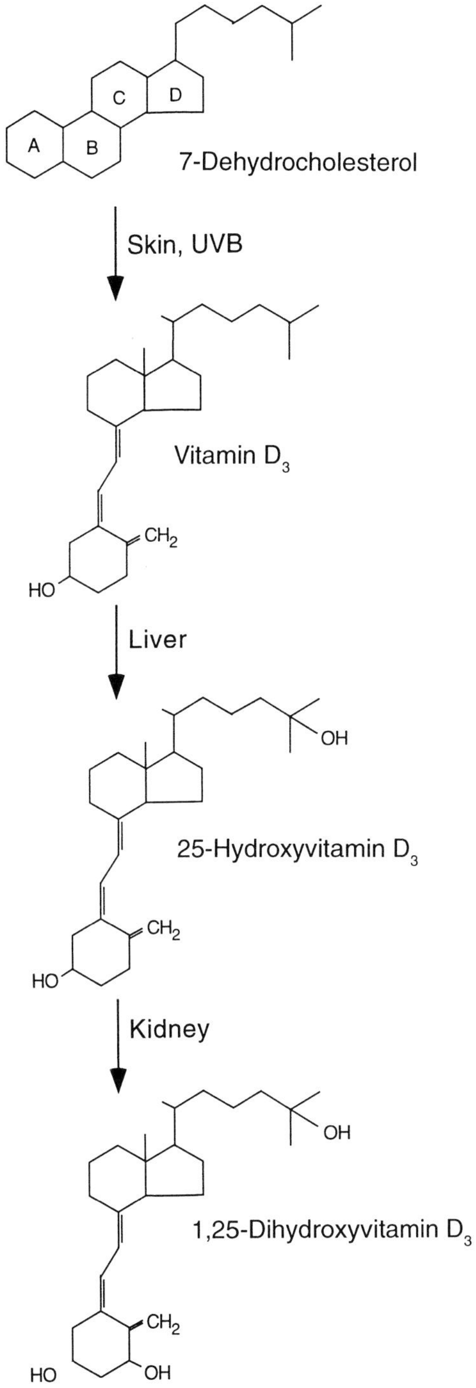

Fig. 1. Metabolic pathway leading to the synthesis of 1,25-(OH)$_2$D$_3$.

An absolute or relative decreased production of 1,25(OH)$_2$D is found in hypoparathyroidism, pseudohypoparathyroidism, renal failure, X-linked hypophosphatemic rickets, oncogenic osteomalacia, and hereditary 1α-hydroxylase deficiency (VDDR I). In 1961, Prader et al. *(13)* described a patient with a deficiency of renal 1α-hydroxylase. This disease is known as VDDR I and more recently was also designated as PDDR *(14)*. PDDR is presumed to be due to mutations in the gene encoding the 1α-hydroxylase. The gene encoding the 1α-hydroxylase enzyme recently has been cloned *(15)*. Linkage analysis has shown that the genetic defect in VDDR I localizes to chromosome 12 at 12q14 *(16,17)*. PDDR (or VDDRI) is an autosomal-recessive disease that is manifested at an early age, presenting with hypotonia, muscle weakness, growth failure, and rickets. Hypocalcemia, elevated PTH levels, increased alkaline phosphatase activity, and low urine calcium excretion are also found. Tetany and convulsions may occur with severe hypocalcemia. Because of the enzyme deficiency, patients with PDDR have normal or elevated 25(OH)D levels and low 1,25(OH)$_2$D levels. Patients with PDDR can be treated with physiological doses of 1,25(OH)$_2$D$_3$ (0.25–2 µg/d), which bypass the deficient enzyme and restore serum calcium concentrations to normal. As discussed below, the low serum levels of 1,25(OH)$_2$D and the effective response to physiological doses of exogenous 1,25(OH)$_2$D$_3$ distinguish PDDR from HVDRR.

24-Hydroxylase

The catabolism of 1,25(OH)$_2$D involves a series of enzymatic steps, the first of which is catalyzed by the enzyme 25-hydroxy D-24-hydroxylase (24-hydroxylase) *(18)*. There is some evidence the 24,25(OH)$_2$D may have some biological activity distinct from 1,25(OH)$_2$D, especially on cartilage cells, and this has recently been supported by experiments with the 24-hydroxylase knockout mouse model *(19)*. This initial 24-hydroxylation and subsequent hydroxylations and oxidations lead to the production of calcitroic acid which is excreted in the urine *(20,21)*. The genes encoding the rat and human 24-hydroxylase (CYP24) have been cloned *(22–24)*. The human 24-hydroxylase gene is located on chromosome 20 at 20q13 *(24)*.

1,25(OH)$_2$D has been shown to regulate 24-hydroxylase gene expression and a vitamin D response element (VDRE) (*see below*) has been identified in the regulatory region of the gene *(25,26)*. Since 24-hydroxylase can be induced by 1,25(OH)$_2$D in many VDR-containing cells, regulation of this gene product has proven to be very useful as a marker of 1,25(OH)$_2$D activity *(27)*.

THE VITAMIN D RECEPTOR

The VDR Gene

An illustration of the VDR and its gene is shown in Fig. 2. In humans, the VDR gene is located on chromosome 12 at 12q13-14 *(17,28,29)*. The structural organization of the human VDR chromosomal gene and its promoter has recently been elucidated by Miyamoto et al. *(30)* The gene encompasses more than 75 kb of DNA, and is composed of 8 coding exons (exons 2–9) and three 5' noncoding exons (exons 1A, 1B, and 1C) *(30)*. The human VDR cDNA contains 4628 nucleotide bases and encodes a protein of 427 amino acids with a predicted molecular mass of 48,000. It can be divided into a DNA binding domain (DBD) containing two zinc fingers and a ligand binding domain (LBD). The first zinc finger module of the DBD is encoded by exon 2, and the second zinc finger

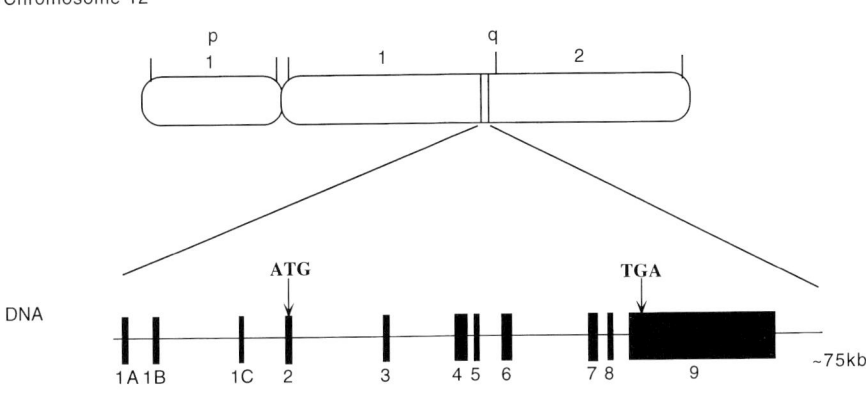

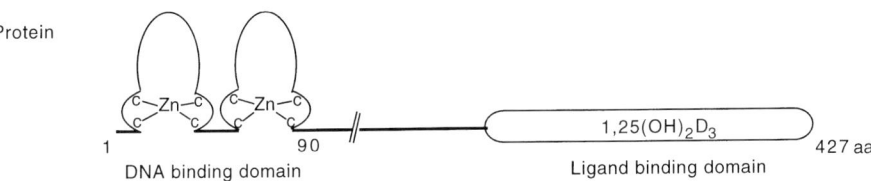

Fig. 2. Chromosomal location, exon–intron organization of the VDR gene, and structural domains of the VDR.

module is encoded by exon 3. The LDB is encoded by exons 6–9, and a hinge region, the region between the DBD and LBD, is encoded by exons 4–6. The chick VDR cDNA was originally cloned by McDonnell et al. *(31)*, which led in the following year to the cloning of the human VDR cDNA by Baker et al. *(32)*.

The system used to number the amino acids in the VDR has been somewhat confusing owing a start codon polymorphism (SCP) *(33)*. This polymorphism alters the putative initiating methionine codon in the sequence reported by Baker et al. *(32)* from ATG to ACG *(33)*. Depending upon the presence or absence of the polymorphism, the VDR would initiate three amino acids downstream at a second ATG. The VDR would thus contain 424 or 427 amino acids. A number of the earlier reports of HVDRR cases based their numbering on 424 amino acids, since the patients and normal subjects had this SCP, which resulted in the loss of the first three amino acids at the N-terminus of the VDR *(3)*. However, since researchers have universally adopted the 427 amino acid numbering system *(32)*, the numbering used in this chapter for mutations previously numbered using 424 amino acids has been adjusted by 3 to correspond to that of the published sequence *(32)*. Interestingly, the SCP has been shown to be associated with low bone mineral density in some populations, but not in others *(34–37)*.

The VDR Protein

The VDR is a ligand-activated nuclear transcription factor *(38)* and a member of the steroid-thyroid-retinoid receptor gene superfamily *(39)*. This gene superfamily includes receptors for the steroid hormones, thyroid hormone, and vitamin A (retinoids), as well

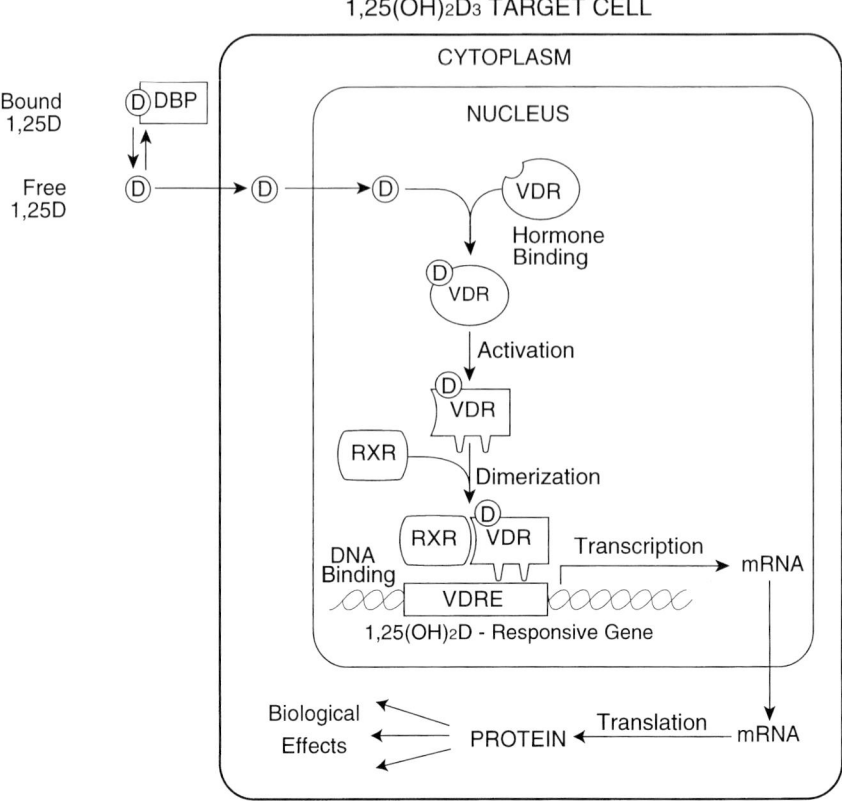

Fig. 3. Schematic representation of 1,25(OH)$_2$D$_3$ action at the cellular level. Abreviations are: D, 1,25(OH)$_2$D$_3$; DBP, vitamin D binding protein; VDR, vitamin D receptor; RXR, retinoid X receptor; and VDRE, vitamin D response element. Reproduced with permission from Feldman D, et al. *(140)*.

as a growing number of orphan receptors for which ligands have not yet been found. The mechanims of action of the VDR is similar to other members of this gene family *(see* Fig. 3). Proteins in this family are structurally similar in that they are composed of a DBD located toward the amino-terminus and an LBD located in the carboxy portion of the protein (Fig. 2). The VDR DBD is located between amino acids 20 and 90 and contains 9 highly conserved cysteine residues, which enable this region to fold into two loops or modules of 12–13 amino acids each *(see* Fig. 4) *(40)*. Each module contains four cysteine residues, which coordinate the binding of one molecule of zinc and allow the formation of a two "zinc finger" structure. The specific interaction of the VDR with DNA is thought to occur through the P-box (amino acids 40–46) located in the first zinc finger module *(41,42)*. In addition, a region located in the second zinc finger module known as the T-box (amino acids 90–101) has been shown to be involved in heterodimerization with RXR and transactivation *(42–44)*.

The architecture of the VDR LBD, which stretches over half of the polypeptide from amino acids 202–427, is more complex. From X-ray crystallographic studies of the LBD in the thyroid receptor (TR) *(45)*, retinoic acid receptor (RAR) *(46)*, and retinoid X receptor (RXR) *(47)*, the LBD is formed by 11–12 α-helices (H1-H12) and two β-sheets (s1-s2) *(see* Fig. 5). These structures form a three-dimensional pocket, which the ligand

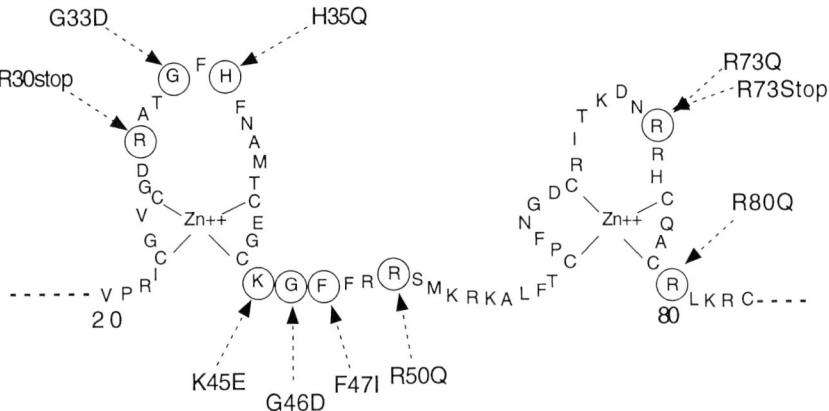

Fig. 4. DBD mutations. Schematic representation of the zinc finger structure of VDR DBD and location of mutations causing HVDRR.

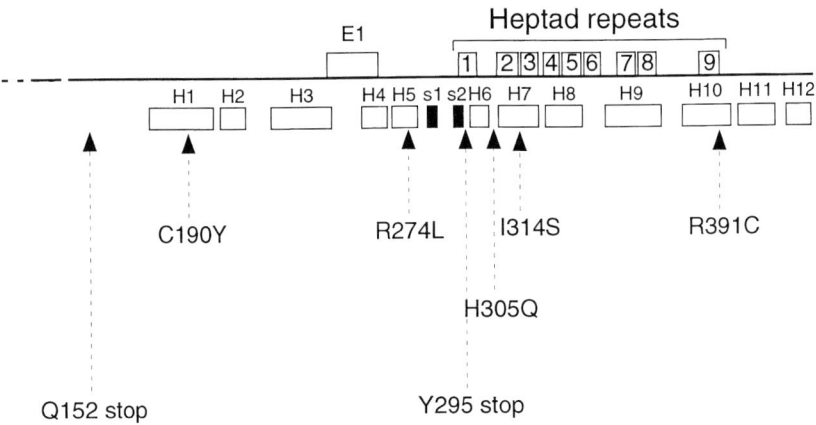

Fig. 5. LBD mutations. Schematic representation of the VDR LBD and location of mutations causing HVDRR.

occupies. The ligand then interacts with specific amino acid residues in the face of the pocket to transmit its signal. The eleventh (or twelfth) α-helix is thought to form a lid that traps the ligand in position.

Like TR, RAR, and RXR, the other members of the vitamin D-thyroid-retinoid subgroup of receptors, the VDR must bind its specific ligand, 1,25(OH)$_2$D, and heterodimerize with RXR in order to initiate transcriptional activity *(48–51)*. RXR is a 55-kDa protein and its ligand is 9-*cis*-retinoic acid *(52,53)*. RXR is found widely distributed in cells and tissues, including tissues that do not express the VDR *(48)*. Within the VDR LBD, there are several structural motifs, including an E1 region and a series of nine heptad repeats that are required for heterodimerization with RXR and for transcriptional activation (*see* Fig. 5) *(49,51)*.

Studies of laboratory-generated mutant VDRs created through deletions or single point mutations have shown that amino acids 382–427 are required for high-affinity hormone binding and heterodimerization with RXR *(51,54)*. A study of conserved cysteine residues in the LBD has shown that Cys288 also plays an important role in binding 1,25(OH)$_2$D$_3$ *(55)*. In addition, Whitfield et al. *(49)* identified two classes of mutants from mutagenesis studies of highly conserved amino acids in the E1 region (amino acids 244–263). One set of mutants binds 1,25(OH)$_2$D normally, but fails to heterodimerize with RXR on VDREs and are transcriptionally inactive. The second class of mutants also binds 1,25(OH)$_2$D normally and forms heterodimers with RXR on VDREs. However, transactivation capacity is still impaired *(49)*. Recently, investigators have shown that substitution of alanine for leucine at amino acid 417 or glutamine at amino acid 420 disrupts transcriptional activation, but not ligand binding, heterodimeric binding, or interaction with transcription factor IIB. The authors suggest that this C-terminal region of the VDR or AF-2 domain is important for VDR interaction with coactivating proteins or other components of the basal transcriptional machinery *(56)*.

The VDR has been shown to undergo a hormone-dependent phosphorylation in intact cells *(57)*. In addition, the VDR has been shown to be phosphorylated on serine residues in vitro by protein kinase C *(58)*, casein kinase-II *(59,60)*, and by the cAMP-dependent protein kinase *(61)*, which strongly suggests that phosphorylation of the VDR may be an essential event required for 1,25(OH)$_2$D-induced gene activation.

MECHANISM OF 1,25(OH)$_2$D ACTION

1,25(OH)$_2$D acts via the VDR to regulate the transcription of vitamin D target genes. The classical actions of 1,25(OH)$_2$D are to promote calcium and phosphate absorption by stimulating transport across the intestinal brush border into the circulation. This process is essential to maintain normal circulating levels of these minerals in the serum, which is required for bone mineralization. However, in recent years it has become clear that VDR and vitamin D responses have been demonstrated in many cell types, leading to a wide array of nonclassical actions unrelated to bone and mineral metabolism. Biochemical and immunocytochemical studies have shown that the VDR is widely distributed in normal human tissues, including intestine, kidney, bone, parathyroids, thyroid, skin, adrenal, liver, breast, pancreas, muscle, and lymphocytes *(62–67)*. In addition, VDR are present in numerous malignant cells *(66,67)*.

The mechanism by which 1,25(OH)$_2$D activates gene transcription through the VDR is a complex multistep process. The 1,25(OH)$_2$D–VDR complex regulates gene expression by interacting with specific nucleotide sequences termed VDREs located in the promoter regions of 1,25(OH)$_2$D-responsive genes. The typical VDRE, the so-called DR3 motif, is an imperfect hexanucleotide direct repeat separated by three nucleotide bases *(44)*. A number of 1,25(OH)$_2$D- regulated genes containing VDREs have been identified, including osteocalcin, osteopontin, calbindin-D$_{9k}$, PTH, type I collagen, and 24-hydroxylase.

A model of 1,25(OH)$_2$D-activated gene transcription is shown in Fig. 3. After being synthesized to the active form in the kidney, 1,25(OH)$_2$D circulates in the blood bound to DBP with a small fraction of free hormone. The free, fat-soluble hormone readily enters target cells through the lipid bilayer of the cell membrane and binds to the VDR in the nucleus. Activation of the VDR is probably initiated by a conformational change in the receptor owing to 1,25(OH)$_2$D binding and specific phosphorylation. The activated 1,25(OH)$_2$D–VDR complex then heterodimerizes with RXR. The heterodimer complex then binds with high affinity and specificity to VDREs in the promoter regions of target genes. After the recruitment of other transcription factors and coactivators *(39,68)*, the

Table 1
Biochemical Profile of Patient with HVDRR[a]

	Normal range	Referral	40 d[b]	80 d[c]	100 d
Ca (mmol/L)	2.2–2.6	1.86	1.77	1.80	1.71
Pi (mmol/L)	1.4–2.2	1.0	1.0	1.0	0.9
ALP (IU/L)	145–320	3056	3991	3800	3609
25 (OH)D (nmol/L)	25–85	30	37.4	250	211
1,25(OH)$_2$D (pmol/L)	40–105	521	953	1830	1560
PTH (pmol/L)	<8	—	34.2	69.9	64.5

[a] Reproduced with permission from Zhu et al. (125).
[b] Treatment: 250 mg elemental calcium q.i.d. and 0.5 µg Rocaltrol b.i.d., 20,000 IU vitamin D$_3$.
[c] Treatment: 250 mg elemental calcium q.i.d. and 5 µg Rocaltrol b.i.d.

VDR-RXR heterodimer enhances the transcription of 1,25(OH)$_2$D-responsive genes. The expression of these new genes leads to protein synthesis and ultimately to a cellular response to the hormone. In some cases, the action of the hormone is inhibitory, which is perhaps best exemplified by the inhibition of PTH gene expression by 1,25(OH)$_2$D *(69)*.

CLINICAL FEATURES OF HVDRR

HVDRR is caused by a generalized resistance to 1,25(OH)$_2$D$_3$ *(5,6,27,70–139)*. The major clinical findings are hypocalcemia and rickets, which is the result of defective mineralization of newly forming bone and preosseous cartilage. Rickets is often severe and exhibited usually within months of birth. Patients suffer from bone pain, muscle weakness, hypotonia, and occasionally convulsions from hypocalcemia. Children are often growth-retarded and in some cases, they develop severe dental caries or exhibit hypoplasia of the teeth *(71,74,75,79,80,98,101)*. Some infants have died from pneumonia caused by poor respiratory movement owing to severe rickets of the chest wall *(74,80,93)*. Many children with HVDRR have sparse body hair, and some have total scalp and body alopecia, including eyebrows and in some cases eyelashes.

An example of the typical biochemistries found in HVDRR cases is shown in Table 1. The clinical findings include low serum levels of calcium and phosphate and elevated serum alkaline phosphatase activity. The patients have elevated serum PTH levels and develop secondary hyperparathyroidism. The serum 1,25(OH)$_2$D$_3$ levels are elevated. This clinical feature distinguishes HVDRR from PDDR, since the serum 1,25(OH)$_2$D$_3$ values in PDDR are depressed. Unlike patient's with PDDR, most patients with HVDRR are resistant to supraphysiologic doses of all forms of vitamin D therapy.

HVDRR follows an autosomal-recessive pattern of inheritance. The recessive nature of the disease is evident from the patient's parents who are heterozygous for the genetic trait, but show no symptoms of the disease and have normal bone development. In many cases, consanguinity is associated with the disease. Males and females are equally affected, and often a family has several affected children.

CELLULAR EVIDENCE OF 1,25(OH)$_2$D RESISTANCE

Studies of VDR from HVDRR patients began shortly after the receptor was demonstrated in skin *(62)*, and it was shown that 1,25(OH)$_2$D$_3$ action could be studied in dermal fibroblasts cultured from human skin biopsies *(110)*. Feldman et al. *(27)* showed that specific [^{3}H]1,25(OH)$_2$D$_3$ binding was absent in cytosolic extracts of cultured fibroblasts from a patient with the HVDRR syndrome. These authors further demonstrated that the fibroblasts were resistant to 1,25(OH)$_2$D$_3$ action by the failure of the cells to induce 24-hydroxylase enzyme activity, which served as a functional marker of 1,25(OH)$_2$D$_3$ responsiveness. Chen et al. *(86)* continued studies on this HVDRR patient and an affected sibling (designated the A family), as well as three additional unrelated HVDRR patients (from the B and C families). They showed that all of the cells from the HVDRR patients lacked [^{3}H]1,25(OH)$_2$D$_3$ binding and failed to respond to 1,25(OH)$_2$D$_3$ with the induction of 24-hydroxylase activity. One unexpected finding was that the fibroblasts from only one parent (the father in the C family) exhibited the heterozygous phenotype of having half of the normal amount of receptor and a half-maximal response to 1,25(OH)$_2$D$_3$. The fibroblasts from the other parents all showed a normal amount of [^{3}H]1,25(OH)$_2$D$_3$ binding and had a normal response to 1,25(OH)$_2$D$_3$ treatment. This may explain the normal phenotype of the heterozygotic parents. A number of HVDRR cases were subsequently examined using cultured skin fibroblasts *(78,80–83)* or bone *(84)*. Some patients had no specific [^{3}H]1,25(OH)$_2$D$_3$ binding *(78,80–84)* similar to the original cases reported by Feldman et al. *(27)*. However, some patients showed normal [^{3}H]1,25(OH)$_2$D$_3$ binding, but their cells were still resistant to 1,25(OH)$_2$D$_3$ treatment *(80,82,84,94)*. Patient's who had normal [^{3}H]1,25(OH)$_2$D$_3$ binding were called "receptor-positive" or "ligand binding-positive," whereas those who showed no specific [^{3}H]1,25(OH)$_2$D$_3$ binding were called "receptor-negative" or "ligand binding-negative." Studies by Griffin and Zerwekh *(82)* and Liberman et al. *(83,84)* confirmed that fibroblasts from HVDRR patient's were resistant to 1,25(OH)$_2$D$_3$ using 24-hydroxylase induction as the marker of hormone responsiveness, but Clemens et al. *(81)* used the loss of growth inhibition by 1,25(OH)$_2$D$_3$ as a marker of resistance. These early studies demonstrated that the HVDRR syndrome was caused by cellular resistance to 1,25(OH)$_2$D$_3$ action and was due to at least two types of defects resulting in ligand binding-positive and ligand binding-negative cells.

In 1982, Pike et al. *(111)* used a monoclonal antibody (MAb) to VDR and a radioligand immunoassay *(112)* and showed that the cytosol from fibroblasts of patients with the ligand binding negative phenotype contained an immunoreactive VDR protein *(113)*. The conclusion drawn from these studies was that the absence of 1,25(OH)$_2$D$_3$ binding was due to a structural abnormality in the LBD of the receptor and not from a lack of expression of the protein.

Fibroblasts from four unrelated individuals with HVDRR, all of whom exhibited the ligand binding-positive phenotype were examined for 1,25(OH)$_2$D$_3$ responses by Gamblin et al. *(88)*. In two cases, complete hormone resistance was seen. In the other two cases, a positive 1,25(OH)$_2$D$_3$ response was observed when the cells were treated with very high concentrations of 1,25(OH)$_2$D$_3$. Interestingly, these patients also showed a calcemic response to high doses of calciferols in vivo. An HVDRR case very similar to the one described above was reported by Castells et al. *(92)*. This patient exhibited very low 1,25(OH)$_2$D$_3$ binding, but did respond to treatment with extremely high doses of 1,25(OH)$_2$D$_3$. In retrospect, these latter cases are probably caused by subtle mutations in the VDR that result in a decreased affinity for hormone. This type of defect can be overcome by treatment with high doses of the hormone.

Two sisters affected with HVDRR and an unaffected sister from a Haitian family (D family) were examined for VDR defects by Hirst et al. *(89)*. Their studies showed that fibroblasts from the affected individuals had normal [^{3}H]1,25(OH)$_2$D$_3$ binding, but were resistant to 1,25(OH)$_2$D$_3$ treatment. Based on the prior observation that the VDR can bind to DNA, these authors used DNA cellulose chromatography to demonstrate that the VDR from the HVDRR fibroblasts that bound to DNA required less salt to elute the VDR from the DNA cellulose than the normal VDR. The authors concluded that the VDR defect was probably located in the DBD, which caused the receptor to have a decreased affinity for DNA. Malloy et al. *(103)* showed a similar DNA binding defect in the VDR from fibroblasts from HVDRR patients (G family), who had normal [^{3}H]1,25(OH)$_2$D$_3$ binding. In addition, they showed that the parents were heterozygous for the VDR having two forms of the receptor, one that had a normal affinity for DNA (wild-type) and one that had a low affinity for DNA (similar to the affected children). In addition, they showed by Western blots that both forms of the receptor had the same molecular weight as the normal VDR. When tested for 1,25(OH)$_2$D$_3$ responsiveness, the parents in the G family showed a normal response. Since the parents were heterozygous for the defective VDR, but showed a normal response to hormone, this demonstrated that the genetic defect was recessive in nature. Liberman et al. *(94)* analyzed four HVDRR patients with ligand binding-positive resistance to 1,25(OH)$_2$D$_3$. Two patients exhibited the same type of DNA binding defect in the VDR as the D and G families. In the other two cases, the authors showed that the VDR had a normal affinity for DNA, but the receptors exhibited a marked decrease in nuclear localization.

Although cultured skin fibroblasts have been the main source for studying HVDRR, a number of other cells have been used to study this disease, including peripheral mononuclear cells *(91)*, phytohemagglutinin (PHA)-stimulated lymphocytes *(97,106)*, myeloid progenitor cells *(99)*, Epstein-Barr virus (EBV) immortalized B-lymphoblasts *(103, 114–116)*, and HTLV-1 virus immortalized T-lymphoblasts *(105)*. In PHA-stimulated lymphocytes, the loss of 1,25(OH)$_2$D$_3$ inhibition of DNA synthesis or the failure to induce 24-hydroxylase activity has been used to diagnose HVDRR rapidly *(97,106)*. It is interesting to note that in contrast to fibroblasts, PHA-stimulated lymphocytes from parents of children with HVDRR exhibit a true heterozygous phenotype, since these cells express intermediate levels of 24-hydroxylase activity *(106)*.

MUTATIONS AFFECTING THE DBD

From the DNA cellulose binding studies of HVDRR cases, it was suspected that the DNA binding defects exhibited by the VDR were likely to be the result of a mutation in the DBD *(89)*. The development of the polymerase chain reaction (PCR) technique by Saiki et al. *(117)* provided a method to amplify gene sequences from small amounts of DNA using a thermostable DNA polymerase. From the biochemical and cellular data obtained from the earlier studies of the HVDRR cases, investigations into the molecular cause of the receptor-positive DNA binding defective phenotype were begun. In 1988, Hughes et al. *(114)* using PCR to amplify each exon of the VDR gene in both the D and G families, identified a unique single-base change in the DBD in each family. In the D family, a G to A transition at nucleotide 304 in exon 3 was identified. This missense mutation resulted in an arginine (CGA) at amino acid residue 73 being replaced by a glutamine (CAA) in the second zinc finger (Arg73Gln) (Fig. 4). The amino acid numbers have been changed by three to be consistent with the 427 amino acid numbering system of Baker et al. *(32)*. In the G family, a G to A transition was identified at nucleotide 184 in exon 2. This missense mutation changed glycine at amino acid residue 33 (GGC) to aspartic acid (GAC) in the first zinc finger (Gly33Asp) (Fig. 4). Only the unique base changes were found in the children with HVDRR, whereas both a normal sequence and the unique base change were found in the parents. The mutations in the VDR found by Hughes et al. *(114)* were the first genetic defects identified in any member of the steroid-thyroid-retinoid receptor superfamily.

The unique base changes found in these families were then recreated in the wild-type cDNA by site-directed mutagenesis and the construct inserted into a eukaryotic expression vector *(114)*. The recreated genes were expressed in COS-1 cells and produced VDRs that had normal [^{3}H]1,25(OH)$_2$D$_3$ binding. DNA cellulose chromatography of the recreated mutant VDRs showed that they had abnormal DNA binding properties similar to the receptors in the HVDRR fibroblasts demonstrating that the single base change gives rise to the DNA binding defective phenotype. The recreated mutant receptors were also tested for their transcriptional activation capacity in CV-1 cells cotransfected with a VDRE reporter plasmid linked to the expression of the chloramphenicol acetyltransferase (CAT) gene *(118)*. Only the normal VDR was able to initiate 1,25(OH)$_2$D$_3$-dependent transcription of the reporter gene; no induction of CAT activity was associated with the mutant VDRs. These data demonstrated that the missense mutations caused the receptors to be inactive transcriptionally and therefore led to the 1,25(OH)$_2$D$_3$ resistance seen in the patients.

Since the original report by Hughes et al. *(114)*, a number of missense mutations have now been identified in the VDR DBD. These mutations and their location in the VDR are summarized in Fig. 4.

Sone et al. *(119)* found a G to A missense mutation in exon 3 in two unrelated patients previously described by Liberman et al. *(83,94)*. This mutation replaces arginine (CGG) with a glutamine (CAG) at amino acid residue 80 located in the second zinc finger (Arg80Gln). The recreated mutant VDR exhibited normal [^{3}H]1,25(OH)$_2$D$_3$ binding, but had a low affinity for DNA and was unable to activate gene transcription *(119)*. This missense mutation Arg80Gln was also identified in two siblings with HVDRR (N family) by Malloy et al. *(120)*. The N family and the families described by Sone et al. *(119)* both had origins in North Africa. However, no relationship between these families could be established.

Saijo et al. *(33)* found a fourth mutation in the VDR DBD. Prior work on fibroblasts from three HVDRR patients from two unrelated families showed normal [^{3}H]1,25(OH)$_2$D$_3$ binding in the affected children. However, the cells showed abnormal nuclear uptake of hormone *(97,104,107)*. A unique G to A base change was identified in exon 3. This missense mutation converts arginine at position 50 to glutamine (Arg50Gln). The mutation was identified in the parents by single-strand conformational polymorphism (SSCP) *(33)*.

Yagi et al. *(121)* identified a C to G missense mutation in the VDR gene from a patient whose fibroblasts exhibited normal [^{3}H]1,25(OH)$_2$D$_3$ binding, but the VDR had a lower affinity for DNA. This mutation, located in exon 2, changed a positively charged histidine at residue 35 to a neutral glutamine (His35Gln). To examine the transactivation capacity of the mutant VDR, the patient's fibroblasts were transiently transfected with a VDRE-CAT reporter plasmid and assayed for 1,25(OH)$_2$D$_3$-induced CAT activity. The fibroblasts were unable to induce gene transcription from the reporter plasmid demonstrating that the cells were resistant to 1,25(OH)$_2$D$_3$. However, when the fibroblasts were cotransformed with the wild-type VDR, they initiated gene transcription in response to hormone treatment indicating that the identified mutation was responsible for the hormone resistance.

An HVDRR patient first reported by Lin and Uttley *(109)* was examined for VDR mutations by Rut et al. *(122)*. In this case, an A to G mutation was found in exon 2. The missense mutation resulted in lysine at amino acid 45 being replaced by glutamic acid (Lys45Glu). In the same report, Rut et al. *(122)* examined the VDR gene in an HVDRR patient described by Simonin et al. *(108)*. They identified a unique T to C base change in exon 2 that resulted in phenylalanine at amino acid 47 being replaced by isoleucine (Phe47Ile). Both recreated mutant VDRs exhibited normal [^{3}H]1,25(OH)$_2$D$_3$ binding, but were transcriptionally inactive *(122)*. In all of the VDR DBD mutations described above, the defects were located in highly conserved amino acids that are common to all of the steroid receptor superfamily genes.

Lin et al. *(123)* examined the VDR gene for mutations in a patient with HVDRR previously described Sakati et al. *(96)*. DNA sequencing of the gene uncovered a unique G to A base change in exon 2. This mutation resulted in a glycine at amino acid 46 being changed to an aspartic acid (Gly46Asp). In contrast to the other DBD mutations described above, the mutation at Gly46 occurs in an amino acid that is not well conserved in the steroid-thyroid-retinoid receptor superfamily. However, Gly46 is conserved among receptors that form heterodimers with RXR proteins, such as TR and RAR. The recreated mutant VDR exhibited normal [^{3}H]1,25(OH)$_2$D$_3$ binding, but showed a reduced affinity for DNA. The mutant VDR was inactive in reporter gene assays, which demonstrated that it was the cause of 1,25(OH)$_2$D$_3$ resistance. The authors demonstrated that the patient was homozygous for the mutation, and the patient's father was a carrier of the mutant allele using PCR and a restriction fragment length polymorphism (RFLP) generated by the mutation.

A previously unreported patient with HVDRR from a Moroccan family was examined for mutations in the VDR gene by Wiese et al. *(124)*. At the cellular level, this patient exhibited the ligand binding negative phenotype, suggesting that the defect was located in the VDR LDB. However, Wiese et al. *(124)* found a mutation in the VDR DBD. A unique C to T base change replaced the codon for arginine (CGA) with an opal termination codon (TGA) (Arg73stop). The Arg73stop mutation occurs in the middle of the second zinc finger deleting the entire LBD and MAb binding sites. Interestingly, the Arg73stop mutation (CGA to TGA) occurs in the same codon that gives rise to the G family mutation (CGA to CAA) (Arg73Gln) *(114)*, but at a different nucleotide base.

Studies of a young boy of French-Canadian origin with HVDRR have been reported by Zhu et al. *(125)*. The patient's fibroblasts lacked specific [^{3}H]1,25(OH)$_2$D$_3$ binding and failed to exhibit 24-hydroxylase mRNA induction after treatment with up to 100 nM 1,25(OH)$_2$D$_3$. Northern blotting showed that the cells expressed a normal-size VDR mRNA, but Western blotting failed to detect any protein. A C to T base substitution was located in exon 2, which changed the codon for arginine (CAG) at amino acid 30 to an opal stop codon (TAG) (Arg30stop). The 29 amino acid polypeptide represents the shortest truncated protein produced by a premature stop mutation in the VDR. The mutation eliminated 398 amino acids including the LBD, the MAb epitope, the second zinc finger module, and a portion of the first zinc finger module. The Arg30stop mutation was also identified in a boy with HVDRR from a family living in Brazil *(126)*.

MUTATIONS AFFECTING THE LBD

To investigate the molecular basis of the ligand binding negative class of defects, investigators have also used PCR to amplify the exons of the LBD to facilitate their studies. The initial molecular analysis of the ligand binding negative phenotype was performed by Ritchie et al. *(115)* in three related families (C, E, and H families) *(27,86, 87,90)*. A single unique base change was identified at nucleotide 970 in exon 8 *(115)*. This single-base change replaced a tyrosine codon (TAC) with an ochre termination codon (TAA) (Tyr295stop) (Fig. 5). The location of the nonsense mutation at amino acid 295 truncates the VDR to a molecular size of 32,000 Dalton, approx 18,000 Daltons shorter than the native protein. The ochre mutation eliminated a major portion of the VDR LBD and resulted in the ligand binding negative phenotype. The recreated mutant VDR containing the ochre mutation exhibited a molecular size of 32,000 Daltons on Western blots, but was unable to bind $[^3H]1,25(OH)_2D_3$. Like the DBD mutants, the truncated VDR failed to activate gene transcription in a VDRE-CAT reporter system demonstrating that this mutation results in a hormone-resistant state.

These three families and four additional related families (F, J, K, and L families) comprised a large kindred in which consanguineous marriages had occurred. As many as eight offspring in this kindred had HVDRR *(127)*. Analysis of $[^3H]1,25(OH)_2D_3$ binding in fibroblasts or EBV-transformed lymphoblasts from the HVDRR-affected children showed that these patients all had the ligand binding negative phenotype *(127)*. However, the truncated VDR could be not demonstrated on Western blots. An analysis of the VDR mRNA by Northern blotting showed that the mutant VDR mRNA was absent in all but one case (the F family) in which it was greatly diminished *(127)*. The absence of a VDR transcript explains the absence of the mutant VDR protein on Western blots. Loss of mRNA transcripts owing to a mutation causing a premature stop in the message have been reported in a number of other genetic diseases *(116)*. In those cases in which it has been studied, the decreased or absent transcript has been attributed to instability of the abnormal mRNA harboring a nonsense mutation. An analysis of an *Rsa*I RFLP created by the mutation showed that all of the patients with HVDRR were homozygous for the Tyr295stop mutation and that their parents were heterozygous for the mutant allele *(127)*.

Two children with HVDRR (A family) described in several early reports *(27,86,87,90)* and shown to exhibit the ligand binding-negative phenotype were examined by Malloy et al. *(127)*. Both patients had the Tyr295stop mutation as demonstrated by *Rsa*I RFLP and DNA sequencing *(127)*. Although this family lived in the same town as the extended kindred described above, they are not related. Interestingly, Wiese et al. *(124)* also found the Tyr295stop mutation in two related patients from Saudi Arabia previously studied by Bliziotes et al. *(101)*.

Since the initial report by Ritchie et al. *(115)*, a number of mutations have now been identified in the VDR LBD. These mutations and their location in the VDR are summarized in Fig. 5.

A Turkish patient with HVDRR shown previously to lack hormone binding and cellular responses to $1,25(OH)_2D_3$ *(102)* was examined for mutations in the VDR gene by Kristjansson et al. *(128)*. They found a single-base change in exon 4 that altered the codon for glutamine (CAG) to an amber termination codon (TAG). This premature stop mutation occurs in the hinge region of the receptor at amino acid 152 (Gln152stop). The mutation deletes 306 amino acids of the LBD and results in the ligand binding-negative phenotype. The truncated VDR exhibited no activity added to $1,25(OH)_2D_3$ in gene activation assays. The Gln152stop mutation was also identified by Wiese et al. *(124)* in an HVDRR patient previously reported by Barsony et al. *(102)*.

A missense mutation in the VDR LBD was described by Rut et al. *(129)* and Kristjansson et al. *(128)*. Preliminary studies of $[^3H]1,25(OH)_2D_3$ binding in fibroblasts from an HVDRR patient from Kuwait *(93)* showed that the patient had a normal complement of VDR. However, the affinity of the receptor for $1,25(OH)_2D_3$ was approx 10-fold lower (K_d $10 \times 10^{-10}M$) than normal controls (K_d $0.7 \times 10^{-10}M$) *(129)*. Resistance to $1,25(OH)_2D_3$ was shown by the failure of the patient's fibroblasts to express 24-hydroxylase activity when treated with the hormone. Genetic analysis uncovered a unique G to T single-base change at nucleotide 821 in exon 7 *(128,129)*. This missense mutation changed a positively charged arginine at amino acid 274 to a neutral leucine (Arg274Leu). Although $[^3H]1,25(OH)_2D_3$ binding studies were not performed, the recreated Arg274Leu mutant VDR did activate gene transcription from a VDRE reporter plasmid. However, the transactivation process required hormone concentrations approx 1000-fold higher than required by the wild-type receptor *(128)*.

A missense mutation in the VDR LBD has been described by Thompson et al. *(130)*. In this case, a unique single-base change was identified in exon 5 that changed cysteine (TGT) to a tryptophan (TGG) at residue 190 (Cys190Trp). Additional studies were not reported.

A patient exhibiting three rare genetic disorders—HVDRR, congenital total lipodystrophy, and persistent müllerian duct syndrome—was described by Van Maldergem et al. *(131)*. The molecular analysis of the HVDRR in this case was reported by Malloy et al. *(132)*. $[^3H]1,25(OH)_2D_3$ binding experiments showed that the patient's fibroblasts had a normal complement of VDR, but the receptor had a slightly lower affinity for $1,25(OH)_2D_3$ than normal fibroblasts. Northern blot analysis showed that induction of 24-hydroxylase mRNA in the patient's fibroblasts required approximately fivefold more $1,25(OH)_2D_3$ compared to control cells. A unique single C to G base change was found in exon 8, resulting in the replacement of histidine at amino acid 305 with glutamine (His305Gln). RFLP analysis showed that the patient and a sibling with HVDRR were homozygous for the mutation, but the parents were heterozygous. In reporter gene assays, the recreated His305Gln mutant VDR exhibited the same hyporesponsiveness to $1,25(OH)_2D_3$ as seen in the patient's fibroblasts.

Two missense mutations in the LBD have been characterized by Whitfield et al. *(133)*. One patient, a girl who was originally seen at 2 yr of age, had the classic symptoms of HVDRR. Fibroblasts from the patient were originally examined by Griffin and Zerwekh *(82)*, who showed that the cells had normal $1,25(OH)_2D_3$ binding, but had defective induction of 24-hydroxylase activity. Nucleotide sequencing of the VDR uncovered a T to G substitution in exon 8, which altered the codon for isoleucine at amino acid 314 to a serine (Ile314Ser). Transactivation experiments showed that high concentrations of $1,25(OH)_2D_3$ were required to achieve normal activity. This patient showed a nearly complete cure when treated with pharmacologic doses of 25-hydroxyvitamin D. The second patient in the study, a young girl with HVDRR and alopecia, was found to have a C to T base change in exon 9, which changed arginine at amino acid 391 to a cysteine (Arg391Cys). Interestingly, in transactivation experiments, the mutant VDR required high concentrations of $1,25(OH)_2D_3$ and increased levels of RXR in order to achieve normal gene transactivation. These studies showed the importance of both $1,25(OH)_2D_3$ binding and heterodimerization with RXR in VDR-mediated gene activation *(133)*.

A single-base change in intron 4 in the VDR gene that caused exon skipping was shown by Hawa et al. *(134)* to be the most probable cause of HVDRR in a young Greek girl. Examination of the VDR message by RT-PCR showed that the patient's RNA sequence diverged from the normal sequence at nucleotide 147. At the point that the sequences diverged, it was noted that the exon 4 sequence was deleted and that the sequence of exon 5 continued in its place. Sequence analysis of the VDR gene uncovered a G to C mutation in the 5'- end of intron 4. This mutation disrupts the consensus sequence for the 5'-donor splice site (normal sequence: GTA/GAGT; mutant sequence: GTA/GACT) and most likely caused exon 4 to be skipped in the processing of the VDR transcript. This exon skipping results in a frame shift that introduces a premature stop codon in the VDR coding sequence. The truncated VDR was unable to bind hormone and failed to induce 24-hydroxylase activity.

There has been one case reported in which a major structural defect in the VDR gene was found to cause HVDRR *(130)*. The defect, found by PCR and Southern blotting, was a deletion in the VDR gene that eliminated exons 7, 8, and 9. This is the only case, thus far reported, in which a partial deletion in the VDR gene has been shown as the cause of HVDRR.

OTHER MUTATIONS

Since the time HVDRR was described as a genetic disorder, it has been suspected that the resistance to $1,25(OH)_2D_3$ exhibited by HVDRR patients was caused by mutations in the VDR. However, although the VDR is the principal factor in the $1,25(OH)_2D_3$ action pathway, target organ resistance to $1,25(OH)_2D_3$ may be caused by mutations in other interacting proteins that participate in the transactivation process. Some of the likely candidates include RXR, or other transcription factors, and coactivators or corepressors. Defects in interacting proteins may inhibit VDR binding to DNA or disrupt the contact between the VDR and the interacting protein.

Hewison et al. *(135)* have described a case of HVDRR that may be caused by a defect in an interacting protein. The patient, a young girl, exhibited all the hallmarks of the disease, including alopecia. Examination of the patient's fibroblasts showed that they expressed a normal-size VDR transcript that had normal binding affinity for $[^3H]1,25(OH)_2D_3$. However, no induction of 24-hydroxylase activity could be detected after treating the fibroblasts with up to 1 µM $1,25(OH)_2D_3$. Although the cells were clearly resistant to $1,25(OH)_2D_3$, the authors could not detect a mutation in the coding region of the VDR gene. The patient's VDR cDNA was reassembled from mRNA from the resistant cells by reverse transcription and PCR. The VDR expressed from the patient's VDR cDNA exhibited a normal transactivation response to $1,25(OH)_2D_3$ in VDRE-CAT reporter assays in CV-1 cells, which indicated that the receptor was normal and that the tissue resistance was not due to a defect in the VDR. Their data suggest that hormone resistance and HVDRR may be caused by mutations in an essential protein other than the VDR that participates in the $1,25(OH)_2D_3$ hormone action pathway.

APPROACHES TO TREATMENT OF HVDRR

Most patients with HVDRR do not respond to treatment with calcitriol or other forms of vitamin D, even at extremely supraphysiologic doses. Many therapies using combinations of calcium and active vitamin D metabolites have been tried to cure the disease.

Vitamin D Therapy

In a few of the earlier reports, some HVDRR patients showed clinical improvement and X-rays showed healing of rickets following administration of pharmacologic doses of vitamin D ranging from 5000 to 40,000 IU/d *(5,6,72)*, 20 to 200 µg of 25(OH)D/d, and 17–20 µg of 1,25(OH)$_2$D$_3$/d *(3)*. Unfortunately, the molecular basis of HVDRR of many of these earlier cases has not been determined. These patients most likely had minor aberrations in the VDR LBD that resulted in a decreased binding affinity for 1,25(OH)$_2$D$_3$. This low affinity for 1,25(OH)$_2$D$_3$ could then be overcome by treatment with high doses of the hormone. It is interesting to note that the HVDRR patients that did respond to treatment usually did not exhibit alopecia.

In general, HVDRR patients who do not have alopecia appear to be more responsive to treatment with vitamin D preparations, whereas those who have alopecia are generally less responsive *(136)*. One HVDRR patient who did not have alopecia responded to treatment with 40,000 U of vitamin D/d *(72)*. This patient was shown to have an I314S mutation in the VDR LBD *(133)*. In a second case, a patient responded to treatment with high doses of 1,25(OH)$_2$D$_3$ (30 µg/d) *(131)*. This patient was shown to have an His305Gln mutation in the VDR LBD, which reduced the affinity of the VDR for 1,25(OH)$_2$D$_3$ *(132)*. These studies suggest that HVDRR patients without alopecia may be caused by defects in the VDR LBD, and these types of defects may sometimes be overcome by treatment with high doses of vitamin D metabolites. On the other hand, one HVDRR patient without alopecia did not respond to treatment with massive doses 1,25(OH)$_2$D$_3$, and his fibroblasts were also unresponsive to hormone treatment *(128)*. In this case, the HVDRR was shown to be caused by an Arg274Leu mutation in the VDR LBD. Interestingly, the recreated Arg274Leu mutant VDR did exhibit transcriptional activity when high doses of hormone were tested *(128)*. A small number of patients with alopecia have been successfully treated with vitamin D metabolites, including vitamin D or 1αOHD *(76,79)*, 25(OH)D plus 1αOHD *(80)*, and 1αOHD or 1,25(OH)$_2$D$_3$ *(73,89,92,100,104)*. Interestingly, in one case where vitamin D and 1,25(OH)$_2$D$_3$ therapies were ineffective, the patient responded to oral phosphorous therapy *(71)*.

Calcium Therapy

Oral calcium administration has sometimes been an effective therapy for treating HVDRR patients. Sakati et al. *(96)* used a high-dose oral calcium therapy of 3–4 g of elemental calcium/d to treat a patient who failed to respond to calciferols. The patient showed clinical improvement within 4 mo of therapy. Long-term iv calcium infusions have also been used to treat HVDRR patients. Balsan et al. *(137)* showed the beneficial effects of iv calcium infusions in a child with HVDRR and alopecia who did not respond to prior treatment with large doses of vitamin D derivatives or oral calcium supplements. The child received high doses of calcium iv during the nocturnal hours over a 9-mo period. Relief from bone pain was observed within the first 2 wk of this therapy, and within 7 mo, the child gained weight and height. Intravenous calcium infusions bypassed the calcium absorption defect in the intestine caused by the disease. Several other studies have used iv calcium infusion to treat children with HVDRR successfully *(101,138,139)*. This approach of iv calcium infusions to bypass the defect in 1,25(OH)$_2$D-mediated intestinal calcium absorption appears to be the most effective means of treating HVDRR patients who fail to respond to vitamin D and oral calcium. After radiological healing of the rickets by iv calcium infusion, high-dose oral calcium therapy has been shown to be an effective means of maintaining normal serum calcium concentrations in some cases *(139)*. In some clinics, if the children fail to respond to oral vitamin D and calcium, the children with HVDRR are now routinely started on this two-step protocol *(139)*.

Spontaneous Healing

Spontaneous healing of the rickets and hypocalcemia has been observed in a few cases of HVDRR *(86,87,89)*. Sometimes the improvement in the disease occurs after long-term relatively ineffective treatment with vitamin D metabolites and mineral replacement, but in other cases, the recovery was noted after the treatment was discontinued *(89)*. Interestingly, fibroblasts taken from HVDRR patients after spontaneous healing occurred continued to exhibit resistance to $1,25(OH)_2D_3$ *(89)*. In one case, spontaneous healing occurred in a patient exhibiting the ligand binding negative phenotype *(86,87)*, which was caused by a Try295stop mutation *(115,116)*. In another case, spontaneous healing occurred in a patient exhibiting ligand binding-positive phenotype *(89)* caused by an Arg73Gln mutation *(114)*. It is interesting to note that in all of the patients who showed spontaneous healing, the alopecia remained unchanged *(86,87,89)*.

CONCLUDING COMMENTS

HVDRR is a rare recessive genetic disorder caused by mutations in the VDR that results in end-organ resistance to $1,25(OH)_2D_3$ action. The major effect of the defective VDR on the vitamin D endocrine system is to decrease intestinal calcium and phosphate absorption, which results in decreased bone mineralization and rickets. Since 1978, more than 40 families exhibiting signs and/or symptoms of HVDRR have been studied. In all cases, the assignment of HVDRR has been based on resistance to vitamin D in combination with high circulating levels of $1,25(OH)_2D_3$. A number of cases have been analyzed for $1,25(OH)_2D_3$ binding and bioactivity, which have shown that the disease was caused by heterogeneous mutations in the VDR gene. A number of cases of HVDRR have not yet been examined for mutations in the VDR gene. Since some of these cases presented late in life, they may have been due to nonhereditary resistance to $1,25(OH)_2D_3$.

Analysis of the syndrome of HVDRR provides many interesting insights into vitamin D physiology and the role of the VDR in mediating $1,25(OH)_2D_3$ action. The VDR has been found in many tissues in the body, widening the scope of potential vitamin D target cells. In addition to maintaining calcium homeostasis, $1,25(OH)_2D_3$ regulates a number of other biological processes *(38,66–68,141,142)*. Important biological actions for vitamin D have been postulated in many of these sites, particularly in the immune and endocrine systems. VDRs have been found in endocrine glands, such as pituitary, pancreas, parathyroid, gonads, and placenta, and $1,25(OH)_2D_3$ has been shown to regulate hormone synthesis and secretion from these glands *(38,66–68,141,142)*. VDRs have also been found in hematolymphopoietic cells, and $1,25(OH)_2D_3$ has been shown to regulate cell differentiation and the production of interleukins and cytokines *(143)*. However, despite the many processes shown to be regulated by $1,25(OH)_2D_3$, children with HVDRR only exhibit symptoms that relate to their calcium deficiency and alopecia. Hochberg et al. *(144)* examined hormone secretion in patients with HVDRR, and found no abnormalities in insulin, TSH, PRL, GH and testosterone levels in serum. Even et al. *(145)* showed that urinary cAMP and renal excretion of potassium, phosphorous, and bicarbonate were normal in HVDRR patients treated with PTH. However, PTH failed to decrease urinary calcium and sodium excretion in these patient to the extent found in the control patients. This suggests that $1,25(OH)_2D_3$ may selectively modulate the renal response to PTH and facilitate the PTH-induced reabsorption of calcium and sodium *(145)*. Although minor aberrations have been noted in the fungicidal activity of neutrophils from HVDRR patients, the patients *(146)* do not exhibit any clinically apparent immunologic defects.

The improvement of rickets by chronic iv calcium infusion or oral calcium raises interesting questions about the role of vitamin D in bone homeostasis. First, correction of hypocalcemia and secondary hyperparathyroidism leads to healing of the rickets as assessed by X-ray and bone biopsy. Thus, although there are many well-defined actions of vitamin D on osteoblasts, the calcium treatment data suggest that $1,25(OH)_2D_3$ action on osteoblasts is not essential in order to form normal bone. The implication is that $1,25(OH)_2D_3$ action is mainly on intestinal mineral absorption to provide calcium and phosphate for bone formation. The same conclusion was reached by Underwood and DeLuca *(147)*, who showed that the development of rickets could be prevented in totally vitamin D-deficient rats by the infusion of calcium and phosphate in the absence of vitamin D.

Second, although $1,25(OH)_2D_3$ is an inhibitor of PTH production in some settings, in the HVDRR children, normalizing serum calcium by iv infusion is sufficient to suppress their PTH overproduction. In addition, iv calcium therapy without phosphate is adequate to correct all of the metabolic abnormalities in children with HVDRR. This suggests that the hypophosphatemia in these patients is mainly the result of secondary hyperparathyroidism.

Third, a number of interesting facts concerning alopecia and HVDRR are worth noting. Since VDR have been found in hair follicles *(62)*, $1,25(OH)_2D_3$ action through the VDR appears to be essential for the differentiation of this structure during embryogenesis. Also, Marx et al. *(136)* have analyzed a number of HVDRR patients, and shown that there is some correlation between the severity of rickets and the presence of alopecia. Patients with alopecia tend to be less responsive to calciferols than those without alopecia. The alopecia or some degree of hair loss appears to be associated with DBD mutations or premature stop mutations, which usually result in complete hormone resistance. Patients with LBD missense mutations in general do not develop alopecia. Alopecia remains unchanged in patients who undergo successful therapy or show spontaneous improvement. In families with a prior history of the disease, the absence of body hair in newborns provides initial evidence for HVDRR. It is of interest to note that alopecia has not been found in other conditions related to vitamin D, including VDDR I and other forms of vitamin D deficiency.

At the time of this writing, 10 point mutations have been found in the VDR DBD, 1 in the hinge region, and 6 in the LBD. A partial deletion encompassing exons 7–9 of the VDR gene has been described in one family. Mutations in the VDR DBD prevent the receptor from activating gene transcription, even though $1,25(OH)_2D_3$ binding is normal. Conversely, missense mutations in the LBD cause a less profound defect, and in some cases, $1,25(OH)_2D_3$ responsiveness was restored to individuals by therapy with high doses of hormone. On the other hand, mutations that introduce premature termination codons, which truncate the VDR, lead to complete hormone resistance.

A prenatal diagnosis of HVDRR is now possible in pregnant women from high-risk families. Cultured cells from chorionic villus samples or amniotic fluid have been used to ascertain whether the fetus has HVDRR using [^{3}H]1,25(OH)$_2$D$_3$ binding, induction of 24-hydroxylase activity, and RFLP analyses *(148,149)*.

A final point is the interesting dilemma regarding the spontaneous improvement in some HVDRR children as they get older. One hypothesis to explain the normalization of the 1,25(OH)$_2$D$_3$ endocrine system in the face of inactive VDRs is that some other transcription factor can substitute for the defective system. Possibly RAR, RXR, or TR can substitute for a nonfunctional VDR and activate the appropriate target genes to reverse the hypocalcemia and restore the bones to normal. This hypothetical explanation remains to be tested.

The biochemical and genetic analysis of the VDR in the HVDRR syndrome has yielded important insights into the structure and function of the receptor in mediating 1,25(OH)$_2$D$_3$ action. Similarly, studies of the affected children with HVDRR continue to provide further insight into the biological role of 1,25(OH)$_2$D$_3$ in vivo. A concerted investigative approach of HVDRR at the clinical, cellular, and molecular level has proven exceedingly valuable in understanding the mechanism of action of 1,25(OH)$_2$D$_3$, and improving the diagnostic and clinical management of this rare genetic disease.

REFERENCES

1. Feldman D, Malloy PJ. Hereditary 1,25-dihydroxyvitamin D resistant rickets: molecular basis and implications for the role of 1,25(OH)$_2$D$_3$ in normal physiology. Mol Cell Endocrinol 1990;72:C57–62.
2. Hughes MR, Malloy PJ, O'Malley BW, Pike JW, Feldman D. Genetic defects of the 1,25-dihydroxyvitamin D3 receptor. J Recept Res 1991;11:699–716.
3. Malloy PJ, Pike JW, Feldman D. Hereditary 1,25-dihydroxyvitamin D resistant rickets. In: Feldman D, Glorieux F, Pike JW, eds. Vitamin D. Academic, San Diego, CA, 1997, pp. 756–788.
4. Albright F, Butler AM, Bloomberg E. Rickets resistant to vitamin D therapy. Am J Dis Child 1937; 54:531–547.
5. Brooks MH, Bell NH, Love L, Stem PH, Orfei E, Queener SF, et al. Vitamin-D-dependent rickets type II. Resistance of target organs to 1,25-dihydroxyvitamin D. N Engl J Med 1978;298:996–999.
6. Marx SJ, Spiegel AM, Brown EM, Gardner DG, Downs RW Jr, Attie M, et al. A familial syndrome of decrease in sensitivity to 1,25-dihydroxyvitamin D. J Clin Endocrinol Metab 1978;47:1303–1310.
7. Holick MF. McCollum Award Lecture, 1994: vitamin D—new horizons for the 21st century. Am J Clin Nutr 1994;60:619–630.
8. Cooke NE, McLeod JF, Wang XK, Ray K. Vitamin D binding protein: genomic structure, functional domains, and mRNA expression in tissues. J Steroid Biochem Mol Biol 1991;40:787–793.
9. Haddad JG, Matsuoka LY, Hollis BW, Hu YZ, Wortsman J. Human plasma transport of vitamin D after its endogenous synthesis. J Clin Invest 1993;91:2552–2555.
10. Okuda KI. Liver mitochondrial P450 involved in cholesterol catabolism and vitamin D activation. J Lipid Res 1994;35:361–372.
11. Henry HL. Vitamin D hydroxylases. J Cell Biochem 1992;49:439.
12. Holick MF. Vitamin D: photobiology, metabolism, and clinical applications. In: DeGroot LJ, Besser M, Burger HG, Loriaux DL, Marshall JC, Odell WD, et al., eds. Endocrinology, vol. 2, 3rd ed. W.B. Saunders, Philadelphia, 1995, pp. 990–1014.
13. Prader VA, Illig R, Heierli E. Eine besondere form der primaren vitamin-D-resistenten rachitis mit hypocalcamie und autosomal-dominantem erbgang: die hereditare pseudo-mangelrachitis. Helvetica Paediatrica Acta 1961;16:452–468.
14. Glorieux FH, St-Arnaud R. Vitamin D pseudodeficiency. In: Feldman D, Glorieux FH, Pike JW, eds. Vitamin D. Academic, San Diego, CA, 1997, pp. 755–764.

15. Takeyama K, Kitanaka S, Sato T, Kobori M, Yanagisawa J, Kato S. 25-Hydroxyvitamin D_3 1α-hydroxylase and vitamin D synthesis. Science 1997;277:1827–1830.
16. Labuda M, Morgan K, Glorieux FH. Mapping autosomal recessive vitamin D dependency type I to chromosome 12q14 by linkage analysis. Am J Hum Genet 1990;47:28–36.
17. Labuda M, Fujiwara TM, Ross MV, Morgan K, Garcia-Heras J, Ledbetter DH, et al. Two hereditary defects related to vitamin D metabolism map to the same region of human chromosome 12q13-14. J Bone Miner Res 1992;7:1447–1453.
18. Okuda K, Usui E, Ohyama Y. Recent progress in enzymology and molecular biology of enzymes involved in vitamin D metabolism. J Lipid Res 1995;36:1641–1652.
19. St-Arnaud R, Glorieux FH. Vitamin D and bone development. In: Feldman D, Glorieux FH, Pike JW, eds. Vitamin D. Academic, San Diego, CA, 1997, pp. 293–304.
20. Makin G, Lohnes D, Byford V, Ray R, Jones G. Target cell metabolism of 1,25-dihydroxyvitamin D_3 to calcitroic acid. Evidence for a pathway in kidney and bone involving 24-oxidation. Biochem J 1989; 262:173–180.
21. Reddy GS, Tserng KY. Calcitroic acid, end product of renal metabolism of 1,25-dihydroxyvitamin D_3 through C-24 oxidation pathway. Biochemistry 1989;28:1763–1769.
22. Ohyama Y, Noshiro M, Okuda K. Cloning and expression of cDNA encoding 25-hydroxyvitamin D_3 24-hydroxylase. FEBS Lett 1991;278:195–198.
23. Guo YD, Strugnell S, Back DW, Jones G. Transfected human liver cytochrome P-450 hydroxylates vitamin D analogs at different side-chain positions. Proc Natl Acad Sci USA 1993;90:8668–8672.
24. Hahn CN, Baker E, Laslo P, May BK, Omdahl JL, Sutherland GR. Localization of the human vitamin D 24-hydroxylase gene (CYP24) to chromosome 20q13.2→q13.3. Cytogenet Cell Genet 1993;62: 192–193.
25. Ohyama Y, Ozono K, Uchida M, Shinki T, Kato S, Suda T, et al. Identification of a vitamin D-responsive element in the 5'-flanking region of the rat 25-hydroxyvitamin D_3 24-hydroxylase gene. J Biol Chem 1994;269:10,545–10,550.
26. Zierold C, Darwish HM, DeLuca HF. Identification of a vitamin D-response element in the rat calcidiol (25-hydroxyvitamin D_3) 24-hydroxylase gene. Proc Natl Acad Sci USA 1994;91:900–902.
27. Feldman D, Chen T, Cone C, Hirst M, Shani S, Benderli A, et al. Vitamin D resistant rickets with alopecia: cultured skin fibroblasts exhibit defective cytoplasmic receptors and unresponsiveness to $1,25(OH)_2D_3$. J Clin Endocrinol Metab 1982;55:1020–1022.
28. Faraco JH, Morrison NA, Baker A, Shine J, Frossard PM. ApaI dimorphism at the human vitamin D receptor gene locus. Nucleic Acids Res 1989;17:2150.
29. Szpirer I, Szpirer C, Riviere M, Levan G, Marynen P, Cassiman JJ, et al. The Spl transcription factor gene (SPI) and the 1,25-dihydroxyvitamin D_3 receptor gene (VDR) are colocalized on human chromosome arm 12q and rat chromosome 7. Genomics 1991;11:168–173.
30. Miyamoto K, Kesterson RA, Yamamoto H, Taketani Y, Nishiwaki E, Tatsumi S, et al. Structural organization of the human vitamin D receptor chromosomal gene and its promoter. Mol Endocrinol 1997;11:1165–1179.
31. McDonnell DP, Mangelsdorf DJ, Pike JW, Haussler MR, O'Malley BW. Molecular cloning of complementary DNA encoding the avian receptor for vitamin D. Science 1987;235:1214–1217.
32. Baker AR, McDonnell DP, Hughes M, Crisp TM, Mangelsdorf DJ, Haussler MR, et al. Cloning and expression of full-length cDNA encoding human vitamin D receptor. Proc Natl Acad Sci USA 1988; 85:3294–3298.
33. Saijo T, Ito M, Takeda E, Huq AH, Naito E, Yokota I, et al. A unique mutation in the vitamin D receptor gene in three Japanese patients with vitamin D-dependent rickets type II: utility of single-strand conformation polymorphism analysis for heterozygous carrier detection. Am J Hum Genet 1991;49: 668–673.
34. Gross C, Eccleshall TR, Malloy PJ, Villa ML, Marcus R, Feldman D. The presence of a polymorphism at the translation initiation site of the vitamin D receptor gene is associated with low bone mineral density in postmenopausal Mexican-American women. J Bone Miner Res 1996;11:1850–1855.
35. Harris SS, Eccleshall TR, Gross C, Dawson-Hughes B, Feldman D. The vitamin D receptor start codon polymorphism (Fok I) and bone mineral density in premenopausal American black and white women. J Bone Miner Res 1997;12:1043–1048.
36. Arai H, Miyamoto K-I, Taketani Y, Yamamoto H, Iemori Y, Morita K, et al. A vitamin D receptor gene polymorphism in the translation initiation codon: Effect on protein activity and relation to bone mineral density in Japanese women. J Bone Miner Res 1997;12:915–921.

37. Eccleshall TR, Garnero P, Gross C, Delmas PD, Feldman D. Lack of correlation between the start codon polymorphism of the vitamin D receptor and bone mineral density in premenopausal French women. J Bone Miner Res 1998;13:31–35.
38. Pike JW. Vitamin D_3 receptors: structure and function in transcription. Ann Rev Nutr 1991;11:189–216.
39. Mangelsdorf DJ, Evans RM. The RXR heterodimers and orphan receptors. Cell 1995;83:841–850.
40. McDonnell DP, Scott RA, Kerner SA, O'Malley BW, Pike JW. Functional domains of the human vitamin D_3 receptor regulate osteocalcin gene expression. Mol Endocrinol 1989;3:635–644.
41. Zilliacus J, Wright AP, Carlstedt-Duke J, Gustafsson JA. Structural determinants of DNA-binding specificity by steroid receptors. Mol Endocrinol 1995;9:389–400.
42. Hsieh JC, Jurutka PW, Selznick SH, Reeder MC, Haussler CA, Whitfield GK, et al. The T-box near the zinc fingers of the human vitamin D receptor is required for heterodimeric DNA binding and transactivation. Biochem Biophys Res Commun 1995;215:1–7.
43. Whitfield GK, Hsieh JC, Jurutka PW, Selznick SH, Haussler CA, MacDonald PN, et al. Genomic actions of 1,25-dihydroxyvitamin D_3. J Nutr 1995;125:1690S–1694S.
44. Freedman LP, Lemon ED. Structural and functional determinants of DNA binding and dimerization by the vitamin D receptor. In: Feldman D, Glorieux FH, Pike JW, eds. Vitamin D. Academic, San Diego, CA, 1997, pp. 127–148.
45. Wagner RL, Apriletti JW, McGrath ME, West BL, Baxter JD, Fletterick RJ. A structural role for hormone in the thyroid hormone receptor. Nature 1995;378:690–697.
46. Renaud JP, Rochel N, Ruff M, Vivat V, Chambon P, Gronemeyer H, et al. Crystal structure of the RAR-gamma ligand-binding domain bound to all-trans retinoic acid. Nature 1995;378:681–689.
47. Bourguet W, Ruff M, Chambon P, Gronemeyer H, Moras D. Crystal structure of the ligand-binding domain of the human nuclear receptor RXR-alpha. Nature 1995;375:377–382.
48. Sone T, Ozono K, Pike JW. A 55-kilodalton accessory factor facilitates vitamin D receptor DNA binding. Mol Endocrinol 1991;5:1587–1586.
49. Whitfield GK, Hsieh JC, Nakajima S, MacDonald PN, Thompson PD, Jurutka PW, et al. A highly conserved region in the hormone-binding domain of the human vitamin D receptor contains residues viral for heterodimerization with retinoid X receptor and for transcriptional activation. Mol Endocrinol 1995;9:1166–1179.
50. Jin CH, Pike JW. Human vitamin D receptor-dependent transactivation in Saccharomyces cerevisiae requires retinoid X receptor. Mol Endocrinol 1996;10:196–205.
51. Jin CH, Kerner SA, Hong MH, Pike JW. Transcriptional activation and dimerization functions in the human vitamin D receptor. Mol Endocrinol 1996;10:945–957.
52. Levin AA, Sturzenbecker LJ, Kazmer S, Bosakowski T, Huselton C, Allenby G, et al. 9-cis retinoic acid stereoisomer binds and activates the nuclear receptor RXR alpha. Nature 1992;355:359–361.
53. Heyman RA, Mangelsdorf DJ, Dyck JA, Stein RE, Eichele G, Evans RM, et al. 9-cis retinoic acid is a high affinity ligand for the retinoid X receptor. Cell 1992;68:397–406.
54. Nakajima S, Hsieh JC, MacDonald PN, Galligan MA, Haussler CA, Whitfield GK, et al. The C-terminal region of the vitamin D receptor is essential to form a complex with a receptor auxiliary factor required for high affinity binding to the vitamin D-responsive element. Mol Endocrinol 1994;8:159–172.
55. Nakajima S, Hsieh JC, Jurutka P, Galligan MA, Haussler CA, Whitfield GK, et al. Examination of the potential functional role of conserved cysteine residues in the hormone binding domain of the human 1,25-dihydroxyvitamin D_3 receptor. J Biol Chem 1996;271:5143–5149.
56. Jurutka PW, Hsieh JC, Remus LS, Whitfield GK, Thompson PD, Haussler CA, et al. Mutations in the 1,25-dihydroxyvitamin D_3 receptor identifying C-terminal amino acids required for transcriptional activation that are functionally dissociated from hormone binding, heterodimeric DNA binding, and interaction with basal transcription factor IIB, in vitro. J Biol Chem 1997;272:14,592–14,599.
57. Pike JW, Sleator NM. Hormone-dependent phosphorylation of the 1,25-dihydroxyvitamin D_3 receptor in mouse fibroblasts. Biochem Biophys Res Commun 1985;131:378–385.
58. Hsieh JC, Jurutka PW, Galligan MA, Terpening CM, Haussler CA, Samuels DS, et al. Human vitamin D receptor is selectively phosphorylated by protein kinase C on serine 51, a residue crucial to its trans-activation function. Proc Natl Acad Sci USA 1991;88:9315–9319.
59. Jurutka PW, Hsieh JC, MacDonald PN, Terpening CM, Haussler CA, Haussler MR, et al. Phosphorylation of serine 208 in the human vitamin D receptor. The predominant amino acid phosphorylated by casein kinase II, in vitro, and identification as a significant phosphorylation site in intact cells. J Biol Chem 1993;268:6791679.

60. Hilliard GM, Cook RG, Weigel NL, Pike JW. 1,25-dihydroxyvitamin D_3 modulates phosphorylation of serine 205 in the human vitamin D receptor: site-directed mutagenesis of this residue promotes alternative phosphorylation. Biochemistry 1994;33:4300–4311.
61. Jurutka PW, Hsieh JC, Haussler MR. Phosphorylation of the human 1,25-dihydroxyvitamin D_3 receptor by cAMP-dependent protein kinase, in vitro, and in transfected COS-7 cells. Biochem Biophys Res Commun. 1993;191:1089–1096.
62. Stumpf WE, Sar M, Reid FA, Tanaka Y, DeLuca HF. Target cells for 1,25-dihydroxyvitamin D_3 in intestinal tract, stomach, kidney, skin, pituitary, and parathyroid. Science 1979;206:1188–1190.
63. Colston K, Hirst M, Feldman D. Organ distribution of the cytoplasmic 1,25-dihydroxycholecalciferol receptor in various mouse tissues. Endocrinology 1980;107:1916–1922.
64. Clemens TL, Garrett KP, Zhou XY, Pike JW, Haussler MR, Dempster DW. Immunocytochemical localization of the 1,25-dihydroxyvitamin D_3 receptor in target cells. Endocrinology 1988;122:1224–1230.
65. Berger U, Wilson P, McClelland RA, Colston K, Haussler MR, Pike JW, et al. Immunocytochemical detection of 1,25-dihydroxyvitamin D receptors in normal in human tissues. J Clin Endocrinol Metab 1988;67:607–613.
66. Walters MR. Newly identified actions of the vitamin D endocrine system. Endocr Rev 1992;13:719–764.
67. Bikle DD. Clinical counterpoint: vitamin D: new actions, new analogs, new therapeutic potential. Endocr Rev 1992;13:765–84.
68. MacDonald PN, Dowd DR, Haussler MR. New insight into the structure and functions of the vitamin D receptor. Semin Nephrol 1994;14:101–118.
69. Demay MB, Kiernan MS, DeLuca HF, Kronenberg HM. Sequences in the human parathyroid hormone gene that bind the 1,25-dihydroxyvitamin D_3 receptor and mediate transcriptional repression in response to 1,25-dihydroxyvitamin D_3. Proc Natl Acad Sci USA 1992;89:8097–8101.
70. Balsan S, Garabedian M, Lieberherr M, Gueris J, Ulmann A. Serum 1,25-dihydroxyvitamin D concentrations in two different types of pseudo-deficiency rickets. In: Norman AW, Bouillon R, Thomasset M, eds. Vitamin D: Basic Research and its Clinical Application. Fourth Workshop on Vitamin D. Walter de Gruyter, New York, 1979, pp. 1143–1149.
71. Rosen JF, Fleischman AR, Finberg L, Hamstra A, DeLuca HF. Rickets with alopecia: an inborn error of vitamin D metabolism. J Pediatr 1979;94:729–735.
72. Zerwekh JE, Glass K, Jowsey J, Pak CY. An unique form of osteomalacia associated with end organ refractoriness to 1,25-dihydroxyvitamin D and apparent defective synthesis of 25-hydroxyvitamin D. J Clin Endocrinol Metab 1979;49:171–175.
73. Fujita T, Nomura M, Okajima S, Furuya H. Adult-onset vitamin D-resistant osteomalacia with the unresponsiveness to parathyroid hormone. J Clin Endocrinol Metab 1980;50:927–931.
74. Liberman UA, Samuel R, Halabe A, Kauli R, Edelstein S, Weisman Y, et al. End-organ resistance to 1,25-dihydroxycholecalciferol. Lancet 1980;1:504–506.
75. Sockalosky JJ, Ulstrom RA, DeLuca HF, Brown DM. Vitamin D—resistant rickets: end-organ unresponsiveness to $1,25(OH)_2D_3$. J Pediatr 1980;96:701–703.
76. Tsuchiya Y, Matsuo N, Cho H, Kumagai M, Yasaka A, Suda T, et al. An unusual form of vitamin D-dependent rickets in a child: alopecia and marked end-organ hyposensitivity to biologically active vitamin D. J Clin Endocrinol Metab 1980;51:685–690.
77. Beer S, Tieder M, Kohelet D, Liberman OA, Vure E, Bar-Joseph G, et al. Vitamin D resistant rickets with alopecia: A form of end organ resistance to 1,25-dihydroxyvitamin D. Clin Endocrinol 1981;14:395–402.
78. Eil C, Liberman UA, Rosen JF, Marx SJ. A cellular defect in hereditary vitamin-D-dependent rickets type II: defective nuclear uptake of 1,25-dihydroxyvitamin D in cultured skin fibroblasts. N Engl J Med 1981;304:1588–1591.
79. Kudoh T, Kumagai T, Uetsuji N, Tsugawa S, Oyanagi K, Chiba Y, et al. Vitamin D dependent rickets: decreased sensitivity to 1,25-dihydroxyvitamin D. Eur J Pediatr 1981;137:307–311.
80. Balsan S, Garabedian M, Liberman UA, Eil C, Bourdeau A, Guillozo H, et al. Rickets and alopecia with resistance to 1,25-dihydroxyvitamin D: two different clinical courses with two different cellular defects. J Clin Endocrinol Metab 1983;57:803–811.
81. Clemens TL, Adams JS, Horiuchi N, Gilchrest BA, Cho H, Tsuchiya Y, et al. Interaction of 1,25-dihydroxyvitamin-D_3 with keratinocytes and fibroblasts from skin of normal subjects and a subject with vitamin-D-dependent rickets, type II: A model for study of the mode of action of 1,25-dihydroxyvitamin D_3. J Clin Endocrinol Metab 1983;56:824–830.

82. Griffin JE, Zerwekh JE. Impaired stimulation of 25-hydroxyvitamin D-24-hydroxylase in fibroblasts from a patient with vitamin D-dependent rickets, type II. A form of receptor-positive resistance to 1,25-dihydroxyvitamin D_3. J Clin Invest 1983;72:1190–1199.
83. Liberman UA, Eil C, Marx SJ. Resistance to 1,25(OH)$_2$D$_3$: Association with heterogeneous defects in cultured skin fibroblasts. J Clin Invest 1983;71:192–200.
84. Liberman UA, Eil C, Hoist P, Rosen JF, Marx SJ. Hereditary resistance to 1,25-dihydroxyvitamin D: defective function of receptors for 1,25-dihydroxyvitamin D in cells cultured from bone. J Clin Endocrinol Metab 1983;57:958–962.
85. Adams JS, Gacad MA, Singer FR. Specific internalization and action of 1,25-dihydroxyvitamin D_3 in cultured dermal fibroblasts from patients with X-linked hypophosphatemia. J Clin Endocrinol Metab 1984;59:556–560.
86. Chen TL, Hirst MA, Cone CM, Hochberg Z, Tietze HU, Feldman D. 1,25-dihydroxyvitamin D resistance, rickets, and alopecia: analysis of receptors and bioresponse in cultured fibroblasts from patients and parents. J Clin Endocrinol Metab 1984;59:383–388.
87. Hochberg Z, Benderli A, Levy J, Vardi P, Weisman Y, Chen T, et al. 1,25-Dihydroxyvitamin D resistance, rickets, and alopecia. Am J Med 1984;77:805–811.
88. Gamblin GT, Liberman UA, Eil C, Downs RWJ, Degrange DA, Marx SJ. Vitamin D dependent rickets type II: Defective induction of 25-hydroxyvitamin D_3-24-hydroxylase by 1,25-dihydroxyvitamin D_3 in cultured skin fibroblasts. J Clin Invest 1985;75:954–960.
89. Hirst MA, Hochman HI, Feldman D. Vitamin D resistance and alopecia: a kindred with normal 1,25-dihydroxyvitamin D binding, but decreased receptor affinity for deoxyribonucleic acid. J Clin Endocrinol Metab 1985;60:490–495.
90. Hochberg Z, Gilhar A, Haim S, Friedman-Birnbaum R, Levy J, Benderly A. Calcitriol-resistant rickets with alopecia. Arch Dermatol 1985;121:646–647.
91. Koren R, Ravid A, Liberman UA, Hochberg Z, Weisman Y, Novogrodsky A. Defective binding and function of 1,25-dihydroxyvitamin D_3 receptors in peripheral mononuclear cells of patients with end-organ resistance to 1,25-dihydroxyvitamin D. J Clin Invest 1985;76:2012–2015.
92. Castells S, Greig F, Fusi MA, Finberg L, Yasumura S, Liberman UA, et al. Severely deficient binding of 1,25-dihydroxyvitamin D to its receptors in a patient responsive to high doses of this hormone. J Clin Endocrinol Metab 1986;63:252–256.
93. Fraher LJ, Karmali R, Hinde FR, Hendy GN, Jani H, Nicholson L, et al. Vitamin D-dependent rickets type II: extreme end organ resistance to 1,25-dihydroxy vitamin D_3 in a patient without alopecia. Eur J Pediatr 1986;145:389–395.
94. Liberman UA, Eil C, Marx SJ. Receptor-positive hereditary resistance to 1,25-dihydroxyvitamin D: chromatography of receptor complexes on deoxyribonucleic acidcellulose shows two classes of mutation. J Clin Endocrinol Metab 1986;62:122–126.
95. Liberman UA, Eil C, Marx SJ. Clinical features of hereditary resistance to 1,25-dihydroxyvitamin D (hereditary hypocalcemic vitamin D resistant rickets type II). Adv Exp Med Biol. 1986;196: 391–406.
96. Sakati N, Woodhouse NJY, Niles N, Harfi H, de Grange DA, Marx S. Hereditary resistance to 1,25-dihydroxyvitamin D: clinical and radiological improvement during high-dose oral calcium therapy. Hormone Res 1986;24:280–287.
97. Takeda E, Kuroda Y, Saijo T, Toshima K, Naito E, Kobashi H, et al. Rapid diagnosis of vitamin D-dependent rickets type II by use of phytohemagglutinin-stimulated lymphocytes. Clin Chim Acta 1986;155:245–250.
98. Laufer D, Benderly A, Hochberg Z. Dental pathology in calcitirol resistant rickets. J Oral Med 1987;42:272–275.
99. Nagler A, Merchav S, Fabian I, Tatarsky I, Hochberg Z. Myeloid progenitors from the bone marrow of patients with vitamin D resistant rickets (type II) fail to respond to 1,25(OH)$_2$D$_3$. Br J Haematol 1987;67:267–271.
100. Takeda E, Kuroda Y, Saijo T, Naito E, Kobashi H, Yokota I, et al. 1 alpha-hydroxyvitamin D_3 treatment of three patients with 1,25-dihydroxyvitamin D-receptor-defect rickets and alopecia. Pediatrics 1987; 80:97–101.
101. Bliziotes M, Yergey AL, Nanes MS, Muenzer J, Begley MG, Viera NE, et al. Absent intestinal response to calciferols in hereditary resistance to 1,25-dihydroxyvitamin D: documentation and effective therapy with high dose intravenous calcium infusions. J Clin Endocrinol Metab 1988;66:294–300.

102. Barsony J, McKoy W, DeGrange DA, Liberman UA, Marx SJ. Selective expression of a normal action of the 1,25-dihydroxyvitamin D_3 receptor in human skin fibroblasts with hereditary severe defects in multiple actions of that receptor. J Clin Invest 1989;83:2093–2101.
103. Malloy PJ, Hochberg Z, Pike JW, Feldman D. Abnormal binding of vitamin D receptors to deoxyribonucleic acid in a kindred with vitamin D-dependent rickets, type II. J Clin Endocrinol Metab 1989; 68:263–269.
104. Takeda E, Yokota I, Kawakami I, Hashimoto T, Kuroda Y, Arase S. Two siblings with vitamin-D-dependent rickets type II: no recurrence of rickets for 14 years after cessation of therapy. Eur J Pediatr 1989;149:54–57.
105. Koeffler HP, Bishop JE, Reichel H, Singer F, Nagler A, Tobler A, et al. Lymphocyte cell lines from vitamin D-dependent rickets type II show functional defects in the 1 alpha,25-dihydroxyvitamin D_3 receptor. Mol Cell Endocrinol 1990;70:1–11.
106. Takeda E, Yokota I, Ito M, Kobashi H, Saijo T, Kuroda Y. 25-Hydroxyvitamin D-24-hydroxylase in phytohemagglutinin-stimulated lymphocytes: intermediate bioresponse to 1,25-dihydroxyvitamin D_3 of cells from parents of patients with vitamin D-dependent rickets type II. J Clin Endocrinol Metab 1990;70:1068–1074.
107. Yokota I, Takeda E, Ito M, Kobashi H, Saijo T, Kuroda Y. Clinical and biochemical findings in parents of children with vitamin D-dependent rickets Type II. J Inherit Metab Dis 1991;14:231-240.
108. Simonin G, Chabrol B, Moulene E, Bollini G, Strouc S, Mattei JF, et al. Vitamin D-resistant rickets type II: apropos of 2 cases. Pediatrie 1992;47:817–820.
109. Lin JP, Uttley WS. Intra-atrial calcium infusions, growth, and development in end organ resistance to vitamin D. Arch Dis Child 1993;69:689–692.
110. Feldman D, Chen T, Hirst M, Colston K, Karasek M, Cone C. Demonstration of 1,25-dihydroxyvitamin D_3 receptors in human skin biopsies. J Clin Endocrinol Metab 1980;51:1463–1465.
111. Pike JW, Donaldson CA, Marion SL, Haussler MR. Development of hybridomas secreting monoclonal antibodies to the chicken intestinal 1 alpha,25-dihydroxyvitamin D_3 receptor. Proc Natl Acad Sci USA 1982;79:7719–7723.
112. Dokoh S, Haussler MR, Pike JW. Development of a radioligand immunoassay for 1,25-dihydroxycholecalciferol receptors utilizing monoclonal antibody. Biochem J 1984;221:129–136.
113. Pike JW, Dokoh S, Haussler MR, Liberman UA, Marx SJ, Eil C. Vitamin D_3—resistant fibroblasts have immunoassayable 1,25-dihydroxyvitamin D_3 receptors. Science 1984;224:879–881.
114. Hughes MR, Malloy PJ, Kieback DG, Kesterson RA, Pike JW, Feldman D, et al. Point mutations in the human vitamin D receptor gene associated with hypocalcemic rickets. Science 1988;242:1702–1705.
115. Ritchie HH, Hughes MR, Thompson ET, Malloy PJ, Hochberg Z, Feldman D, et al. An ochre mutation in the vitamin D receptor gene causes hereditary 1,25-dihydroxyvitamin D_3-resistant rickets in three families. Proc Natl Acad Sci USA 1989;86:9783–9787.
116. Malloy PJ, Hochberg Z, Tiosano D, Pike JW, Hughes MR, Feldman D. The molecular basis of hereditary 1,25-dihydroxyvitamin D_3 resistant rickets in seven related families. J Clin Invest 1996;86: 2071–2079.
117. Saiki RK, Gelfand DH, Stoffel S, Scharf SJ, Higuchi R, Horn GT, et al. Primer-directed enzymatic amplification of DNA with a thermostable DNA polymerase. Science 1988;239:487–491.
118. Sone T, Scott RA, Hughes MR, Malloy PJ, Feldman D, O'Malley BW, et al. Mutant vitamin D receptors which confer hereditary resistance to 1,25-dihydroxyvitamin D_3 in humans are transcriptionally inactive in vitro. J Biol Chem 1989;264:24,230–20,234.
119. Sone T, Marx SJ, Liberman UA, Pike JW. A unique point mutation in the human vitamin D receptor chromosomal gene confers hereditary resistance to 1,25-dihydroxyvitamin D_3. Mol Endocrinol 1990; 4:623–631.
120. Malloy PJ, Weisman Y, Feldman D. Hereditary 1 alpha,25-dihydroxyvitamin D-resistant rickets resulting from a mutation in the vitamin D receptor deoxyribonucleic acid-binding domain. J Clin Endocrinol Metab 1994;78:313–316.
121. Yagi H, Ozono K, Miyake H, Nagashima K, Kuroume T, Pike JW. A new point mutation in the deoxyribonucleic acid-binding domain of the vitamin D receptor in a kindred with hereditary 1,25-dihydroxyvitamin D-resistant rickets. J Clin Endocrinol Metab 1993;76:509–512.
122. Rut AR, Hewison M, Kristjansson K, Luisi B, Hughes MR, O'Riordan JL. Two mutations causing vitamin D resistant rickets: modelling on the basis of steroid hormone receptor DNA-binding domain crystal structures. Clin Endocrinol 1994;41:581–590.

123. Lin NU-T, Malloy PJ, Sakati N, Al-Ashwal A, Feldman D. A novel mutation in the deoxyribonucleic acid-binding domain of the vitamin D receptor gene causes hereditary 1,25-dihydroxyvitamin D resistant rickets. J Clin Endocrinol Metab 1996;81:2564–2569.
124. Wiese RJ, Goto H, Prahl JM, Marx SJ, Thomas M, al-Aqeel A, et al. Vitamin D-dependency rickets type II: truncated vitamin D receptor in three kindreds. Mol Cell Endocrinol 1993;90:197–201.
125. Zhu WJ, Malloy PJ, Chabot G, Delvin E, Feldman D. Hereditary 1,25-dihydroxyvitamin D resistant rickets due to an opal mutation causing premature termination of the vitamin D receptor. J Bone Miner Res 1998;13:259–264.
126. Mechica JB, Leite MOR, Mendonca BE, Frazzatto EST, Borelli A, Latronico AC. A novel nonsense mutation in the first zinc finger of the vitamin D receptor causing hereditary 1,25-dihydroxyvitamin D_3-resistant rickets. J Clin Endocrinol Metab 1997;82:3892–3894.
127. Malloy PJ, Hughes HR, Pike JW, Feldman D. Vitamin D receptor mutations and hereditary 1,25-dihydroxyvitamin D resistant rickets. In: Norman AW, Bouillon R, Thomasset M, eds. Vitamin D: Gene Regulation, Structure–Function Analysis, and Clinical Application. Eighth Workshop on Vitamin D. Walter de Gruyter, New York, 1991, pp. 116–124.
128. Kristjansson K, Rut AR, Hewison M, O'Riordan JL, Hughes MR. Two mutations in the hormone binding domain of the vitamin D receptor cause tissue resistance to 1,25 dihydroxyvitamin D_3. J Clin Invest 1993;92:12–16.
129. Rut AR, Hewison M, Rowe P, Hughes M, Grant D, O'Riordan JLH. A novel mutation in the steroid binding region of the vitamin D receptor (VDR) gene in hereditary vitamin D resistant rickets (HVDRR). In: Norman AW, Bouillon R, Thomasset M, eds. Vitamin D: Gene Regulation, Structure–Function Analysis, and Clinical Application. Eighth Workshop on Vitamin D. Walter de Gruyter, New York, 1991, pp. 94,95.
130. Thompson E, Kristjansson K, Hughes M. Molecular scanning methods for mutation detection: application to the 1,25-dihydroxyvitamin D receptor. Abstracts from the Eighth Workshop on Vitamin D, Paris, France, 1991,p. 6.
131. Van Maldergem L, Bachy A, Feldman D, Bouillon R, Maassen J, Dreyer M, et al. Syndrome of lipoatrophic diabetes, vitamin D resistant rickets, and persistent miillerian ducts in a Turkish boy born to consanguineous parents. Am J Med Genet 1996;64:506–513.
132. Malloy PJ, Eccleshall TR, Gross C, Van Maldergem L, Bouillon R, Feldman D. Hereditary vitamin D resistant rickets caused by a novel mutation in the vitamin D receptor that results in decreased affinity for hormone and cellular hyporesponsiveness. J Clin Invest 1997;95:297–304.
133. Whitfield GK, Selznick SH, Haussler CA, Hsieh JC, Galligan MA, Jurutka PW, et al. Vitamin D receptors from patients with resistance to 1,25-dihydroxyvitamin D3: point mutations confer reduced transactivation in response to ligand and impaired interaction with the retinoid X receptor heterodimeric partner. Mol Endocrinol 1996;10:1617–1631.
134. Hawa NS, Cockerill FJ, Vadher S, Hewison M, Rut AR, Pike JW, et al. Identification of a novel mutation in hereditary vitamin D resistant rickets causing exon skipping. Clin Endocrinol 1996;45:85–92.
135. Hewison M, Rut AR, Kristjansson K, Walker RE, Dillon MJ, Hughes MR, et al. Tissue resistance to 1,25-dihydroxyvitamin D without a mutation of the vitamin D receptor gene. Clin Endocrinol 1993;39:663–670.
136. Marx SJ, Bliziotes MM, Nanes M. Analysis of the relation between alopecia and resistance to 1,25-dihydroxyvitamin D. Clin Endocrinol 1986;25:373–381.
137. Balsan S, Garabedian M, Larchet M, Gorski AM, Cournot G, Tau C, et al. Long-term nocturnal calcium infusions can cure rickets and promote normal mineralization in hereditary resistance to 1,25-ihydroxyvitamin D. J Clin Invest 1986;77:1661–1667.
138. Weisman Y, Bab I, Gazit D, Spirer Z, Jaffe M, Hochberg Z. Long-term intracaval calcium infusion therapy in end-organ resistance to 1,25-dihydroxyvitamin D. Am J Med 1987;83:984–990.
139. Hochberg Z, Tiosano D, Even L. Calcium therapy for calcitriol-resistant rickets. J Pediatr 1992;121:803–808.
140. Feldman D, Malloy PJ, Gross C. Vitamin D: metabolism and action. In: Marcus R, Feldman D, Kelsey J, eds. Osteoporosis. Academic, San Diego, CA, 1996, pp. 205–235.
141. Reichel H, Koeffler HP, Norman AW. The role of the vitamin D endocrine system in health and disease. N Engl J Med 1989;320:980–991.
142. Darwish H, DeLuca HF. Vitamin D-regulated gene expression. Crit Rev Eukaryot Gene Expr 1993;3:89–116.

143. Manolagas SC, Yu XP, Girasole G, Bellido T. Vitamin D and the hematolymphopoietic tissue: a 1994 update. Semin Nephrol 1994;14:129–143.
144. Hochberg Z, Borochowitz Z, Benderli A, Vardi P, Oren S, Spirer Z, et al. Does 1,25-dihydroxyvitamin D participate in the regulation of hormone release from endocrine glands? J Clin Endocrinol Metab 1985;60:57-61.
145. Even L, Weisman Y, Goldray D, Hochberg Z. Selective modulation by vitamin D of renal response to parathyroid hormone: a study in calcitriol-resistant rickets. J Clin Endocrinol Metab 1996;81:2836–2840.
146. Etzioni A, Hochberg Z, Pollak S, Meshulam T, Zakut V, Tzehoval E, et al. Defective leukocyte fungicidal activity in endorgan resistance to 1,25-dihydroxyvitamin D. Pediatr Res 1989;25:276–279.
147. Underwood JL, DeLuca HF. Vitamin D is not directly necessary for bone growth and mineralization. Am J Physiol 1984;246:E493–498.
148. Weisman Y, Jaccard N, Legum C, Spirer Z, Yedwab G, Even L, et al. Prenatal diagnosis of vitamin D-dependent rickets, type II: response to 1,25-dihydroxyvitamin D in amniotic fluid cells and fetal tissues. J Clin Endocrinol Metab 1990;71:937–943.
149. Weisman Y, Malloy PJ, Krishnan AV, Jaccard N, Feldman D, Hochberg Z. Prenatal diagnosis of calcitriol resistant Rickets (CRR) by 1,25(OH)$_2$D$_3$ binding, 24-hydroxylase induction and RFLP analysis. Abstracts from the Ninth Workshop on Vitamin D, Orlando, 1994, p. 106.

5
Decreased Responsiveness to Extracellular Ca^{2+} Owing to Abnormalities in the Ca^{2+}_o-Sensing Receptor

Edward M. Brown, MD, Olga Kifor, MD, and Mei Bai, PHD

CONTENTS

INTRODUCTION
THE CAR: A KEY ELEMENT IN NORMAL MINERAL ION HOMEOSTASIS
SYNDROMES WITH GENERALIZED RESISTANCE TO CA^{2+}_o
ACQUIRED TISSUE-SPECIFIC RESISTANCE TO CA^{2+}_o IN PRIMARY
  AND UREMIC HYPERPARATHYROIDISM
SUMMARY
ACKNOWLEDGMENTS
REFERENCES

INTRODUCTION

Calcium (Ca^{2+}) ions serve a large variety of intra- and extracellular roles *(1,2)*. Cytosolic free calcium ions (Ca_i) are a key intracellular second messenger and cofactor for enzymes *(2,3)*. Ca^{2+}_i controls diverse cellular functions, such as muscular contraction, glycogen metabolism, secretion, cellular motility, and proliferation *(2)*. Basal levels of Ca_i are maintained at ~100 nM, which is nearly four orders of magnitude lower than the extracellular ionized calcium concentration (Ca^{2+}_o) (~1 mM). During cellular activation and intracellular signaling, Ca^{2+}_i can undergo large fluctuations as a result of release of Ca^{2+} from its intracellular stores and/or uptake of extracellular Ca^{2+} through various Ca^{2+}-permeable influx pathways in the plasma membrane. Ca^{2+}_o, in contrast, remains virtually invariant, varying over a range of only a few percent under normal circumstances *(1,4–6)*. Extracellular calcium ions are also essential for numerous critical functions, such as promoting skeletal integrity, clotting of the blood, and controlling neuromuscular excitability.

From: *Contemporary Endocrinology: Hormone Resistance Syndromes*
Edited by: J. L. Jameson © Humana Press Inc., Totowa, NJ

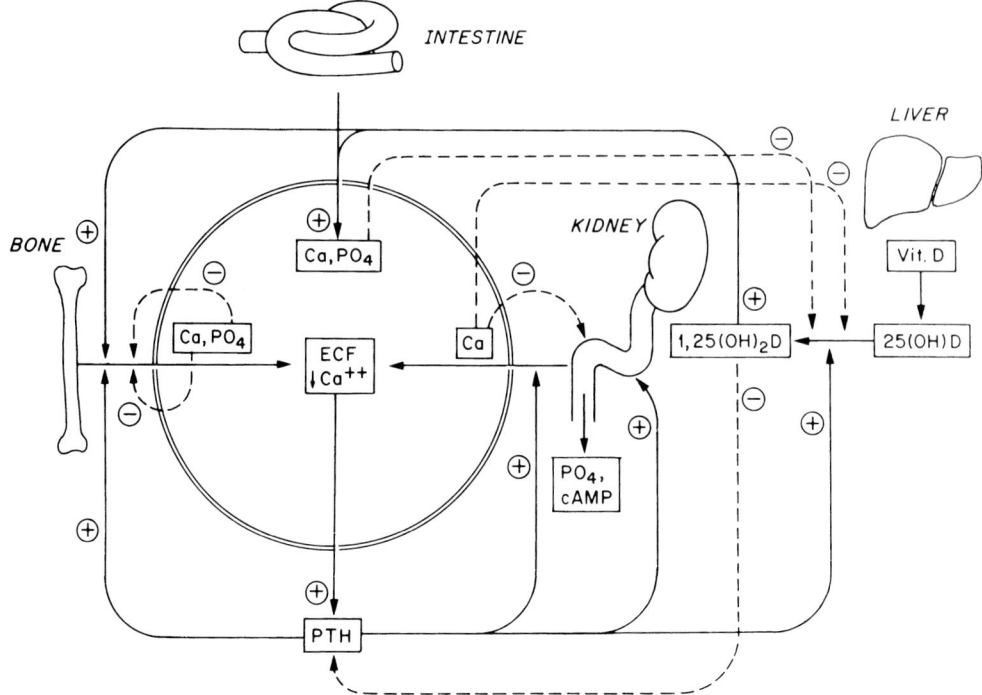

Fig. 1. Schematic representation of the regulatory system maintaining Ca^{2+}_o homeostasis. The solid arrows and lines indicate effects of PTH and $1,25(OH)_2D_3$; dotted arrows and lines show examples of how extracellular Ca^{2+} or phosphate ions directly act on target tissues. Abbreviations are as follows: Ca^{2+}, calcium; PO_4, phosphate; ECF, extracellular fluid; PTH, parathyroid hormone; $1,25(OH)_2D$, 1,25-dihydroxyvitamin D; $25(OH)D$, 25-hydroxyvitamin D; minus signs show inhibitory effects, whereas plus signs show stimulatory actions. Reproduced with permission from ref. *6a*.

Ca^{2+}_o is maintained nearly invariant by means of a homeostatic system that has two central elements (Fig. 1) *(1,4–6)*. The first of these are Ca^{2+}_o-sensing cells that secrete calciotropic hormones (i.e., parathyroid cells, renal proximal tubular cells, and thyroidal C-cells). When the extracellular calcium concentration changes, these cells modulate their hormonal secretion (e.g., of parathyroid hormone [PTH], 1,25-dihydroxyvitamin D [1,25{OH}$_2$D], and calcitonin [CT], respectively) in order to normalize Ca^{2+}_o. These calciotropic hormones can then modulate the second element of the system, tissues that regulate the fluxes of mineral ions into or out of the extracellular fluid (e.g., bone, kidney and intestine—*see* Fig. 1) in order to normalize Ca^{2+}_o. The cloning and characterization of a cell-surface, G-protein-coupled, Ca^{2+}_o-sensing receptor (CaR) *(7)* has advanced our knowledge of mineral ion homeostasis on two fronts: first, it has elucidated considerably how Ca^{2+}_o-sensing cells recognize changes in the Ca^{2+}_o signal. Second, it has made it possible to establish that inherited forms of human hyper- or hypocalcemia can result from mutations in the CaR that decrease or increase, respectively, its activity *(8,9)*. This chapter will review the clinical and molecular features of patients with disorders in which abnormalities in the CaR reducing its activity lead to Ca^{2+}_o-resistance, following a brief description of the mechanisms by which the CaR functions as a central part of the mechanism through which the level of Ca^{2+}_o is "set" by the mineral ion homeostatic system.

THE CaR:
A KEY ELEMENT IN NORMAL MINERAL ION HOMEOSTASIS
Cloning of the CaR

Expression cloning in *Xenopus laevis* oocytes provides a useful approach to cloning phosphatidylinositol- (PI) coupled receptors when no molecular probes for the receptor in question are available. Parathyroid cells are known to respond to elevated levels of Ca^{2+}_o with activation of phospholipase C (PLC) *(10,11)*, as well as with transient followed by sustained increases in Ca_i *(12)* and inhibition of adenylate cyclase that is pertussis toxin-sensitive *(13)*. These responses are characteristic of several of the "calcium-mobilizing" members of the superfamily of G-protein-coupled receptors that couple positively and negatively, respectively, to PLC and adenylate cyclase *(1)*. In addition, injection of *X. laevis* oocytes with mRNA extracted from bovine parathyroid glands rendered the oocytes responsive to Ca^{2+}_o *(14,15)*. It was, therefore, feasible to utilize a similar methodology to screen a bovine parathyroid cDNA library and isolate a full-length clone of CaR *(7)*. Subsequent use of hybridization-based screening approaches has made it possible to clone additional CaRs from human parathyroid (Fig. 2) *(16,16a)* and kidney *(17)*, rat kidney *(18)* and brain *(19)*, a rat C-cell-derived cell line (rMTC44-2 cells) *(20)*, rabbit kidney *(21)*, and chicken parathyroid *(22)*. These latter CaRs are all highly homologous to the bovine parathyroid CaR, and are thought to represent species and tissue homologs of the same ancestral gene.

The CaR has a large amino-terminal extracellular domain (ECD) that comprises over 600 amino acids and contains important determinants of the binding of Ca^{2+}_o and other polycationic agonists to the receptor *(23)*. A generally hydrophobic middle region of the receptor contains seven predicted, membrane-spanning segments and identifies the CaR as a member of the superfamily of G-protein-coupled receptors (GPCR). The CaR is thought to couple through its intracellular loops as well as its carboxyl- (C) terminal tail to the various G-proteins that mediate its biological actions (probably $G_{q/11}$ for activating PLC *(24)* and G_i for inhibiting adenylate cyclase *(13,24,25)*. Thus far, only a single isoform of the CaR has been isolated, in contrast to another family of the GPRCs that is structurally homologous to the CaR, the metabotropic glutamate receptors (mGluRs), which includes numerous family members, each of which couples principally to one effector system (e.g., mGluRs 1 and 5 activate PLC, for example) *(26)*.

Tissue Distribution and Physiological Roles of the CaR

Numerous tissues express transcripts for the CaR, many of which do not play obvious roles in systemic calcium homeostasis *(7,16,18,19)*. These include the parathyroid glands, C-cells, kidney, lung, intestine, and numerous regions of brain. The use of *in situ* hybridization and immunohistochemistry with anti-CaR antibodies has permitted more precise localization of the CaR in several of these tissues *(18,19)*. This information, combined with biochemical and physiological data delineated later, has significantly illuminated how the CaR regulates several of the tissues expressing it.

In the parathyroid, the CaR activates at least three phospholipases—PLC, phospholipase A_2 (PLA$_2$) and phospholipase D (PLD) *(27)*—and may also inhibit adenylate cyclase *(1,13)*. It is not currently known with certainty, however, which of these intracellular signaling pathways is the most important one through which the CaR inhibits PTH secretion, rather than stimulating secretion, as is usually the case following activation of

PLC by GPCRs in other hormone-secreting cells. Indeed, in C-cells, the same CaR appears to mediate the stimulatory effects of extracellular calcium on calcitonin secretion *(20,28,29)*. The marked impairment of high Ca^{2+}_o-induced inhibition of PTH release in patients with homozygous inactivating CaR mutations *(30,31)* and in mice that are homozygous for targeted disruption of the CaR gene *(32)* (*see* Neonatal Severe Hyperparathyroidism [NSHPT] and Mouse Models of FHH and NSHPT *below*) strongly support a key role for the CaR in the regulation of PTH secretion by Ca^{2+}_o. In addition, the CaR may contribute to the high Ca^{2+}_o-induced decrease in preproPTH mRNA levels, since so-called calcimimetic CaR activators that act allosterically on the CaR to increase its affinity for Ca^{2+}_o mimic the actions of elevating the extracellular calcium concentration to inhibit this parameter *(33)*. Finally, the marked chief cell hyperplasia often encountered in both NSHPT *(34,35)* and in mice homozygous for "knockout" of the CaR gene *(32)* provide indirect evidence for a role of the receptor in suppressing parathyroid cellular proliferation.

The CaR is expressed within most segments of the nephron (i.e., the glomerulus, proximal convoluted and straight tubules, medullary thick ascending limb [MTAL] and cortical thick ascending limb [CTAL], distal convoluted tubule [DCT], cortical collecting duct [CCD], outer medullary [OMCD], and inner medullary collecting ducts [IMCD]) *(18,36)*. CaR transcripts and protein are present at the highest levels in the CTAL in which the CaR is located on the basolateral surface of the cells, where it likely senses systemic/blood Ca^{2+}_o. The CaR in CTAL probably regulates tubular handling of both Ca^{2+} and Mg^{2+}, decreasing their reabsorption when Ca^{2+}_o is high and reducing it when Ca^{2+}_o is low *(37)*. The CaR and PTH receptors are both expressed in CTAL and DCT, thereby presumably enabling mutually antagonistic interactions between the actions of the two receptors on Ca^{2+}_o reabsorption in this nephron segment *(38)*. The role of the CaR in DCT remains to be elucidated, but it may also regulate calcium reabsorption in this location. In contrast to the CTAL, the CaR in the IMCD is situated principally on the apical/luminal surface of the tubular epithelial cells *(39)*, thereby permitting sensing of urinary Ca^{2+}_o.

Perfusion of the IMCD in vitro with high levels of Ca^{2+}_o directly inhibits vasopressin-stimulated water flow *(39)*, an action that may be mediated by the CaR. This action, possibly combined with the inhibition of NaCl transport in the MTAL during hypercalcemia *(40)*, may provide an explanation for the impaired urinary concentrating capacity frequently encountered in hypercalcemic individuals *(41)*. It is also possible that this interaction between the systems regulating water and mineral ion homeostasis provides a "trade-off" that reduces the risk of overly elevated levels of urinary Ca^{2+}_o in the IMCD during antidiuresis *(37,39)*.

Although much has already been learned about the role of the CaR in tissues involved in regulating mineral ion metabolism, the function(s) of the receptor in those that are not, such as the brain and lung, remains to be elucidated. It is likely, however, that in such locations the CaR responds to local rather than systemic changes in Ca^{2+}_o *(19,42)*. For example, there can be marked decreases in Ca^{2+}_o within the interstitial fluid surrounding brain cells during increases in cellular activity as a result of the influx of large amounts of calcium via various neuronal calcium channels *(43)*. In addition, it is possible that Mg^{2+}_o *(44)* or other endogenous CaR ligands, such as spermine *(45)*, also function as CaR agonists in the brain or elsewhere where their concentrations are in a range where they can modulate CaR activity.

SYNDROMES WITH GENERALIZED RESISTANCE TO Ca^{2+}_o

Familial Hypocalciuric Hypercalcemia (FHH)

CLINICAL FEATURES OF FHH

In 1972, Foley et al. described the benign clinical features of a hypercalcemic syndrome that they named familial benign hypercalcemia (FBH) *(46)*. Although several other types of familial hypercalcemia had been identified previously, this report first clearly delineated the characteristic clinical features of individuals with this apparently benign form of hypercalcemia. Subsequent investigators have identified and studied a large number of families with this condition over the ensuing 25 years, particularly Marx and colleagues *(47)* and Heath and coworkers *(48)*. Because of the striking alterations in renal Ca^{2+} handling in affected members of families with this condition, Marx et al. called it FHH *(49)*. Many but not all authors refer to the condition by this name, although in some cases it is called FBH or familial benign hypocalciuric hypercalcemia (FBHH).

FHH is a rare genetic syndrome with a prevalence considerably lower than that of primary hyperparathyroidism (PHPT), which, as described in more detail below, is a more common form of tissue-specific resistance to Ca^{2+}_o *(50,51)*. FHH is an autosomal-dominantly inherited condition, and is characterized by lifelong and usually asymptomatic hypercalcemia that is of mild to moderate severity (usually <12 mg/dL) *(47–49)*. Despite their hypercalcemia, affected patients have few if any of the symptoms and complications that are characteristic of other hypercalcemic disorders. These latter symptoms comprise most commonly gastrointestinal abnormalities (e.g., anorexia, nausea, and constipation), renal complications (i.e., impaired renal function, defective urinary concentrating ability, nephrolithiasis and/or nephrocalcinosis), and mental disturbances *(4,52)*. In some families, the degree of hypercalcemia in affected individuals is greater than generally observed in other FHH kindreds (i.e., in the range of 12–14 mg/dL *[47]*). Even in these kindreds with unusually elevated levels of Ca^{2+}_o, however, there is usually a remarkable dearth of the symptomatology commonly found in hypercalcemia. Although early studies described nonspecific manifestations of hypercalcemia, such as fatigue, in some families with FHH *(47)*, subsequent series have not confirmed this finding, suggesting that it might have been the result of ascertainment bias *(48,53)*.

Affected members of some FHH kindreds have suffered from pancreatitis or chondrocalcinosis (calcification of the cartilage in joints) (47), leading to the suggestion that these manifestations might be *bone fide* complications of this syndrome. More recent studies, however, have found that pancreatitis is no more common in affected than in unaffected family members of kindreds with FHH or in the general population. Moreover, most patients with FHH who have had pancreatitis also had other factors predisposing to this condition (e.g., gallstones or alcoholism) *(53)*. Studies carried out subsequently have likewise failed to demonstrate that chondrocalcinosis is more common in patients with FHH than in the general population *(48,53)*. Although one study *(48)* described an apparently elevated incidence of gallstones in FHH, most large reviews of kindreds with this condition have not found this to be a complication of the disorder.

BIOCHEMICAL FEATURES OF FHH

The level of hypercalcemia in FHH is similar to that encountered in modern-day series of patients with PHPT that is of mild to moderate severity. Both the serum total and ionized calcium concentrations are elevated to a similar extent *(47,48)*. Individuals with FHH frequently have some degree of reduction in their serum phosphate concentrations, although the level of phosphate generally remains within the normal range and is reduced to a lesser extent than in patients with PHPT *(48)*. The serum magnesium concentration in FHH tends to be in the upper part of the normal range or mildly elevated. Elevation in serum magnesium may be more common in those families with FHH that exhibit greater degrees of elevation in serum calcium concentration. Indeed there is a positive correlation between the levels of serum calcium and magnesium in FHH, in contrast to this relationship in PHPT, where there is a negative correlation between these parameters *(47)*.

A characteristic biochemical feature of patients with FHH is their inappropriately "normal" circulating levels of PTH in spite of their overt hypercalcemia *(53)*, particularly when PTH is measured using assays that are specific for the intact hormone *(54,55)*. Occasionally, however, PTH levels are in the lower portion of the normal range or are frankly elevated *(56)*. In the latter circumstance, it can be difficult in individual cases to distinguish patients with FHH from those who have mild primary hyperparathyroidism on the basis of serum levels of calcium and PTH alone, particularly in the approximately 5–10% of cases of PHPT in which the intact PTH level falls within the upper part of the normal range. Patients with FHH likewise exhibit deranged Ca^{2+}_o-regulated PTH secretion during induced hyper- and/or hypocalcemia, and show elevations in their "set point" (the level of Ca^{2+}_o half-maximally suppressing circulating PTH levels) *(57,58)*. That is, to reduce the level of PTH to any given extent, individuals with FHH require a level of serum calcium concentration that is 10–15% higher than that lowering PTH to the same extent in normal subjects. This finding documents that there is a mild to moderate degree of resistance of the parathyroid glands to the inhibitory action of Ca^{2+}_o on PTH secretion in this condition. Pathological parathyroid glands from patients with PHPT have a similar, but somewhat more severe abnormality in set point *(59)* as well as additional defects in the control of PTH secretion, including increases in minimal and maximal secretory rates at high and low Ca^{2+}_o, respectively *(58,59)*. In view of their frequently normal levels of circulating intact PTH, it is not surprising that in FHH there is generally normal histology of the parathyroid glands or very mild parathyroid hyperplasia *(60,61)*.

In some cases, patients with FHH have had total or partial parathyroidectomy performed because they were thought to have PHPT. Patients with FHH display a distinctly unusual clinical course following surgical intervention that afforded further clues that this disorder differed fundamentally from typical primary hyperparathyroidism. Hypercalcemia recurred within a few days or weeks in 21 of 27 patients with FHH who underwent from one to four neck explorations each *(53)*. Of the remaining patients, only two were rendered permanently normocalcemic without additional treatment *(53)*. In fact, long-term normocalcemia was only achieved in those patients who were rendered totally aparathyroid and were then treated with vitamin D (5 of the total of 27 patients). In contrast, recurrent hypercalcemia after removal of a parathyroid adenoma is distinctly unusual, whereas in primary parathyroid hyperplasia, recrudescence of hypercalcemia generally takes place after several years if it occurs at all *(4,52)*.

Patients with FHH usually have serum 25(OH)D and 1,25(OH)$_2$D levels in the normal range *(62–64)*, and their intestinal calcium absorption is normal or slightly reduced. There can be a blunted homeostatic response to reduced dietary calcium intake in FHH, with smaller than anticipated increases in gastrointestinal Ca^{2+} absorption and 1,25(OH)$_2$D levels *(64)*. Patients with PHPT, in contrast, have elevated levels of these two parameters in a substantial fraction of cases *(52)*. Although indices of bone turnover (e.g., urinary hydroxyproline excretion) can be slightly increased *(65,66)*, bone mineral density is most commonly within the normal range in patients with FHH *(48,65,67)*. Moreover, there is no increase in the risk of fractures. A recently reported FHH kindred from Oklahoma had several affected family members with osteomalacia, but the latter form of bone disease is very uncommon in other kindreds with FHH *(68)*. In addition, the form of FHH in this Oklahoman kindred is genetically distinct from that present in most families with this syndrome *(see below)*.

An additional characteristic biochemical finding in FHH is abnormally avid renal tubular reabsorption of Ca^{2+} and Mg^{2+}, despite the presence of hypercalcemia *(49,56)*. The renal calcium-to-creatinine clearance ratio is the parameter of renal Ca^{2+} handling that is most commonly used to demonstrate this abnormality; it is <0.01 in about 80% of persons with FHH. In PHPT, in contrast, this clearance ratio is >0.01 in most patients, whereas in other hypercalcemic disorders, the accompanying suppression of PTH reduces renal tubular calcium reabsorption further, and the calcium to creatinine clearance ratio is correspondingly higher. Thus, a low calcium-to-creatinine clearance ratio when combined with a normal level of PTH and the presence of an autosomal-dominant inheritance of mild, asymptomatic hypercalcemia usually makes the diagnosis of FHH a relatively straightforward one.

In occasional FHH kindreds, there is hypercalciuria and even actual renal stone disease in some family members *(69)*. It is not known at present whether these families represent a variant of FHH or the coexistence of an additional cause(s) of deranged renal calcium handling that increases Ca^{2+} excretion despite the expected hypocalciuric effect of the FHH gene(s). Of interest, the enhanced renal tubular calcium reabsorption persists even after total parathyroidectomy *(62,70)*. Thus, the abnormal renal Ca^{2+} handling does not depend on the presence of PTH, but represents an entirely separate derangement in renal Ca^{2+}_o sensing/handling. In one study, the abnormal renal Ca^{2+} handling was localized to the thick ascending limb of the nephron (TAL) *(70)*, whereas the results of another study suggested that the defect resided within the proximal tubule *(71)*. Recent studies on the mechanisms though which the CaR regulates renal handling of calcium and magnesium suggest that the CTAL represents a key locus for the abnormal divalent cation handling in FHH *(see below) (37)*.

Although glomerular filtration rate and renal blood flow can be reduced in patients with many forms of hypercalcemia, these two parameters of renal function are generally well preserved in FHH, providing additional indirect evidence for altered renal responsiveness to Ca^{2+}_o in this disorder *(47)*. Moreover, persons with FHH are able to concentrate their urine normally *(72)*, which contrasts with the reduced maximal urinary concentrating capacity that can be present in other forms of hypercalcemia and sometimes leads to frank nephrogenic diabetes insipidus *(41)*.

Therefore, the clinical and biochemical features of FHH provided strong evidence that this disorder represented some form of inherited abnormality in Ca^{2+}_o sensing and/or handling by the parathyroid, kidney, and perhaps other tissues. For example, patients with FHH lack the mental and gastrointestinal symptoms commonly found in other forms of hypercalcemia. Because of its benign clinical course and the difficulty in obtaining a surgical "cure" in this condition (an accomplishment of doubtful value in a patient who is otherwise asymptomatic), there is currently a consensus that surgical intervention should be avoided in FHH *(53)*. It is of considerable importance, therefore, to differentiate

FHH from PHPT in order to avoid futile and unnecessary exploration of the neck in the former.

GENETICS OF FHH

FHH is inherited in an autosomal-dominant fashion and has a penetrance approaching 100% *(47,48,53)*. Moreover, the biochemical abnormalities can be detected immediately after birth in affected infants *(73)*. In some families, the degree of hypercalcemia is so mild that it may be difficult to document its presence without repeated measurements of serum calcium concentration. Even in these families, however, the serum calcium concentrations of affected family members are clearly distinguishable from those of unaffected members *(55)*. The disease gene was initially mapped to the long arm of chromosome 3 (band q21-24) through linkage analysis by Chou et al. in four large families with FHH *(55)*. Linkage analysis also permitted formal proof that individuals with FHH have one abnormal and one normal allele of the FHH gene, and are, therefore, heterozygous for the disease gene *(55,74)*. In subsequent studies, 90% or more of families with FHH that are sufficiently large for genetic analysis have a disease gene that is linked to the locus on chromosome 3q *(73,75)*. One family, however, exhibited linkage of a disorder with a phenotype indistinguishable from that of FHH to the short arm of chromosome 19 (band 19p13.3) *(75)*, confirming that FHH is genetically heterogeneous. Moreover, the FHH kindred from Oklahoma described above with atypical features (i.e., osteomalacia in a few affected members and a tendency toward progressive increases in serum PTH in older affected members) was not linked to either the locus on chromosome 3 or to that on chromosome 19 *(68,76)*. Neonatal severe hyperparathyroidism (NSHPT), a severe form of hyperparathyroidism encountered in infants, sometimes occurs in FHH kindreds. Some of these cases of NSHPT represent the homozygous form of FHH that is linked to chromosome 3 *(74,77)*. However, as described in more detail below, neonatal hyperparathyroidism of varying degrees of severity can also be seen in some cases in infants who are heterozygous for the disease gene on chromosome 3 *(73,78)*. NSHPT will be discussed in detail later in Neonatal Severe Hyperparathyroidism (NSHPT).

IDENTIFICATION AND FUNCTIONAL CHARACTERIZATION OF CaR MUTATIONS IN FHH3Q

Because Ca^{2+}_o-sensing by parathyroid and kidney (and probably other tissues) are abnormal in FHH, the newly cloned CaR represented an obvious candidate gene. Pollak et al. *(9)* showed that the CaR gene, like the FHH gene, was present on the long arm of chromosome 3 and then employed a ribonuclease A (RNase A) protection assay to search for mutations in the CaR in three separate families with FHH that was linked to chromosome 3. They documented the presence of a unique missense mutation in each of these families (e.g., a change in a single nucleotide substituting a new amino acid for the one that is normally coded for). In these three families, the mutations were: arg185gln,* glu297lys, and arg795trp. These three mutations could not be found in the genomic DNA of 50 normal subjects (Fig. 2). Subsequently, additional CaR mutations have been identified in about two-thirds of the families in which the FHH gene is linked to chromosome 3 *(73,79–82)*. Each family usually has its own unique mutation, although in occasional cases, the same mutation is present in families that appear to be unrelated. In most cases,

---

*This mutation was inadvertently reported as arg186glu *(9)*. Moreover, note that the numbering for amino acid residues in the bovine CaR *(9)* is numerically one greater than that for the human CaR (e.g., *73*) because of the presence of an extra residue in the ECD of the bovine CaR.

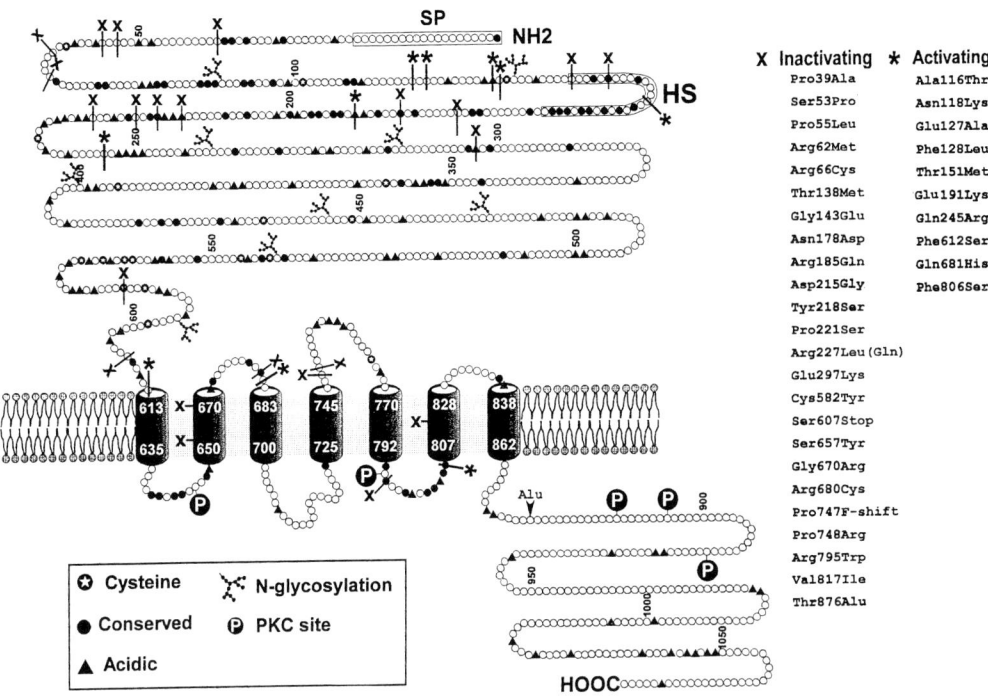

Fig. 2. Schematic representation of the topology of the Ca^{2+}_o-sensing receptor cloned from human parathyroid, indicating locations of activating and inactivating mutations. Abbreviations are the following: SP = signal peptide; HS = hydrophobic segment. Also indicated are the positions of missense and nonsense mutations causing either familial hypocalciuric hypercalcemia (FHH) or autosomal-dominant hypocalcemia; the latter are represented with the three-letter amino acid code. The normal amino acid is indicated before and the mutant amino acid following the number of the appropriate codon. From ref. *16a* with permission.

these are missense mutations, which cluster within several distinct regions of the CaR gene (Fig. 2). These regions include:

1. The first half of the extracellular domain.
2. A portion of the ECD immediately preceding TM1 (the first membrane-spanning segment).
3. Within the transmembrane domains, the intracellular loops (ICLs) or the extracellular loops (ECLs).

Only about two-thirds of the families that have the FHH gene linked to chromosome 3, however, have identifiable mutations in the coding sequence of the CaR. In the remaining families, mutations are presumably present within introns or in upstream or downstream regulatory domains of the CaR that interfere with the normal expression of the gene. Several apparently benign polymorphisms are present within the C-terminal tail of the CaR. These amino acid variations are not associated with overt hypo- or hypercalcemia and are present within a substantial proportion of the population (~10–30%) *(80)*.

Several additional types of mutations have recently been described in FHH. In one family, there is a nonsense mutation just proximal to the first membrane-spanning segment (e.g., ser607stop), which introduces a stop codon *(73)*, resulting in a truncated and presumably biologically inactive receptor lacking all of the CaR's membrane-spanning domains. Another mutation comprises both a deletion of one nucleotide and a transversion (e.g., change from one nucleotide to another) of an adjacent nucleotide within codon 747. This modifies the downstream reading frame, thereby resulting in premature termination of the CaR protein after residue 776 in the transmembrane domains *(73)*. Finally, in a Nova Scotian family, there is insertion of a 383-bp Alu repetitive sequence at codon 876 *(79)*. The Alu element contains an unusually long poly (A/T) tract and is in an orientation opposite to that of the CaR gene. The Alu element also contains stop signals within all three reading frames, thereby predicting a truncated CaR protein with a long stretch of phenylalanines within its C-terminal tail that is encoded by the repeating UUU codons within the mutant CaR's messenger RNA (mRNA) as a result of translation of the poly (A/T) tract. Interestingly, the Alu repetitive element has approximately doubled in size in a subsequent generation of this family *(83)*.

Recent studies have utilized mammalian cells to express mutated CaRs that have been engineered to include many of the FHH mutations identified to this point *(84)*. Figure 3 shows the effects of several point mutations on high Ca^{2+}_o-evoked elevations in Ca^{2+}_i in transiently transfected human embryonic kidney (HEK293) cells. Several of these mutations, including arg185gln and arg795trp, markedly reduce the apparent affinity and/or maximal biological activity of the mutated CaRs. Others (e.g., thr138met or arg62met) only diminish apparent affinity modestly, without changing the maximal response *(84)*. Mutant CaRs that show the most dramatic decreases in their biological activities can also show reduced quantities of the mature, fully glycosylated form of the receptor protein on the cell surface. Of interest, the degree of elevation in the serum calcium concentration of heterozygous individuals in these families (e.g., with FHH) is generally mild. For example, the family harboring the nonsense mutation at codon 607 has serum calcium concentrations that are at or only slightly above the upper limit of normal *(73)*. In addition, as described in more detail later, mice that have been generated with targeted disruption of one CaR allele likewise have very mild hypercalcemia *(32)*. Mutations that profoundly impair the biological activity of the receptor in spite of showing an apparently normal pattern of protein expression on western analysis, in contrast, can be associated with more severe hypercalcemia. For instance, affected members of families with the arg795trp or arg185gln mutations *(84)* exhibit serum calcium concentrations averaging approx 2 and about 3 mg/dL higher, respectively, than in unaffected family members of these kindreds. In fact, although coexpression of most FHH mutations with the wild-type receptor (e.g., mimicking the heterozygous state of FHH patients in vivo) had little or no apparent effect on the normal receptor's function, both arg795trp and arg185gln (Fig. 4) shifted the EC_{50} (effective agonist concentration eliciting half of the maximal response) of the wild-type CaR to the right. The latter two mutant CaRs probably interfere with the function of the normal receptor in some fashion, potentially through one of the following mechanisms:

1. The normal receptor reaches the cell surface in reduced quantities.
2. There is a decrease in the concentration of G protein(s) that is available to the normal receptor, because the mutant CaR forms an inactive complex with this G-protein(s).
3. The normal and mutant receptors form inactive complexes on the cell surface.

At present available data do not allow differentiation among these various possibilities.

Transient transfection has also been performed with cDNAs that code for two mutant CaRs containing the deletion and transversion in codon 747 *(85)* or the alu repetitive element inserted at codon 876 *(86)*. In both cases, the transfected mutant CaRs showed

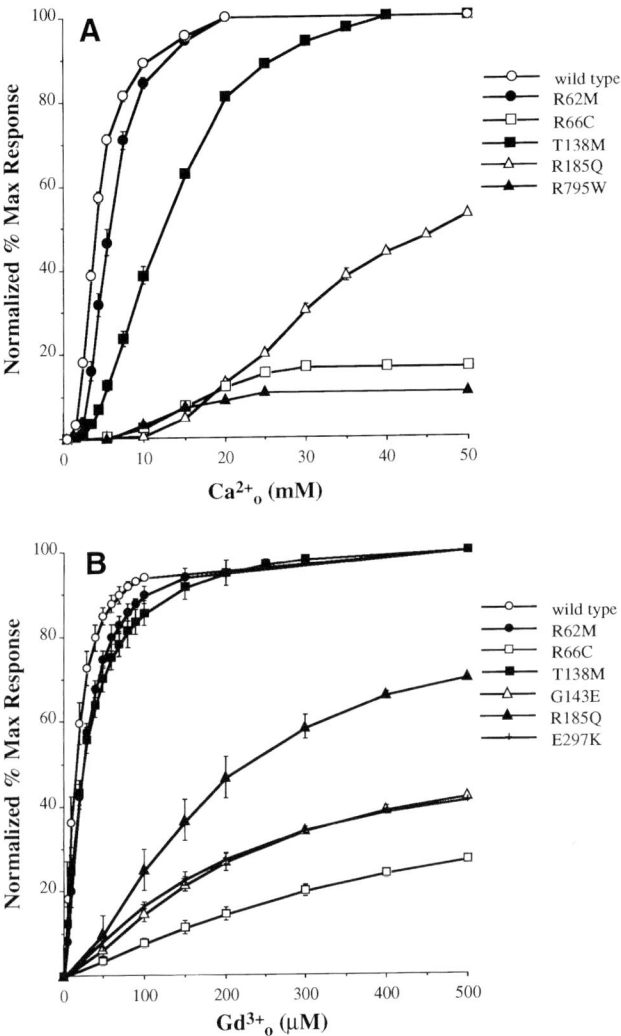

Fig. 3. Expression of CaRs with FHH mutations in HEK293 cells. The results illustrate the effects of varying levels of Ca^{2+}_o or Gd^{3+}_o on Ca^{2+}_i in HEK293 cells transiently transfected with wild-type CaR or the indicated mutant CaRs. Results are normalized to indicate percent of the maximal response of the wild-type CaR. Reproduced from ref. *84* with permission.

essentially no biological activity. Truncated proteins were indeed produced in vitro as assessed by Western blot analysis that were of the sizes anticipated based on the premature stop codons that were predicted to be encountered as a result of these two mutations.

The mutations that fall within the TMs of the CaR likely interfere with the receptor's activation by decreasing its cell-surface expression and/or interfering with conformational changes in the transmembrane domains and/or intracellular loops that are necessary for activation of the G-protein(s) with which the CaR interacts. As noted above, the single FHH family that has been described to date with a mutation in an ICL (e.g., arg795trp, which resides within the third intracellular loop), markedly impairs the activation of phospholipase C in spite of the fact that near-normal quantities of the mature CaR protein appear to be expressed *(84)*. The third intracellular loop of most GPCRs is thought

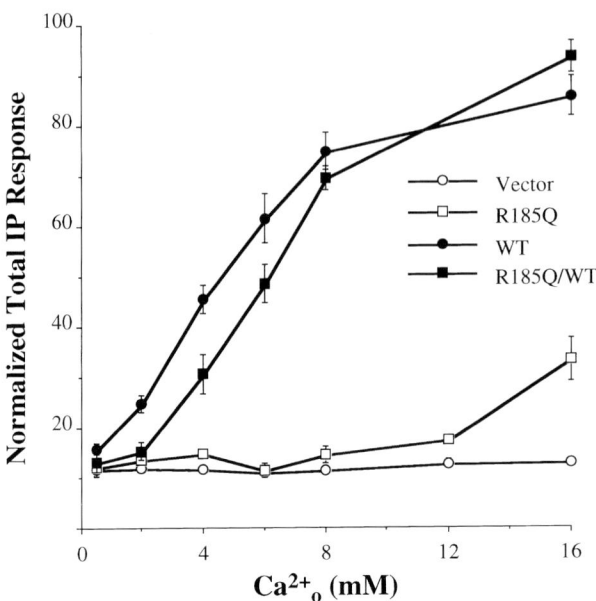

Fig. 4. Coexpression of mutant (arg185gln) and wild-type CaRs in HEK293 cells. The results shown the high Ca^{2+}_o-evoked increases in total cellular inositol phosphates in HEK 293 cells transiently transfected with empty vector (e.g., not containing any CaR cDNA), wild-type CaR, a mutant CaR bearing arg185gln, or both the mutant and wild-type CaRs. Note the "dominant negative" effect of the CaR bearing arg185gln when cotransfected with the wild-type CaR, thereby right shifting the high Ca^{2+}_o-evoked activation of the wild-type receptor and presumably contributing to the more severe hypercalcemia noted in the neonate with *de novo* heterozygous FHH resulting from this mutation. From ref. 78 with permission.

to play a crucial role in determining the specificity of their coupling to G-proteins *(87,88)*, although the second ICL of the mGluRs may be the most important in this regard *(89)*.

The following tentative conclusions can be drawn about structure–function relationships for the CaR from the studies just described. Some of the missense mutations in the ECD produce a range of alterations in the CaR's apparent affinity for Ca^{2+}_o (as well as for the trivalent CaR agonist, Gd^{3+}_o) without bringing about any clear-cut alterations in the expression of the mature CaR protein (as determined from Western analysis) *(84)*. Thus, the altered amino acid might modify the affinity of the CaR directly or indirectly for Ca^{2+}_o. The eventual elucidation of the ECD's three-dimensional structure will be necessary to elucidate the precise nature of the interactions of the protein with its polycationic ligands. The Hill coefficient for the activation of the expressed CaR by Ca^{2+}_o is approximately three, indicating that there may be at least three interacting binding sites for Ca^{2+}_o *(84)*. The Hill coefficient for stimulation of the receptor by Gd^{3+}_o, in contrast, is about one; thus it is possible that trivalent cations interact with the CaR differently than does Ca^{2+}_o. Of interest, Hammerland and coworkers *(23)* have shown that a mutant CaR from which essentially the entire ECD had been removed was no longer activated by Ca^{2+}_o, but still showed a relatively robust activation by Gd^{3+}_o. This result suggests the possibility of a binding site for Gd^{3+}_o (but perhaps not for Ca^{2+}_o) within the CaR's extracellular loops or even TMs, or that binding of Ca^{2+}_o to this site by itself is not sufficient to initiate signal transduction. The results observed with the CaR deletion mutant could also explain why mutant calcium-sensing receptors identified in some families with FHH do not exhibit any responses to Ca^{2+}_o, but are stimulated reasonably well by Gd^{3+}_o *(84)*.

We draw the following conclusions about FHH from the information just reviewed:

1. It is genetically heterogeneous, but more than 90% of families with this disorder likely have the form of the condition that is linked to chromosome 3.
2. Approximately two-thirds of persons with FHH linked to chromosome 3q have inactivating mutations that lie within the CaR's coding region, and most families have their own unique mutations. Most of the mutations reside within the ECD and probably reduce the apparent affinity of the receptor for Ca^{2+}_o or interfere with its biosynthesis or expression on the cell surface. Some mutations within TMs or ICLs may disrupt the processes that are necessary for initiation of productive signal transduction. All mutations likely interfere with the CaR's responsiveness to Ca^{2+}_o, thereby rendering FHH patients mildly to moderately "resistant" to the normal actions of Ca^{2+}_o on CaR-expressing tissues.
3. The remaining one-third of families with the form of FHH that is linked to chromosome 3, but who do not have mutations that can be identified within the CaR's coding region, may have mutations in promoter or enhancer sequences of the receptor's gene.
4. The abnormal parathyroid dynamics in persons with FHH confirm that the receptor plays a key role in the regulation of PTH secretion by Ca^{2+}_o. Moreover, the direct regulation of renal Ca^{2+}, Mg^{2+}, and water handling by Ca^{2+}_o (i.e., the overly avid renal tubular reabsorption of these divalent cations in spite of the associated hypercalcemia and the apparent lack of the normal inhibitory effect of high Ca^{2+}_o on maximal urinary concentrating capacity *[72]*) provide in vivo evidence that supports an important role for the CaR in regulating several key aspects of renal function.
5. In occasional families disease genes for a phenotypically similar disorder map to additional, as-yet-undefined loci on chromosome 19p and elsewhere. The eventual identification of these genes could provide important insights into the nature of additional Ca^{2+}_o-sensing receptors or other molecular components required for normal functioning of the CaR and its signaling pathways.
6. Since the disorder is genetically heterogeneous and about one-third of patients with FHH_{3q} do not have mutations that can be detected within the CaR's coding region, the diagnosis of FHH will probably continue to be based on the clinical documentation of mild, PTH-dependent hypercalcemia accompanied by relative hypocalciuria (e.g., a calcium-to-creatinine clearance ratio <0.01) that is autosomal-dominantly inherited. Actual mutational analysis may provide useful information in specific settings, such as in differentiating FHH from PHPT in patients with no relatives available for genetic screening or in those patients who have *de novo* CaR mutations.
7. Finally, because the great majority of patients with FHH have a benign clinical course, parathyroidectomy should be performed very rarely, if at all.

Neonatal Severe Hyperparathyroidism (NSHPT)

CLINICAL AND BIOCHEMICAL FEATURES OF NHSPT

NSHPT often presents with dramatic clinical manifestations, which include severe symptomatic hypercalcemia in association with hyperparathyroid bone disease that is often diagnosed in the immediate postnatal period, but can present as late as the age of 6 mo; in these cases, the condition is appropriately termed NSHPT *(53)*. NSHPT, because of its dramatic presentation, was described well before FHH *(90)*. A recent review of 49 cases of NSHPT indicated that most presented at birth or shortly thereafter, usually within the first week of life *(53)*. Affected infants commonly exhibited failure to thrive, anorexia, constipation, respiratory distress, and hypotonia. Additional clinical findings included chest wall deformity and, less commonly, craniotabes, dysmorphic facies, and rectovaginal or anovaginal fistulas *(35,91–94)*. Respiratory complications can arise as a result of thoracic deformity and sometimes include a flail chest syndrome owing to multiple rib fractures that can produce substantial morbidity *(53,91)*.

The hypercalcemia in NSHPT is commonly severe, on the order of 14 to 20 mg/dL, and levels as high as 30.8 mg/dL have been described *(94)*. In spite of this marked hypercalcemia, relative hypocalciuria is observed in some cases when it has been measured, even if there is no family history of FHH *(95)*. When they have been determined, serum magnesium concentrations have been elevated well above normal on occasion *(82)*. Serum levels of PTH are generally frankly high, usually being elevated 5- to 10-fold, although the increase can sometimes be modest *(30,96)*. Skeletal radiographs often reveal marked undermineralization, and there can be fractures of the long bones and ribs as well as subperiosteal erosions, metaphyseal widening, and occasionally, rickets *(91,97)*. Bone histology shows typical osteitis fibrosa cystica *(93)*. At the time of parathyroidectomy, the parathyroid glands are usually frankly enlarged, and their combined weights may be many times those of the parathyroid glands of children at this age; they typically show chief cell or water clear-cell hyperplasia. Occasionally, if glandular enlargement is less apparent, the low content of fat in parathyroid glands of children can complicate interpretation of the histological findings. Usually, however, the presence of nonsuppressed levels of PTH in the setting of severe hypercalcemia makes the diagnosis of neonatal hyperparathyroidism obvious *(53,93,98)*. There have been no cases reported in which a parathyroid adenoma caused neonatal severe hyperparathyroidism.

Sometimes, the degree of hypercalcemia in neonatal hyperparathyroidism is less marked, in the range of 11–12 mg/dL. Such cases might more appropriately be called neonatal hyperparathyroidism (NHPT), particularly in the absence of hyperparathyroid bone disease, which is generally considered to be a prerequisite for the diagnosis of NSHPT. Furthermore, cases have been described, particularly in the more recent literature, in which the disorder ran a self-limited course, reverting to a milder form of hypercalcemia resembling FHH at 6–7 mo of age, even with conservative medical therapy *(86,91)*. The recent identification of CaR mutations not only in FHH, but also in NSHPT and less severe cases of NHPT will probably widen the spectrum of this disorder further as additional milder cases are uncovered. A particularly interesting case in this regard was recently identified—a woman of 35 yr who was the product of a consanguineous marriage of two individuals with FHH. The serum calcium concentrations of the parents were in the upper part of the normal range *(82)*. This individual did not present as NSHPT, and her homozygous state was only identified in adulthood by mutational analysis. In spite of her serum calcium concentrations being in the range of 15–17 mg/dL, she was asymptomatic, although she exhibited mild mental retardation. Interestingly, her serum magnesium concentration was elevated to a similar degree (approx 50% above the upper limit of the normal range), and her intact PTH level was just at the upper end of normal *(82)*. Despite her striking hypercalcemia, her renal function appeared to be normal.

In reports published prior to 1982 *(53)*, NSHPT frequently had a fatal outcome without prompt and aggressive medical therapy followed by surgical intervention. This has not been invariably true in the more recently reported clinical experience, probably because wider recognition of the broader clinical spectrum of the disorder, combined with improved medical treatment of severe hypercalcemia, has enabled successful medical management in an increasing number of cases over the past 15 yr. In symptomatic cases, initial management should include vigorous hydration as well as the use of antiresorptive therapy (e.g., bisphosphonates, such as pamidronate) and respiratory support if needed. If the infant's condition is especially severe or deteriorates despite aggressive medical treatment, total parathyroidectomy with or without autotransplantation of a portion of one of the excised parathyroid glands is indicated within the first month of life *(31,53,96)*. If autotransplantation is not performed or if the graft is not viable, lifelong management of the ensuing hypoparathyroidism with vitamin D (generally with 1,25[OH]$_2$D), and oral calcium supplements may be needed to prevent symptomatic hypocalcemia *(30,93)*. There is usually dramatic clinical improvement following parathyroidectomy. There can likewise be rapid healing of the skeletal lesions, even though the hypercalcemia will usually recur rapidly following autotransplantation or if less than a total parathyroidectomy has been carried out *(53,93)*. Similar clinical improvement has been described more recently in cases of NSHPT following medical management *(53)*, suggesting that the natural history of NSHPT or NHPT can include biochemical improvement in the degree of hyperparathyroidism, a point to be returned to later.

THE GENETICS OF NSHPT

Early descriptions of familes with FHH showed that children with the clinical picture of NSHPT could coexist in these kindreds *(47,77,93,99)*. In 15 families with FHH, three infants in two separate kindreds had NSHPT *(47,77,93)*, suggesting that some cases of NSHPT could represent the homozygous form of FHH. In an additional family with two children affected by NSHPT, the parents were related and exhibited hypocalciuria as well as mild increases in their serum ionized calcium concentrations (despite the fact that their total calcium concentrations were normal) *(64)*, again providing evidence that NSHPT can be the homozygous form of FHH. Pollak and coworkers subsequently demonstrated in an additional study of 11 families in which the abnormal gene mapped to the long arm of chromosome 3q *(74)* that consanguineous unions of affected individuals in four of these families produced children with NSHPT. The pattern of inheritance of genetic markers that were closely linked to the FHH gene in these four families provided strong evidence that NSHPT can represent the homozygous form of the same disease. Subsequent studies in which FHH arose from mutations in the CaR gene confirmed that the inheritance of two abnormal copies of the CaR gene can produce NSHPT *(9,79,81)*. Because they have no normal copies of the CaR gene, these infants exhibit much more severe biochemical and clinical manifestations than is found in those with the heterozygous state (FHH) owing to severe resistance of CaR-expressing tissues, particularly the parathyroid glands, to Ca$^{2+}_o$.

NHPT AS A RESULT OF HETEROZYGOUS CaR MUTATIONS

Not all cases of NSHPT and NHPT, however, represent the homozygous form of FHH. In fact, most cases described in the literature occur sporadically or in FHH families where there is only one affected parent *(35,100,101)*. What are possible explanations for such cases? Children with NSHPT or NHPT might conceivably be compound heterozygotes and harbor two abnormal CaR alleles, each with a different mutation—one producing obvious hypercalcemia in one parent, but the other being so mild that hypercalcemia was not readily detected biochemically in the other parent. Alternatively, a mutation might be present in one allele of the CaR gene as well as in another gene that can cause an FHH-like clinical picture (i.e., the one on chromosome 19p *[75]* or the locus on neither chromosome 3 nor 19 *[76]*). No documentation of any such compound heterozygotes, however, has been provided to date.

Another possible explanation that has been suggested for the development of NHPT in a child with a single abnormal allele of the CaR arising from an FHH father, but a normal mother is the impact of normal maternal Ca^{2+} homeostasis on the parathyroid glands of the affected fetus *in utero (77)*. The placenta actively transports calcium from mother to fetus, generating a level of Ca^{2+}_o in cord blood that is about 10% higher than the maternal calcium concentration *(77)*. Therefore, a normal mother would expose fetal parathyroid glands with even modestly abnormal Ca^{2+}_o-sensing owing to FHH to a level of Ca^{2+}_o that would be recognized as relatively hypocalcemic. The latter would then "overstimulate" the fetal parathyroid glands, producing an additional degree of "secondary" fetal/neonatal hyperparathyroidism that would be superimposed on the abnormal Ca^{2+}_o-sensing already present as a result of the presence of a heterozygous CaR mutation. This possibility is supported by the fact that cases of NHPT can occur with an autosomal-dominant pattern of inheritance when the father had FHH and the mother appeared to have normal mineral ion homeostasis *(53,94)*. In the postnatal period, the "secondary" hyperparathyroidism would eventually subside, producing a clinical state with the clinical and biochemical features of FHH. In most described cases, however, children with FHH born to a normal mother do not have hypercalcemia that is more severe than that in infants born to an affected mother (although serum calcium concentrations and PTH levels are not available in the immediate postnatal period in most such infants). In addition, there are no obvious differences between mice heterozygous for targeted disruption of the CaR gene that are born to normal mothers and to those that are heterozygous for disruption of the CaR gene (*see* Mouse Models of FHH and NSHPT) *(32)*.

Recent studies have also shown that infants can present with the clinical appearance of NSHPT or NHPT as a result of *de novo* heterozygous CaR mutations *(73)* (i.e., the presence a single *de novo* CaR mutation in the offspring of normal parents). Two such infants presented with hyperparathyroid bone disease, but with hypercalcemia that was less severe than that seen in NSHPT owing to homozygous CaR mutations *(73)*. We have recently documented an additional case of *de novo* heterozygous NSHPT in which the infant harbored the same arg185gln mutation that was described previously and is associated with a greater degree of elevation of the serum calcium concentration than in most FHH kindreds *(9,78,93)*. In this latter case, the relatively large discrepancy between the set points of the mother's and infant's parathyroid glands may have produced more severe hyperparathyroidism in the prenatal period, leading to hyperparathyroid bone disease in the infant at the time of birth *(78)*.

Therefore, the biochemical and clinical findings in NHPT, which in homozygous cases represents the "knockout" of the CaR gene, suggest the following conclusions about the function of the CaR in humans: (1) These cases highlight the receptor's importance in fetal and neonatal calcium metabolism. (2) They demonstrate that, in addition to its involvement in Ca^{2+}_o-regulated PTH secretion, the CaR likely functions directly or indirectly to inhibit parathyroid cellular proliferation, because there is substantial hyperplasia of the parathyroid glands in NSHPT.

Mouse Models of FHH and NSHPT

Recently, Ho and coworkers *(32)* utilized targeted disruption of the CaR gene to produce mice heterozygous or homozygous for gene inactivation of the CaR gene, which, therefore, provide animal models of FHH and NSHPT, respectively. Introducing DNA coding for the neomycin resistance gene within the third exon of the CaR gene caused complete absence of detectable CaR protein in the parathyroid gland and kidneys of the mice with homozygous CaR "knockout" and approx 50% reductions in the receptor protein in the heterozygous mice *(32)*. The heterozygous mice have a normal appearance, are fertile, and have a normal life-span. Their serum calcium concentrations average 10.4 mg/dL, about 10% higher than in their normal litter mates. The heterozygotes also have slight (~10–15%), but significant increases in their serum magnesium concentrations. Serum PTH levels are also slightly elevated, being about 50% higher in the heterozygous than in the normal mice, and urinary calcium concentration is reduced slightly compared to that of the normal mice. Skeletal X-rays are normal. Therefore, mice that are heterozygous for targeted disruption of the CaR gene exhibit many of the same phenotypic and biochemical features present in individuals with FHH.

Interestingly, the heterozygous mice show little, if any, upregulation of the CaR protein arising from the remaining normal gene in spite of the total lack of CaR production from the inactivated allele. As a result, the levels of expression of the CaR protein in parathyroid and kidney are about half of those of the wild-type animals *(32)*. This ~50% decrease in CaR protein expression in the parathyroid produces a mild (~10%) elevation in the apparent set point for Ca^{2+}_o-regulated PTH release, which is similar in magnitude to that seen in FHH families with mutations that would produce a totally inactive CaR from the mutant allele (e.g., ser607stop). Therefore, the results in the heterozygous mice provide further support for the concept that the abnormal CaR allele in FHH frequently acts as a null mutation (i.e., it leads to the total absence of the gene product or to one that is totally inactive), and the abnormal parathyroid function results principally from a decrease in the number of normally functioning CaRs *(47,48,53)*. This pathophysiology is analogous to that which has recently been postulated to be present in parathyroid adenomas (*see* Acquired Tissue-Specific Resistance to Ca^{2+}_o in Primary and Uremic Hyperparathyroidism), which apparently do not harbor CaR mutations in the great majority of cases *(102)*, but show, on average, about a 60% reduction in CaR immunoreactivity *(50,51)*.

Mice that are homozygous for inactivation of the CaR gene, in contrast, although of nearly normal size at birth, grow substantially more slowly than their normal or heterozygous littermates during the first few weeks postnatally—if they survive this long *(32)*. This growth retardation may occur, in part, because they compete poorly with their more vigorous littermates for maternal milk. The homozygotes also exhibit severe hypercalcemia, averaging 14.8 mg/dL. Their serum magnesium concentrations, in contrast, are slightly, but not significantly higher than those of the heterozygous mice. As with FHH and NSHPT, this elevation in serum magnesium concentration in the setting of reduced levels of functional CaRs provides indirect evidence that the CaR also functions as a magnesium-sensing receptor that contributes to the "setting" of Mg^{2+}_o *(44)*. Serum PTH levels are about 10-fold higher than those in the normal mice, an increase that is comparable to that seen with NSHPT. In spite of their severe hypercalcemia, the calcium concentration in the urine is lower than that in normal mice *(32)*. Skeletal X-rays showed numerous abnormalities, with reductions in apparent mineral density, kyphoscoliosis, and bowing of the long bones. Most homozygous mice die within the first 2 wk postnatally, with only occasional ones surviving for 3 or 4 wk. Therefore, the biochemical and clinical features of mice that are homozygous for targeted disruption of the CaR gene exhibit a number of similarities to the human disorder, NSHPT. Much work needs to be carried out, however, employing this animal model to study further abnormalities in Ca^{2+}_o sensing in tissues that normally express the CaR, including those that are seemingly uninvolved in calcium metabolism (e.g., the brain).

ACQUIRED, TISSUE-SPECIFIC RESISTANCE TO CA^{2+}_o IN PRIMARY AND UREMIC HYPERPARATHYROIDISM

A large number of studies have demonstrated the presence of abnormal regulation of PTH secretion by Ca^{2+}_o in primary hyperparathyroidism as well as in some cases of uremic secondary hyperparathyroidism, particularly those in which frank hypercalcemia supervenes *(59,103)*. We investigated the possibility that this abnormal Ca^{2+}_o-regulated PTH secretion, like that in FHH, might arise from CaR mutations *(102)*. In nearly 40 parathyroid tumors, however, including cases of parathyroid adenomas, primary parathyroid hyperplasia, uremic secondary/tertiary hyperparathyroidism, and parathyroid carcinoma, no mutations could be detected in the CaR gene *(102)*. In subsequent studies, we *(51)* and others *(50)* have utilized immunohistochemistry *(50,51)* as well as *in situ* hybridization *(50)* with CaR-specific antisera and RNA probes, respectively, to determine the levels of expression of the CaR mRNA and protein in various types of hyperparathyroidism. These studies demonstrated a substantial reduction (e.g., 50% or more in most cases) of both the CaR protein *(50,51)* and mRNA *(50)*.

Thus, currently available data indicate that there is a reduction in the number of normally functioning CaRs on the cell surface of the parathyroid cell in both FHH and PHPT, although this abnormality results from different molecular mechanisms (e.g., CaR mutations in the former and yet-to-be-defined mechanisms in the latter). It is not currently known with certainty whether this reduction in cell-surface CaR immunoreactivity is the major cause of the abnormal Ca^{2+}_o-sensing by pathological parathyroid glands in PHPT and severe uremic hyperparathyroidism. Nevertheless, the apparent correlation between the greater increase in parathyroid set point in PHPT *(59,103)* than in most cases of FHH *(57)* and the greater degree of reduction in cell-surface CaR expression in the former *(50,51)* than that thought to be present in the latter *(78,84,86)* (as well as in mice heterozygous for CaR "knockout" *[32]*) provides indirect evidence for the importance of cell-surface CaR expression in determining the set point of the parathyroid. An important difference between PHPT and FHH or NSHPT, however, is that in hyperparathyroidism, the Ca^{2+}_o resistance is limited to the pathological parathyroid gland(s), whereas in the latter two conditions, it is generalized (e.g., involved all tissues expressing the CaR) *(104)*.

The mechanism(s) underlying the reductions in CaR expression in primary and uremic hyperparathyroidism remains to be elucidated. It is of interest, however, that the development of tertiary (i.e., hypercalcemic) hyperparathyroidism in patients on dialysis for chronic renal insufficiency is often accompanied by the development of the same genetic lesions in the parathyroid gland that are encountered in primary hyperparathyroidism (i.e., deletion or inactivation of a putative tumor-suppressor gene on chromosome 11) *(105,106)*. It is possible, therefore, that the reduction in CaR expression in hyperparathyroidism is a secondary phenomenon, resulting from other alterations in parathyroid function that lead to the hyperplasia of parathyroid cells characteristic of all forms of primary and uremic hyperparathyroidism. In addition, in occasional cases of primary or secondary hyperparathyroidism, there may loss of heterozygosity (i.e., loss of chromosomal markers from one of the two chromosomes encoding a given gene) in the region of the CaR gene, possibly reducing expression of the CaR by a gene dosage effect *(107)*.

SUMMARY

The cloning of the CaR has documented directly that several of the cell types that recognize and respond to even minute perturbations in Ca^{2+}_o do so through a receptor-mediated process that is similar to that through which a variety of cells respond to numerous hormones, neurotransmitters, and other extracellular messengers—a cell-surface Ca^{2+}_o-sensing receptor. Thus Ca^{2+}_o can act as an extracellular, first messenger in addition to serving its more widely recognized role as an intracellular second messenger. Of the tissues that express the CaR, several are important components of the mineral ion homeostatic system and have been known for many years to sense Ca^{2+}_o. The presence of the CaR on several types of kidney cells, however, provides strong indirect evidence that several of the poorly understood actions of Ca^{2+}_o on renal function are CaR-mediated. The latter actions include increased renal excretion of Ca^{2+} and Mg^{2+} owing to the effects of hypercalcemia on CaRs in the CTAL and probably the DCT, which act along with the concomitant reduction in renal calcium reabsorption that results from high Ca^{2+}_o-induced suppression of PTH release. The impaired maximal urinary concentrating capacity present in some hypercalcemic individuals is probably also a manifestation of a functionally significant coordination of the homeostatic systems that govern renal calcium and water handling. The purpose of integrating and coordinating these homeostatic systems is probably to decrease the risk of renal stone formation and resultant damage to the kidney during disposal of calcium loads. Human syndromes that result from Ca^{2+}_o "resistance" owing to loss-of-function CaR mutations provide further support for the CaR's role in controlling renal function. A great deal remains to be learned, however, about this receptor's role in regions of the body, such as brain, where it likely responds to local and not systemic changes in Ca^{2+}_o. Developing drugs that activate or inhibit the CaR will be of great potential utility for the treating conditions in which the receptor is either under- or overactive. Clinical trials are presently under way that are assessing the efficacy of "calcimimetic" CaR activators for treating primary and secondary hyperparathyroidism *(108)*. Finally, there may be further receptors for Ca^{2+}_o *109–112)*, which could conceivably be encoded by the additional genetic loci that produce the clinical syndrome of FHH, or for other ions (indeed the CaR probably functions as a physiologically significant Mg^{2+}_o-receptor *[44]*).

ACKNOWLEDGMENTS

The authors gratefully acknowledge generous grant support from the following: the USPHS DK41415 (to EMB), 46422 (to EMB), 48330 (to EMB), 52005 (to EMB), and 09436 (to MB), as well as the St. Giles Foundation and NPS Pharmaceuticals, Inc., Salt Lake City, UT (to EMB).

REFERENCES

1. Brown E. Extracellular Ca^{2+} sensing, regulation of parathyroid cell function, and role of Ca^{2+} and other ions as extracellular (first) messengers. Physiol Rev 1991;71:371–411.
2. Pietrobon D, Di Virgilio F, Pozzan T. Structural and functional aspects of calcium homeostasis in eukaryotic cells. Eur J Biochem 1990;120:599–622

3. Berridge M, Irvine R. Inositol trisphosphate, a novel second messenger in cellular signal transduction. Nature 1984;312:315–321.
4. Aurbach G, Marx S, Spiegel A. Parathyroid hormone, calcitonin, and the calciferols. In: Wilson J, Foster D, eds. Textbook of Endocrinology, 7th ed, W. B. Saunders, Philadelphia, 1985, pp. 1137–1217.
5. Parfitt A. Bone and plasma calcium homeostasis. Bone 1987;8(Suppl 1):1–8.
6. Kurokawa K. The kidney and calcium homeostasis. Kidney Int. 1994;45(Suppl 44):S97–S105.
6a. Brown EM, Pollak M, Hebert SC. Cloning and characterization of extracellular Ca^{2+}-sensing receptors from parathyroid and kidney: molecular physiology and pathophysiology of Ca^{2+}-sensing. The Endocrinologist 1994;4:419–426.
7. Brown E, Gamba G, Riccardi D, Lombardi D, Butters R, Kifor O, et al. Cloning and characterization of an extracellular Ca^{2+}-sensing receptor from bovine parathyroid. Nature 1993;366:575–580.
8. Pollak M, Brown E, Estep H, McLaine P, Kifor O, Park J, et al. Autosomal dominant hypocalcaemia caused by Ca^{2+}-sensing receptor gene mutation. Nature Genet 1994;8:303–307.
9. Pollak M, Brown E, Chou Y, Hebert S, Marx S, Steinmann B, et al. Mutations in the human Ca^{2+}-sensing receptor gene cause familial hypocalciuric hypercalcemia and neonatal severe hyperparathyroidism. Cell 1993;75:1297–1303.
10. Kifor O, Kifor I, Brown EM. Effects of high extracellular calcium concentrations on phosphoinositide turnover and inositol phosphate metabolism in dispersed bovine parathyroid cells. J Bone Miner Res 1992;7:1327–1335.
11. Shoback D, Membreno L, McGhee J. High calcium and other divalent cations increase inositol trisphosphate in bovine parathyroid cells. Endocrinology 1988;123:382–389.
12. Nemeth EF, Wallace J, Scarpa A. Stimulus-secretion coupling in bovine parathyroid cells. Dissociation between secretion and net changes in cytosolic Ca^{++}. J Biol Chem 1986;261:2668–2674.
13. Chen C, Barnett J, Congo D, Brown E. Divalent cations suppress 3',5'-adenosine monophosphate accumulation by stimulating a pertussis toxin-sensitive guanine nucleotide-binding protein in cultured bovine parathyroid cells. Endocrinology 1989;124:233–239.
14. Racke F, Hammerland L, Dubyak G, Nemeth E. Functional expression of the parathyroid cell calcium receptor in *Xenopus* oocytes. FEBS Lett 1993;333:132–136.
15. Chen T, Pratt S, Shoback D. Injection of bovine parathyroid poly(A)+ RNA into Xenopus oocytes confers sensitivity to high extracellular calcium. J Bone Miner Res 1994;9:293–300.
16. Garrett J, Capuano I, Hammerland L, Hung B, Brown E, Hebert S, et al. Molecular cloning and functional expression of human parathyroid calcium receptor cDNAs. J Biol Chem 1995;270:12,919–12,925.
16a. Brown EM, Bai M, Pollak M. Familial benign hypocalcemic hypercalcemia and other syndromes of altered responsiveness to extracellular calcium. In: Krane JM, Avioli LV, eds. Metabolic Bone Diseases, 3rd ed., Academic, San Diego, CA, 1998.
17. Aida K, Koishi S, Tawata M, Onaya T. Molecular cloning of a putative Ca^{2+}-sensing receptor cDNA from human kidney. Biochem Biophys Res Commun 1995;214:524–529.
18. Riccardi D, Park J, Lee W, Gamba G, Brown E, Hebert S. Cloning and functional expression of a rat kidney extracellular calcium/polyvalent cation-sensing receptor. Proc Natl Acad Sci USA 1995;92:131–135.
19. Ruat M, Molliver M, Snowman A, Snyder S. Calcium sensing receptor: molecular cloning in rat and localization to nerve terminals. Proc Natl Acad Sci USA 1995;92:3161–3165.
20. Garrett JE, Tamir H, Kifor O, Simin RT, Rogers KV, Mithal A, et al. Calcitonin-secreting cells of the thyroid express an extracellular calcium receptor gene. Endocrinology 1995;136:5202–5211.
21. Butters R, Chattopadhyay N, Nielsen P, Smith C, Mithal A, Kifor O, et al. Cloning and characterization of a calcium-sensing receptor from the hypercalcemic New Zealand white rabbit reveals unaltered responsiveness to extracellular calcium. J Bone Miner Res 1997;12:568–579.
22. Diaz R, Hurwitz S, Chattopadhyay N, Pines M, Yang Y, Kifor O, et al. Cloning, expression and tissue localization of the calcium-sensing receptor in the chicken (Gallus domesticus). Am J Physiol 1997; 273:R1008–R1016.
23. Hammerland LG, Krapcho KJ, Alasti N, Garrett JE, Capuano IV, Hung BCP, et al. Cation binding determinants of the calcium receptor revealed by functional analysis of chimeric receptors and a deletion mutant. J Bone Miner Res 1995;10:S156 (Abstract).
24. Varrault A, Pena M, Goldsmith P, Mithal A, Brown E, Spiegel A. Expression of G protein alpha-subunits in bovine parathyroid. Endocrinology 1995;136:4390–4396.
25. Rogers K, Dunn C, Hebert S, Brown E, Nemeth E. Pharmacological comparison of bovine parathyroid, human parathyroid, and rat kidney calcium receptors expressed in HEK 293 cells. J Bone Miner Res 1995;10:S483 (Abstract T516).

26. Nakanishi S. Metabotropic glutamate receptors: synaptic transmission, modulation and plasticity. Neuron 1994;13:1031–1037.
27. Kifor O, Diaz R, Butters R, Brown E. The Ca^{2+}-sensing receptor mediates activation of phospholipases C, A_2, and D by high extracellular Ca^{2+} in bovine parathyroid cells and human embryonic kidney (HEK293) cells. J. Bone. Mineral Res. 1997;12:715–725.
28. Fried R, Tashjian AJ. Unusual sensitivity of cytosolic free Ca^{2+} to changes in extracellular Ca^{2+} in rat C-cells. J Biol Chem 1986;261:7669–7674.
29. Scherubl H, Schultz G, Hescheler J. Electrophysiological properties of rat calcitonin-secreting cells. Mol Cell Endocrinol 1991;82:293–301.
30. Marx S, Lasker R, Brown E, Fitzpatrick L, Sweezey N, Goldbloom R, et al. Secretory dysfunction in parathyroid cells from a neonate with severe primary hyperparathyroidism. J Clin Endocrinol Metab 1986;62:445–449.
31. Cooper L, Wertheimer J, Levey R, Brown E, LeBoff M, Wilkinson R, et al. Severe primary hyperparathyroidism in a neonate with two hypercalcemic parents: management with parathyroidectomy and heterotopic autotransplantation. Pediatrics 1986;78:263–268.
32. Ho C, Conner DA, Pollak M, Ladd DJ, Kifor O, Warren H, et al. A mouse model for familial hypocalciuric hypercalcemia and neonatal severe hyperparathyroidism. Nature Genet 1995;11:389–394.
33. Garrett J, Steffey M, Nemeth E. The calcium receptor agonist R-568 suppresses PTH mRNA in cultured bovine parathyroid cells. J Bone Miner Res 1995 10(Suppl 1):387 (Abstract).
34. Randall C, Lauchlan S. Parathyroid hyperplasia in an infant. Am J Dis Child 1963;105:364–367.
35. Spiegel A, Harrison H, Marx S, Brown E, Aurbach G. Neonatal primary hyperparathyroidism with autosomal dominant inheritance. J Pediatr 1977;90:269–272.
36. Riccardi D, Lee W-S, Lee K, Segre G, Brown E, Hebert S. Localization of the extracellular Ca^{2+}-sensing receptor and PTH/PTHrP receptor in rat kidney. Am J Physiol 1996;271:F951–F956.
37. Brown E, Hebert S. A cloned Ca^{2+}-sensing receptor: A mediator of direct effects of extracellular Ca^{2+} on renal function? J Am Soc Nephrol 1995;6:1530–1540.
38. Chabardes D, Imbert M, Clique A, Montegut M, Morel F. PTH-sensitive adenyl cyclase activity in different segments of the rabbit nephron. Pflugers Arch 1975;354:229–239.
39. Sands J, Naruse M, Baum M, Hebert S, Brown E, Harris W. An apical extracellular calcium/polyvalent cations-sensing receptor (CaR) regulates vasopressin-elicited water permeability in rat kidney inner medullary collecting duct. J Clin Invest 1997;99:1399–1405.
40. Hebert S, Andreoli T. Control of NaCl transport in the thick ascending limb. Am J Physiol 1984; 246:F745–F756.
41. Suki W, Eknoyan G, Rector F Jr, Seldin D. The renal diluting and concentrating mechanism in hypercalcemia. Nephron 1969;6:50–61.
42. Brown E, Vassilev P, Hebert S. Calcium as an extracellular messenger. Cell 1995;83:679–682.
43. Heinemann U, Lux HD, Gutnick MJ. Extracellular free calcium and potassium during paroxysmal activity in cerebral cortex of the rat. Exp Brain Res 1977;27:237–243.
44. Strewler G. Famialial hypocalciuric hypercalcemia—from the clinic to the calcium sensor. West J Med 1994;160:579–580.
45. Quinn S, Ye C, Diaz R, Kifor OMB, Vassilev P, Brown E. The calcium-sensing receptor: a target for polyamines. Am J Physiol 1997;273:C1315–C1323.
46. Foley Jr T, Harrison H, Arnaud C, Harrison H. Familial benign hypercalcemia. J Pediatr 1972;81:1060–1067.
47. Marx SJ, Attie MF, Levine MA, Spiegel AM, Downs RW Jr, Lasker RD. The hypocalciuric or benign variant of familial hypercalcemia: clinical and biochemical features in fifteen kindreds. Medicine (Baltimore) 1981;60:397–412.
48. Law W Jr, Heath H III. Familial benign hypercalcemia (hypocalciuric hypercalcemia). Clinical and pathogenetic studies in 21 families. Ann Int Med 1985;105:511–519.
49. Marx S, Speigel A, Brown E, Koehler J, Gardner D, Brennan M, et al. Divalent cation metabolism. Familial hypocalciuric hypercalcemia versus typical primary hyperparathyroidism. Am J Med 1978; 65:235–242.
50. Gogusev J, Duchambon P, Hory B, Giovannini M, Goureau Y, Sarfati E, et al. Depressed expression of calcium receptor in parathyroid gland of patients with hyperparathyroidism. Kidney Int 1997;51:328–336.
51. Kifor O, Moore FJ, Wang P, Goldstein M, Vassilev P, Kifor I, et al. Reduced immunostaining for the extracellular Ca^{2+}-sensing receptor in primary and uremic secondary hyperparathyroidism. J Clin Endocrinol Metab 1996;81:1598–1606.

52. Stewart A, Broadus A. Mineral metabolism. In: Felig P, Baxter J, Broadus A, Frohman, L, eds. Endocrinology and Metabolism, 2nd ed, McGraw-Hill, New York, 1987, pp. 1317–1453.
53. Heath D. Familial hypocalciuric hypercalcemia. In: Bilezikian J, Marcus R, Levine MA, eds. The Parathyroids, Raven, New York, 1994, pp. 699–710.
54. Gunn I, Wallace J. Urine calcium and serum ionized calcium, total calcium and parathyroid hormone concentrations in the diagnosis of primary hyperparathyroidism and familial benign hypercalcaemia. Ann Clin Biochem 1992;29:52–58.
55. Chou Y-H, Brown E, Levi T, Crowe G, Atkinson A, Arnqvist H, et al. The gene responsible for familial hypocalciuric hypercalcemia maps to chromosome 3q in four unrelated families. Nature Genet 1992;1: 295–300.
56. Heath H III. Familial benign (hypocalciuric) hypercalcemia. a troublesome mimic of primary hyperparathyroidism. Endocrinol Metab Clin North Am 1989;18:723–740.
57. Auwerx J, Demedts M, Bouillon R. Altered parathyroid set point to calcium in familial hypocalciuric hypercalcaemia. Acta Endocrinologica (Copenh) 1984;106:215–218.
58. Khosla S, Ebeling PR, Firek AF, Burritt MM, Kao PC, Heath H III. Calcium infusion suggests a "setpoint" abnormality of parathyroid gland function in familial benign hypercalcemia and more complex disturbances in primary hyperparathyroidism. J Clin Endocrinol Metab 1993;76:715–720.
59. Brown E. Four parameter model of the sigmoidal relationship between parathyroid hormone release and extracellular calcium concentration in normal and abnormal parathyroid tissue. J Clin Endocrinol Metab 1983;56:572–581.
60. Law W Jr, Carney J, Heath H III. Parathyroid glands in familial benign hypercalcemia (familial hypocalciuric hypercalcemia). Am J Med 1984;76:1021–1026.
61. Thogeirsson U, Costa J, Marx S. The parathyroid glands in familial hypocalciuric hypercalcemia. Hum Pathol 1981;12:229–237.
62. Davies M, Adams P, Lumb G, Berry J, Loveridge N. Familial hypocalciuric hypercalcemia: evidence for continued enhanced renal tubular reabsorption of calcium following total parathyroidectomy. Acta Endocrinol 1984;106:499–504.
63. Kristiansen J, Rodbro P, Christiansen C, Brochner Mortensen J, Carl J. Familial hypocalciuric hypercalcemia II: Intestinal calcium absorption and vitamin D metabolism. Clin Endocrinol 1985;23: 511–515.
64. Law Jr W, Bollman S, Kumar R, Heath H III. Vitamin D metabolism in familial benign hypercalcemia (hypocalciuric hypercalcemia) differs from that in primary hyperparathyroidism. J Clin Endocrinol Metab 1984;58:744–747.
65. Kristiansen J, Rodbro P, Christiansen C, Johansen J, Jensen J. Familial hypocalciuric hypercalcemia. III: Bone mineral metabolism. Clin Endocrinol (Oxford) 1987;26:713–716.
66. Menko F, Bijouvet O, Fronen J, et al. Familial benign hypercalcemia: study of a large family. Q J Med 1983;206:120–140.
67. Abugassa S, Nordenstrom J, Jarhult J. Bone mineral density in patients with familial hypocalciuric hypercalecmia (FHH). Eur J Surg 1992;158:397–402.
68. McMurtry C, Schranck F, Walkenhorst D, Murphy W, Kocher D, Teitelbaum S, et al. Significant developmental elevation in serum parathyroid hormone levels in a large kindred with familial benign (hypocalciuric) hypercalcemia. Am J Med 1992;93:247–258.
69. Pasieka J, Andersen M, Hanley D. Familial benign hypercalcemia: hypercalciuria and hypocalciuria in affected members of a small kindred. Clin Endocrinol 1990;33:429–433.
70. Attie M, Gill JJ, Stock J, Spiegel A, Downs RJ, Levine M, et al. Urinary calcium excretion in familial hypocalciuric hypercalcemia. Persistence of relative hypocalciuria after induction of hypoparathyroidism. J Clin Invest 1983;72:667–676.
71. Kristiansen J, Brochner-Mortensen J, Pedersen K. Renal tubular reabsorption of calcium in familial hypocalciuric hypercalcaemia. Acta Endocrinol (Copenh) 1986;112:541–546.
72. Marx SJ, Attie MF, Stock JL, Spiegel AM, Levine MA. Maximal urine-concentrating ability: familial hypocalciuric hypercalcemia versus typical primary hyperparathyroidism. J Clin Endocrinol Metab 1981;52:736–740.
73. Pearce S, Trump D, Wooding C, Besser G, Chew S, Heath D, et al. Calcium-sensing receptor mutations in familial benign hypercalcaemia and neonatal hyperparathyroidism. J Clin Invest 1995;96:2683–2692.
74. Pollak M, Chou Y, Marx S, Steinmann B, Cole D, Brandi M, et al. Familial hypocalciuric hypercalcemia and neonatal severe hyperparathyroidism. effects of mutant gene dosage on phenotype. J Clin Invest 1994;93:1108–1112.

75. Heath III H, Jackson C, Otterud B, Leppert M. Genetic linkage analysis of familial benign (hypocalciuric) hypercalcemia: evidence for locus heterogeneity. Am J Hum Genet 1993;53:193–200.
76. Trump D, Whyte M, Wooding C, Pang J, Pearce S, Kocher D, et al. Linkage studies in a kindred from Oklahoma, with familial benign (hypocalciuric) hypercalcaemia (FBH) and developmental elevations in serum parathyroid hormone levels, indicate a third locus for FBH. Hum Genet 1995;96:183–187.
77. Marx S, Fraser D, Rapoport A. Familial hypocalciuric hypercalcemia. Mild expression of the gene in heterozygotes and severe expression in homozygotes. Am J Med 1985;78:15–22.
78. Bai M, Pearce S, Kifor O, Trivedi S, Stauffer U, Thakker R, Brown E, et al. In vivo and in vitro characterization of neonatal hyperparathyroidism resulting from a de novo, heterozygous mutation in the Ca^{2+}-sensing receptor gene—normal maternal calcium homeostasis as a cause of secondary hyperparathyroidism in familial benign hypocalciuric hypercalcemia. J Clin Invest 1997;99:88–96.
79. Janicic N, Pausova Z, Cole DEC, Hendy GN. Insertion of an Alu sequence in the Ca^{2+}-sensing receptor gene in familial hypocalciuric hypercalcemia and neonatal severe hyperparathyroidism. Am J Hum Genet 1995;56:880–886.
80. Heath H III, Odelberg S, Jackson C, Teh B, Hayward N, Larsson C, et al. Clustered inactivating mutations and benign polymorphisms of the calcium receptor gene in familial benign hypocalciuric hypercalcemia suggest receptor functional domains. J Clin Endocrinol Metab 1996;81:1312–1317.
81. Chou Y-H, Pollak M, Brandi M, Toss G, Arnqvist H, Atkinson A, et al. Mutations in the human Ca^{2+}-sensing receptor gene that cause familial hypocalciuric hypercalcemia. Am J Hum Genet 1995;56:1075–1079.
82. Aida K, Koishi S, Inoue M, Nakazato M, Tawata M, Onaya T. Familial hypocalciuric hypercalcemia associated with mutation in the human Ca^{2+}-sensing receptor gene. J Clin Endocrinol Metab 1995;80:2594–2598.
83. Janicic N, Pausova Z, Cole D, Hendy G. De novo expansion of an Alu insertion mutation of the Ca^{2+}-sensing receptor gene in familial hypocalciuric hypercalcemia and neonatal severe hyperparathyroidism. J Bone Miner Res 1995 10(Suppl. 1):S191.
84. Bai M, Quinn S, Trivedi S, Kifor O, Pearce S, Pollak M, et al. Expression and characterization of inactivating and activating mutations of the human Ca^{2+}_o-sensing receptor. J Biol Chem 1996;271:19,537–19,545.
85. Pearce S, Bai M, Quinn S, Kifor O, Brown E, Thakker R. Functional characterization of calcium-sensing receptor mutations expressed in human embryonic kidney cells. J Clin Invest 1996;98:1860–1866.
86. Bai MNJ, Trivedi S, Quinn S, Cole D, Brown E, Hendy G. Markedly reduced activity of mutant calcium-sensing receptor with an inserted Alu element from a kindred with familial hypocalciuric hypercalcemia and neonatal sevre hyperparathyroidism. J Clin Invest 1997;99:1917–1925.
87. Bockaert J. G proteins, G protein-coupled receptors: structure, function and interactions. Curr Opinion Neurobiol 1991;1:32–42.
88. Jackson T. Structure and function of G protein coupled receptors. Pharmacol Ther 1991;50:425–442.
89. Pin J-P, Gomeza T, Joly C, Bockaert J. The metabotropic glutamate receptors: their second intracellular loop plays a critical role in the G-protein coupling specificity. Biochem Soc Trans 1994;23:910–96.
90. Landon J. Parathyroidectomy in generalized osteitis fibrosa cystica. J Pediatr 1932;1:544–560.
91. Eftekhari F, Yousefzadeh D. Primary infantile hyperparathyroidism: clinical, laboratory, and radiographic features in 21 cases. Skeletal Radiol 1982;8:201–208.
92. Steinmann B, Gnehm H, Rao V, Kind H, Prader A. Neonatal severe primary hyperparathyroidism and alkaptonuria in a boy born to related parents with familial hypocalciuric hypercalcemia. Helv Paediatr Acta 1984;39:171–186.
93. Marx S, Attie M, Spiegel A, Levine M, Lasker R, Fox M. An association between neonatal severe primary hyperparathyroidism and familial hypocalciuric hypercalcemia in three kindreds. N Engl J Med 1982;306:257–284.
94. Gaudelus J, Dandine M, Nathanson M, Perelman R, Hassan M. Rib cage deformity in neonatal hyperparathyroidism (letter). Am J Dis Child 1983;137:408–409.
95. Mallette L. The functional and pathologic spectrum of parathyroid abnormalities in hyperparathyroidism. In: Bilezikian J, Marcus R, Levine, MA, eds. The Parathyroids. Raven, New York, 1994, pp. 423–455.
96. Fujimoto Y, Hazama H, Oku K. Severe primary hyperparathyroidism in a neonate having a parent with hypercalcemia: treatment by total parathyroidectomy and simultaneous heterotopic autotransplantation. Surgery 1990;108:933–938.

97. Grantmyre E. Roentgenographic features of "primary" hyperparathyroidism in infancy. J Can Assoc Radiol 1973;24:257–260.
98. Fujita T, Watanabe N, Fukase M, Tsutsumi M, Fukami T, Imai Y, et al. Familial hypocalciuric hypercalcemia involving four members of a kindred including a girl with severe neonatal primary hyperparathyroidism. Mineral Electrolyte Metab 1983;9:51–54.
99. Matsuo M, Okita K, Takemine H, Fujita T. Neonatal primary hyperparathyroidism in familial hypocalciuric hypercalcemia. Am J Dis Child 1982;136:728–731.
100. Harris S, D'Ercole A. Neonatal hyperparathyroidism: the natural course in the absence of surgical intervention. Pediatrics 1989;83:53–56.
101. Page L, Haddow J. Self-limited neonatal hyperparathyroidism in familial hypocalciuric hypercalcemia. J Pediatr 1987;111:261–264.
102. Hosokawa Y, Pollak M, Brown E, Arnold A. Mutational analysis of the extracellular Ca^{2+}-sensing receptor gene in human parathyroid tumors. J Clin Endocrinol Metab 1995;80:3107–3110.
103. Akerstrom G, Rastad J, Ljunghall S, Ridefelt P, Juhlin C, Gylfe E. Cellular physiology and pathophysiology of the parathyroid glands. World J Surg 1991;15:672–680.
104. Chattopadhyay N, Mithal A, Brown E. The calcium receptor: a new handle on the diagnosis and treatment of parathyroid disorders. Endocr Rev 1996;17:289–307.
105. Arnold A, Brown M, Urena P, Gaz R, Sarafati E, Drueke T. Monoclonality of parathyroid tumors in chronic renal failure and in primary parathyroid hyperplasia. J Clin Invest 1995;95:2047–2053.
106. Rosenberg C, Kim H, Shows T, Kronenberg H, Arnold A. Rearrangement and overexpression of DS11S287E, a candidate oncogene on chromosome 11q13 in benign parathyroid adenoma. Oncogene 1991;6:449–453.
107. Farnebo F, Teh B, Dotzenrath C, Wassif W, Svensson A, White I, et al. Differential loss of heterozygosity in familial, sporadic and uremic hyperparathyroidism. Hum Genet 1997;99:342–349.
108. Nemeth E. Ca^{2+} receptor-dependent regulation of cellular function. News Physiol Sci 1995;10:1–5.
109. Malgaroli A, Meldolesi J, Zambone-Zallone A, Teti A. Control of cytosolic free calcium in rat and chicken osteoclasts. The role of extracellular calcium and calcitonin. J Biol Chem 1989;264:14,342–14,349.
110. Lundgren S, Hjalm G, Hellman P, Ek B, Juhlin C, Rastad J, et al. A protein involved in calcium sensing of the human parathyroid and placental cytotrophoblast cells belongs to the LDL-receptor protein superfamily. Exp Cell Res 1994;212:344–350.
111. Saito A, Pietromonaco S, Loo A, Farquhar M. Complete cloning and sequencing of rat gp330/"megalin", a distinctive member of the low density lipoprotein receptor gene family. Proc Natl Acad Sci USA 1994;91:9725–9729.
112. Zaidi M, Datta H, Patchell A, Moonga B, MacIntyre I. "Calcium-activated" intracellular calcium elevation: a novel mechanism of osteoclast regulation. Biochem Biophys Res Commun 1989;183:1461–1465.

6 Resistance to TSH

Peter Kopp, MD

CONTENTS

HORMONE RESISTANCE
SPORADIC CONGENITAL HYPOTHYROIDISM
MOLECULAR CHARACTERISTICS OF THE TSH RECEPTOR
INTRACELLULAR SIGNALING IN THYROID FOLLICULAR CELLS
MUTATIONS IN SEVEN-TRANSMEMBRANE RECEPTORS
RESISTANCE TO TSH
ACTIVATING MUTATIONS IN THE TSH RECEPTOR
TSH RESISTANCE: SUMMARY AND SIGNIFICANCE
ACKNOWLEDGMENTS
REFERENCES

HORMONE RESISTANCE

The concept of hormone resistance evolved from the observation that pseudohypoparathyroidism is caused by peripheral unresponsivness to bioactive parathyroid hormone *(1)*. This pathogenic principle has now been recognized to result in a wide array of endocrine disorders. Hormone resistance may be complete or partial. In partial resistance, elevated hormone levels may compensate for the reduced target tissue sensitivity and, thus, result in sufficient stimulation of the cellular signaling pathways necessary for a normal metabolic response. The molecular defects causing hormone resistance include mutations in hormone receptors or in their postreceptor signaling pathways. Antibodies directed against hormones or their receptors, as well as the absence of target cells, can result in a similar phenotype.

The diagnosis of hormone resistance is made by the demonstration of an elevated bioactive hormone level coexisting with deficient hormone action. This constellation may be mimicked by detection of an elevated immunoreactive, but bioinactive hormone. Because hormone resistance is often caused by loss-of-function mutations, phenotypic alterations usually will arise only if both alleles are defective. The mode of inheritance of many of these disorders is thus autosomal-recessive. More rarely, a defective allele may exert a dominant-negative function over the wild-type allele. Consequently, the disease will be transmitted in an autosomal-dominant manner, as exemplified by resistance to thyroid hormone *(2,3)* (*see* Chapter 7).

From: *Contemporary Endocrinology: Hormone Resistance Syndromes*
Edited by: J. L. Jameson © Humana Press Inc., Totowa, NJ

SPORADIC CONGENITAL HYPOTHYROIDISM

The cloning of many of the key genes involved in the regulation of the hypothalamic–pituitary–thyroid axis, combined with careful clinical and biochemical characterizations, made it possible to define the molecular basis of many thyroid disorders, among them being some of the defects resulting in sporadic forms of congenital hypothyroidism. The differential diagnosis of congenital hypothyroidism includes syndromes of hormone resistance to thyrotropin-releasing hormone (TRH) at the level of the pituitary gland, to thyroid-stimulating hormone (TSH) in thyroid follicular cells, and to the action of thyroid hormones in peripheral tissues (Table 1, Fig.1).

In iodine-sufficient regions, permanent sporadic congenital hypothyroidism affects about 1/3000 to 1/4000 newborns *(4–6)*. The majority of newborns with sporadic congenital hypothyroidism appear normal at birth, and fewer than 10% will be diagnosed as hypothyroid based solely on clinical signs. Because diagnosis may be difficult and early treatment is essential in order to avoid permanent mental retardation, neonatal screening programs have been established in most developed countries *(7–9)*.

A heterogeneous group of developmental abnormalities, thyroid dysgenesis, accounts for 80–90% of sporadic congenital hypothyroidism (Table 1, Fig. 1) *(10,11)*. It includes thyroid agenesis, ectopic thyroid tissue, or cysts of the thyroglossal duct. In these patients, radioiodine scanning will be negative (agenesis) or demonstrate an abnormal location or structure of the gland (ectopy, hemiagenesis). The etiology of these developmental defects, which only rarely recur within a family, remains elusive.

Genetic defects in the gene products involved in the regulation of the hypothalamic–pituitary–thyroid axis are rare. They often display an autosomal-recessive mode of inheritance and account for 10–20% of patients with congenital hypothyroidism. They can be broadly classified into two major categories based on the absence or presence of a goiter (Table 1, Fig. 1). Agoitrous hypothyroidism indicates that an element of the TRH–TSH–thyroid axis is defective, thus impeding stimulation of thyroid cell growth and function. The goitrous forms point to a defect in one of the steps of thyroid hormone synthesis within the thyrocyte.

Thus far, there have been no reports on patients with documented defects in the TRH gene. Targeted disruption of the TRH gene in mice led to an overtly hypothyroid phenotype *(12)*. Remarkably, the TSH levels were elevated in these mice, but displayed diminished biological activity. Similar biochemical constellations with elevated TSH with reduced bioactivity have been found in some individuals with central hypothyroidism *(13–15)*. A defect in the TRH receptor in the pituitary thyrotroph results in resistance to TRH. Recently, the first patient with isolated central hypothyroidism owing to a compound heterozygous defect in the TRH receptor gene was reported *(16)*. Clinically, the patient presented with short stature, delayed bone age, and reduced intelligence. The basal TSH levels were normal and T4 was below normal. Intravenous administration of TRH did not result in an increase of TSH and prolactin.

Other causes of TSH deficiency include tumors, infiltrative diseases, or congenital malformations of the hypothalamus or the pituitary gland, and are typically associated with deficiencies of other hypophyseal hormones *(17)*. Very rarely, familial panhypopituitarism and familial pituitary agenesis are inherited in an autosomal- recessive manner *(18)*. The molecular defects underlying these genetic forms of panhypopituitarism are mostly unknown. Very recently, mutations in the homeodomain protein Prop1 have been

Table 1
Causes of Permanent Sporadic Congenital Hypothyroidism[a]

	Molecular Defect	Inheritance	Chromosome	MIM#
Thyroid dysgenesis				
Thyroid agenesis	Unknown	—		218700
Athyreosis				
Hemiagenesis				
Ectopic thyroid gland	Unknown	—		225250
Lingual				
Suprahyoid				
Infrahyoid				
Familial dyshormonogenesis				
Without goiter				
Hypothalamic–pituitary axis				
Familial hypopituitarism	Prop1, unknown	AR, X	?, X	601538, 262600, 262710, 262710, 312000
TRH	Unknown	AR	3p	275120
TRH receptor	Inactivating mutations	AR	8q23	188545
Pit1	Inactivating mutations	AR, AD	3p11	173110
TSHβ-subunit	Inactivating mutations	AR	1p13	188540
TSH resistance				
TSH receptor	Inactivating mutations	AR	14q31	275200
G$_s$α (pseudohypoparathyroidism)	Inactivating mutations	AR, AD	20q13.2	139320
Other postreceptor defects	Not identified	AR, AD	?	
With goiter				
Sodium-iodine symporter	Inactivating mutations	AR	19p12-13.2	601843
Thyroperoxidase	Inactivating mutations	AR	2p25	274500
Hydrogenperoxide generation	Not identified	?	?	
Thyroglobulin	Inactivating mutations	AR,(AD)	8q24	188450
Dehalogenase	Not identified	?	?	274800
Pendred syndrome	Inactivating mutations	AR		274600
TTF1	Not identified	AR (?)	14q13	600635
Pax8	Inactivating mutations	AD	2q12-q14	1674125

[a] Abbreviations: TRH (receptor), thyrotropin-releasing hormone (receptor); Prop1, transcriptionf factor prophet of Pit1; Pit1, pituitary transcription factor 1; TSHβ, thyrotropin β subunit; TSH (receptor), thyrotropin (receptor); Gα, stimulatory G-protein subunit α; TTF1, thyroid transcription factor 1; Pax8, homebox transcription factor 8; AR, autosomal-recessive; AD, autosomal-dominant; MIM#, Mendelian Inheritance in man catalog #.

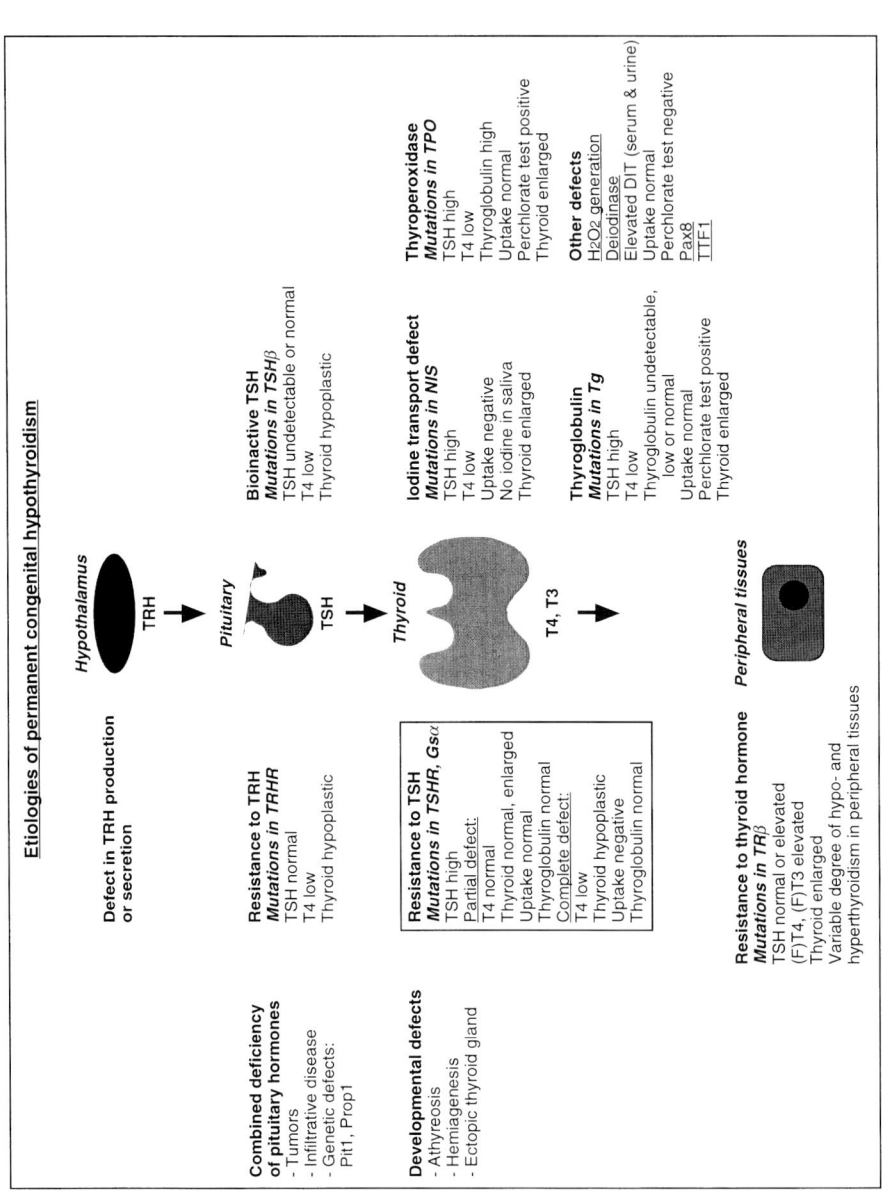

Fig. 1. Etiologies of permanent congenital hypothyroidism. TRH(R) = TSH-releasing hormone (receptor); Prop1 = prophet of Pit1 homeodomain transcription factor; TSH(R) = thyroid-stimulating hormone (receptor); $G_s\alpha$ = stimulatory G-protein subunit α; NIS = sodium-iodide symporter; TPO = thyroperoxidase (gene); Tg = thyroglobulin (gene); (F)T4 = (free) thyroxine; (F)T3 = free triiodothyronine; DIT = diiodothyronine; Pax8 = transcription factor pax8; TTF1 = thyroid transcription factor 1.

reported in such families *(19)*. Two well-characterized genetic defects that cause TSH deficiency include mutations in the Pit-1 gene, and in the gene encoding the β-subunit of TSH. Pit-1 is a pituitary-specific transcription factor involved in the development and function of somatotrophs, lactotrophs, and thyrotrophs *(20)*. Mutations in the Pit-1 gene were initially identified in strains of dwarf mice (Jackson mouse, Snell mouse) *(21)*, and a variety of Pit-1 mutations have now also been described in humans *(22,23)*. If the mutation affects the DNA binding domain of the transcription factor, the defect is inherited as an autosomal-recessive trait. Pit-1 mutations that lead to a loss of the transactivation properties of the protein, while retaining the ability to bind DNA, display a dominant-negative effect and are thus inherited in an autosomal-dominant manner. Isolated hereditary TSH deficiency with secondary congenital hypothyroidism may be caused by autosomal-recessive mutations in the TSHβ chain *(24–27)*. In these patients, the TSH levels are low or even unmeasurable.

In addition to mutations in the TSHβ subunit that lead to a loss of biological activity, the response to TSH may be impaired at the level of the thyroid gland. This can be owing to defects in the TSH receptor itself or its signaling cascade, and forms the focus of this chapter. Depending on the degree of reduced function, a diminution in TSH receptor signaling will lead to overt hypothyroidism or an elevated TSH with normal T4 levels, a constellation referred to as euthyroid hyperthyrotropinemia *(28–30)*.

Complete or partial unresponsivness to TSH can theoretically be caused by any element in the mediation of TSH action. Total insensitivity to TSH is predicted to result in a small hypoplastic thyroid gland, and reduced synthesis and secretion of thyroid hormones. In partial resistance, the size of the gland and the thyroid hormone levels are expected to be normal at the expense of an elevated TSH in order to overcome the insensitivity. If the defect involves only the thyroid gland, the responsible gene should be expressed in a thyroid-specific manner. It is obvious that the TSH receptor fulfills these criteria, and consequently, it is one of the primary candidates to harbor such a defect.

A quantitative or qualitative reduction in the second component of the TSH-dependent signaling cascade, the $G_s\alpha$-subunit, has been recognized to cause pseudohypoparathyroidism (PHP) *(31,32)*. In PHP Ia, the insensitivity to ligand stimulation results not only in resistance to parathyroid hormone (PTH), but it may include unresponsiveness to TSH and the gonadotropins *(see* Chapter 3).

Defects in one of the multiple steps involved in hormone synthesis within the thyroid follicular cell itself are typically associated with goitrous enlargement of the thyroid because of the intact trophic effects of TSH on cell growth (Table 1, Fig. 1) *(33)*.

MOLECULAR CHARACTERISTICS OF THE TSH RECEPTOR

TSH exerts its stimulating effect on thyroid follicular cells through the TSH receptor, a member of the large family of G-protein-coupled seven-transmembrane receptors (GPCR) (Fig. 2). In addition to its central role in thyroid physiology, the TSH receptor is a key element in several pathophysiological entities. It is the target of stimulating antibodies that cause Graves' hyperthyroidism, whereas blocking antibodies may lead to autoimmune hypothyroidism *(34)*. In addition, as recognized during the last 5 yr, the TSH receptor can be activated or inactivated by gain- or loss-of-function mutations (Fig. 3, Table 2) *(35–37)*.

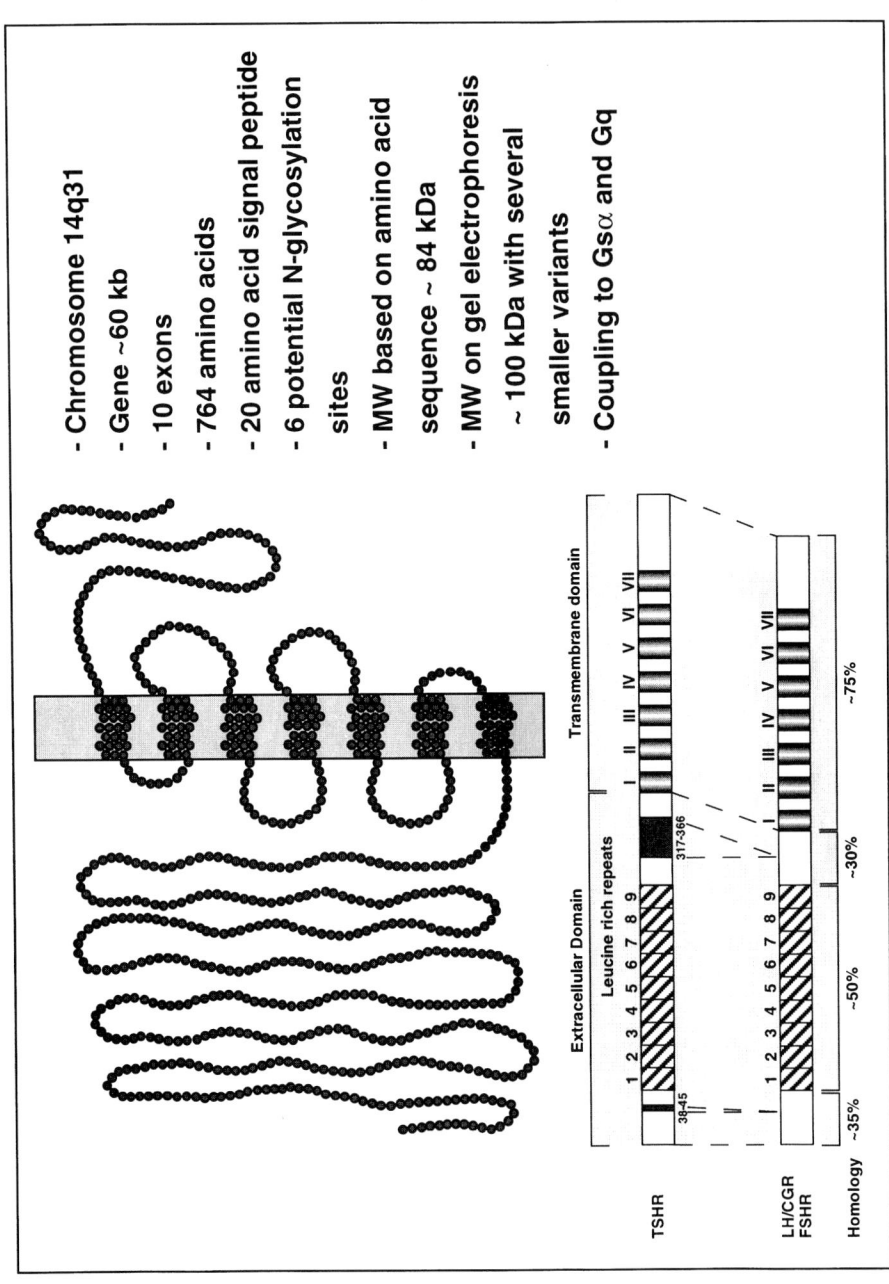

Fig. 2. The TSH receptor is a member of the G-protein-coupled seven-transmembrane receptors. The single-copy human gene is located on chromosome 14q13 and spans more than 60 kb. The first 9 exons of its 10 exons encode the extracellular domain, and exon 10 codes for the carboxy-terminal part of the extracellular domain, the whole transmembrane domain, and the intracellular carboxy-terminus. The open reading frame consists of 2295 nucleotides and encodes a protein of 764 amino acids. The first 20 amino acids form a signal peptide sequence. The homology with the gene structure of the closely related receptors for LH/CG and FSH is shown schematically.

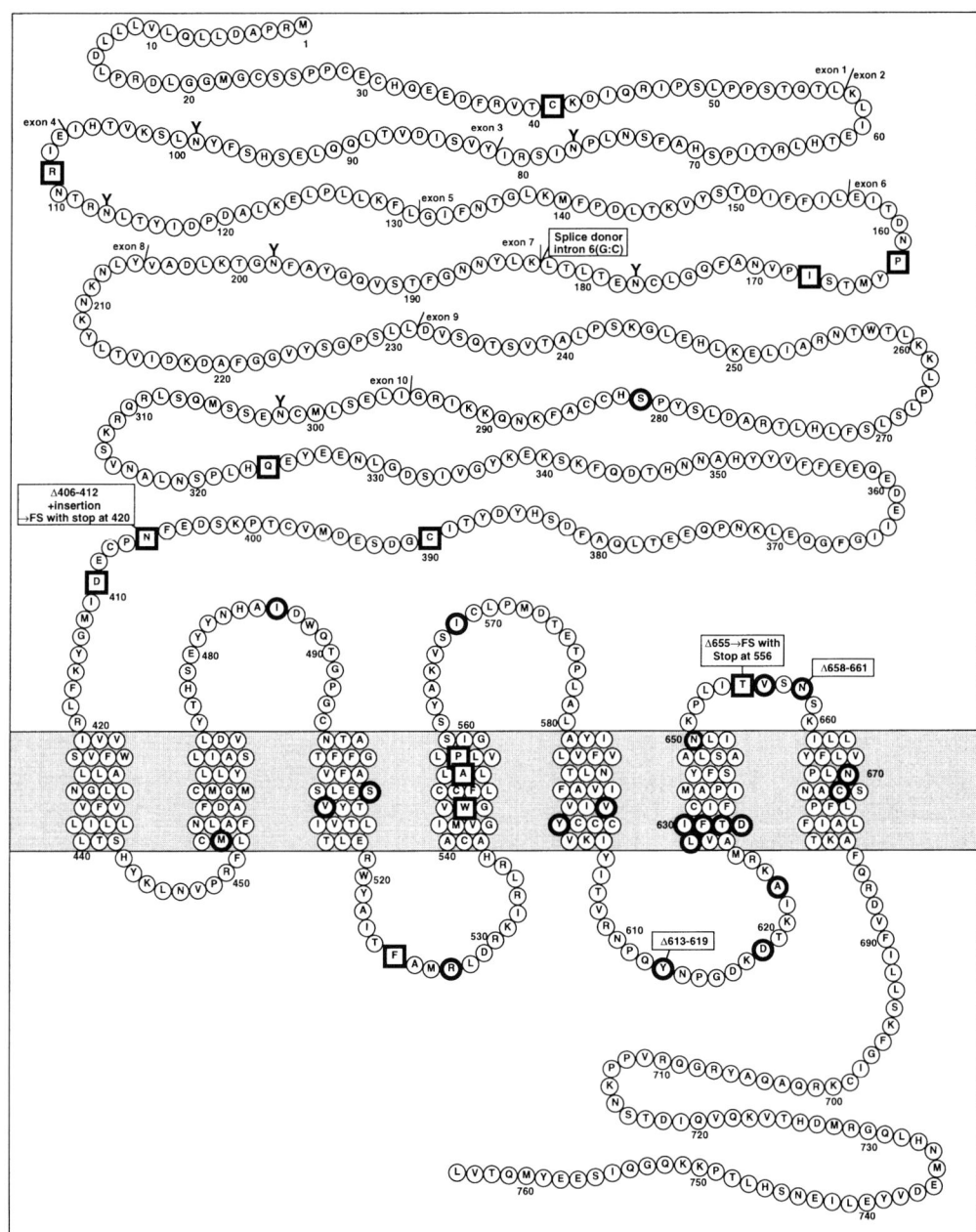

Fig. 3. Naturally occurring activating (circles) and inactivating (squares) mutations in the TSH receptor. The Y symbols indicate potential aspargine-linked glycosylation sites.

Together with the receptors for the other glycoprotein hormones follicle-stimulating hormone (FSH) and luteinizing hormone/choriogonadotropin (LH/CG), the TSH receptor forms a distinct subfamily of GPCRs defined by a large amino-terminal extracellular domain involved in binding of the hormone *(38–40)*. The receptors share a high degree of homology in the transmembrane domain of about 70%. The main differences are found in the extracellular domain with a 39% degree of homology with the FSH receptor, and 45% with the LH/CG receptor *(38–40)* (*see* Chapter 10).

Table 2
Naturally Occurring Mutations in the TSH Receptor

Activating mutations	Activating mutations	Activating mutations	Activating mtuations
	Somatic mutations in toxic adenomas and differentiated thyroid carcinomas*	Familial germline mutations	Sporadic germline mutations
Cys41Ser	Ser281Asn/Thr/Ile		
Arg109Gln	Met453Thr		Met453Thr
Pro162Ala	Ile486Phe/Met		
Ile167Asn		Ser505Arg	Ser505Asn
Gln324X		Val509Ala	
Cys390Trp		Arg528His	
del406-412 +insertion: FS and stop at 420	Ile568Thr		
Asp410Asn			Val 597Leu
Phe525Leu	Tyr601Asn		
Trp546X	Del 613-619		
Ala553Thr	Asp619Gly		
Pro556Leu (*hyt/hyt* mouse)	Ala623Ile/Ser/Val	Ala623Val	
Splice donor intron 6(G;C)	Leu629Phe	Leu629Phe	
delAC codon 655: FS and stop at 656	Ile630Leu		
	Phe631Leu/Cys		Phe631Leu
	Thr632Ile/Ala*		Thr632Ile
	Asp633Ala/Glu/Tyr/His*		
		Asn650Tyr	
	Val 656Phe		
	del658-661		
		Asn670Ser	
		Cys672Tyr	
Refs. 28–30,46,47,80,93, 105,106,112,113,115	Refs. 35,62,85, 128–137,160–162	Refs. 140,144–146	Refs. 145, 153–157

The first TSH receptor cDNA to be cloned was from the dog *(39,41)*. Subsequently, cDNAs have been cloned in various species, including human, rat, mouse, cow, sheep, and cat *(42–50)*. Species comparisons reveal a very high degree of homology of more than 90%. The single-copy human gene is located on chromosome 14q13 *(51,52)*. Structurally, it spans more than 60 kb and it contains 10 exons *(53)* (Figs. 2 and 3). The first 9 exons encode the extracellular domain, and exon 10 codes for the carboxy-terminal part of the extracellular domain, the whole transmembrane domain, and the intracellular carboxy-terminus. The promoter of the TSH receptor gene contains binding sites for the thyroid transcription factor 1 (TTF1 or T/ebp), and the cAMP response element binding protein (CREB), among other *cis*-acting sequences *(54,55)*.

The TSH receptor gene is predominantly expressed in the thyroid gland where several mRNA species have been described, the most important encompassing 4.6 kb. By reverse transcriptase polymerase chain reaction (RT-PCR), the mRNA has also been detected in adipocytes and retroorbital tissue *(56,57)*. The significance of the latter observation in the development of Graves' ophthalmopathy remains controversial *(58,59)*. The open reading frame consists of 2295 nucleotides and encodes a protein of 764 amino acids. The first 20 amino acids form a signal peptide sequence. The human extracellular domain contains six potential asparagine-linked glycosylation sites, some of which are important for normal receptor function *(60)*. The TSH receptor differs from the LH/CG receptor by the presence of two unique insertions of 8 and 50 amino acids in the extracellular domain *(40)*. Based on the putative homology with the ribonuclease inhibitor, a three-dimensional model of the extracellular domain has been proposed *(61)*. It is thought that eight leucine-rich repeats form a horseshoe-shaped concave surface composed of parallel β-sheets, which interact with the ligand *(61,62)*.

In vitro studies with thyroid cells in culture and mammalian cells transfected with the TSH receptor suggest that the receptor is not only present as a single chain on the cell surface, but also as a two-subunit form *(63,64)*. An extracellular α-subunit seems to be linked by disulfide bonds to the membrane-spanning β-subunit after cleavage of a connecting peptide *(65)*. The functional significance of this phenomenon is presently unknown. Shedding of the the extracellular α-subunit or the connecting peptide may be implicated in the development of autoantibodies *(66)*.

As occurs with other GPCRs, the TSH receptor undergoes homologous desensitization after exposure to TSH *(38)*. It is likely that this regulation is in part mediated by G-protein receptor kinases, which phosphorylate the receptor *(67,68)*. Subsequently, β-arrestin may bind to the phosphorylated receptor and result in its uncoupling from $G_s\alpha$. Transcriptional downregulation of the TSH receptor is thought to be mediated by a splice variant of cAMP response element modulator (CREM), the transcriptional repressor inducible cAMP early repressor (ICER) *(69)*.

INTRACELLULAR SIGNALING IN THYROID FOLLICULAR CELLS

After the binding of TSH, the receptor undergoes a structural change resulting in the activation of G-proteins. The TSH receptor couples primarily to $G_s\alpha$, which activates adenylyl cyclase. The ensuing increase in cAMP leads to phosphorylation of protein kinase A and activation of targets in the cytosol and the nucleus, e.g., the transcription factor CREB. The TSH-dependent cAMP cascade is the major regulator of growth, differentiation, and hormone secretion of thyroid follicular cells *(70)*. At least in humans, the TSH receptor also stimulates the phospholipase C-dependent inositol phosphate pathway at higher doses of TSH by coupling to Gq *(71–74)*. This pathway activates iodination and thyroid hormone synthesis. In addition, there is evidence that the TSH receptor may be coupled to other G-proteins *(75)*. In contrast to the receptors for LH/CG and FSH, the TSH receptor displays constitutive basal activity, i.e., readily measurable spontaneous activity in the absence of ligand.

The predominant role of the cAMP pathway for thyroid cell growth and function has been further illustrated by activating mutations in the TSH receptor and by several transgenic models with chronic overstimulation of this pathway *(36,76–78)*.

MUTATIONS IN SEVEN TRANSMEMBRANE RECEPTORS

The molecular pathogenesis of many endocrine and non-endocrine diseases has been elucidated by the demonstration of inactivating and activating mutations in GPCRs *(31,32,79)*. An alteration in the genetic code may cause loss of function by several distinct mechanisms *(32)*. Mutations may result in a quantitative or qualitative reduction of mRNA or protein synthesis. If protein synthesis is still possible, amino acid substitutions may alter protein folding and thus the ability of the gene product to reach its cellular location *(80)*. Mutations can also impair binding of a ligand *(47)* or disrupt interaction with other proteins. Under normal conditions, GPCRs are thought to change spontaneously between conformations that favor or impede G protein coupling. Binding of a ligand stabilizes the receptor in its active conformation. Conversely, an alteration in the amino acid sequence may favor an inactive conformation. This model is supported by the identification of inverse agonists, ligands that compete with agonists for receptor binding, but result in diminished G-protein activation *(81)*. Because both alleles have to be present in an inactive form to result in a clinical phenotype, the inheritance of these disorders is often autosomal-recessive. Alternatively, a mutated allele may act as a dominant-negative and thus impede proper function of the normal gene product. In these cases, one will encounter an autosomal-dominant mode of transmission. The detection of gain of function mutations was prompted by the discovery that site-directed mutagenesis of critical residues in the third intracellular loop of the α_{1b}-adrenergic receptor can lead to constitutive activation in the absence of ligand *(82)*. Subsequently, naturally occurring activating mutations were discovered in rhodopsin *(83)*, the receptors for LH and TSH *(84,85)*, and followed by the detection of activating mutations in many other GPCRs *(32)*. These mutations result in constitutive activation even in the absence of ligand. They are by definition dominant and can be transmitted in an autosomal-dominant mode.

Naturally occurring mutations in these receptors provide insights into the molecular etiology, the structure–function relationship of the proteins, and may become important for diagnostic and therapeutic purposes.

RESISTANCE TO TSH

Several cases suggestive of resistance to TSH, but not consistent with PHP Ia, have been recognized in clinical studies (Table 3) *(86–92)*. The molecular correlate, loss-of-function mutations in the TSH receptor gene, has only be identified recently in a subset of such patients (Table 4) *(28–30,80,93)*. In other patients with resistance to TSH, the TSH receptor gene was found to be normal, indicating that additional defects in the signal cascade remain to be discovered (Table 5) *(94,95)*.

Clinical Studies on TSH Resistance

The first case suggestive of resistance to TSH at the level of thyroid follicular cells was reported by Stanbury et al. in 1968 *(86)*. This patient, an 8-yr-old boy with congenital hypothyroidism, was the offspring of first cousins. He had high serum levels of biologically active TSH, but his thyroid did not respond to TSH in vivo or in vitro, findings consistent with end-organ resistance (Table 3). A patient with congenital hypothyroidism described by Job et al. had a normally located gland with absent increase in radioiodine uptake after administration of TSH *(87)*. A 19-yr-old male, the son of first-degree cousins, with congenital hypothyroidism, severely retarded development, and a bone age of 7 yr was studied by Medeiros-Neto et al. *(88)*. In this case, TSH was shown to increase after TRH stimulation in vivo and to be biologically active in vitro. Thyroid tissue from this patient was obtained by biopsy for in vitro studies. The basal cAMP levels were signifiantly lower in comparison to normal thyroid tissue. After incubation with bovine TSH, there was no increase in cAMP accumulation and only a minimal rise in ^{131}iodine incorporation. In addition, there was no detectable thyroglobulin. The exact site of the TSH resistance could not be further defined *(88)*.

A very similar case was studied by Codaccioni et al. *(90)*. The proband was the product of a consanguineous marriage, and congenital hypothyroidism was first noted at the age

Table 3
Clinical Descriptions of TSH Unresponsiveness with Unknown Molecular Defect[a]

	Stanbury et al.	Job et al.	Medeiros-Neto et al.	Codaccioni et al.	Aarseth et al.	Takamatsu et al.		Mimouni et al.
Reference	86	87	88	90	91	92	92	116
Sex	Male	Male	Male	Male	Male	Sister	Brother	11 patients propositus
Age at diagnosis	2 yr evaluation at 8 yr	4 yr	19 yr	18 mo; evaluation at 17 yr	16 mo; evaluation at 53 yr	infancy evaluation at 26 yr	5 yr evaluation at 29 yr	3.8 yr
Consanguinity	Yes	No	Yes	Yes	No	Yes		No
Familial occurrence	Possible	No	Possible	No	Yes	Yes		Yes
Mode of inheritance	AR	?	AR	AR	AR ?	AR		AD
Goiter Scintigraphy	Absent ^{131}iodine normal location	Absent ^{131}iodine normal location	Absent ^{131}iodine normal location	Absent ^{131}iodine normal location	Absent ^{131}iodine normal location	Absent ^{123}iodine normal location	Absent ^{123}iodine normal location	Absent ^{123}iodine normal location
Uptake at 24 h	27–37% 19% after TSH	13% 17% after TSH	41% 49% after TSH	31% 34% after TSH 38% after dibutyryl cAMP	20% 26% after TSH increase aafter dibutyryl cAMP infusion	9%	5%	11%
TSH mU/L	145	Elevated	52 192 after TRH	385	450	125	220	100
T4 nmol/L[b] FT4 pmol/L	9–12.9	41.2	36	<12.9	24.5	<3 pmol/L	<3 pmol/L	8 pmol/L
Tg µg/L	Not detectable in tissue	nd	<5 absent in tissue	<1	<5	<5	<5	nd
ClO$_4$/SCN discharge	SCN 19% ClO$_4$ normal	nd	ClO$_4$ normal	SCN normal	ClO$_4$ normal	nd	nd	nd
Additional tests	Absent response to TSH in vivo and in vitro	nd	No increase in cAMP after bTSH administration; TSH bioactivity normal	TSH binding and bioassay normal; no increase in cAMP after bTSH administration, but after NaF	Increase in radioiodine uptake after dibutyryl cAMP infusion	TSH bioactivity normal; no TSH-receptor mutations in subsequent studies (ref. 97)		TSH bioactivity normal

[a] Abbreviations: TSH, thyrotropin; T4, thyroxin; ClO$_4$, perchlorate; SCN, thiocyanate; AR, autosomal recessive; AD, autosomal dominant; nd, not done.
[b] T4 values in most original publications in µg/dL (T4 in µg/dL → nmol/L = × 12.87).

Table 4
TSH Resistance Due to Mutations in the TSH Receptor Gene[a]

	Sunthornthep-varakul et al.	De Roux et al.	De Roux et al.	De Roux et al.	De Roux et al.	Clifton-Bligh et al.	Biebermann et al.	Abramowicz et al.
Reference	28	29	29	29	29	30	112, 113	93
Diagnosis	EHT	EHT	EHT	EHT	EHT	EHT	CH	CH
Amino acid substitution	Pro162Ala Ile167Asn	Cys41Ser Phe525Leu	Gln324Stop Asp410Asn	Cys390Trp Trp546Stop	Pro162Ala	Arg109Gln Trp546Stop	Cys390Trp + Deletion and insertion resulting in 420X	Ala553Thr
Sex	Three sisters	Female	Female	Male	Female	Male	Female	Male Female
Age at diagnosis	2 neonatal 1 age four	Neonatal	Neonatal in index patient	Neonatal	Neonatal	Neonatal	Neonatal	Neonatal
Consanguinity	No	No	No	No	Yes	No	No	Yes
Familial occurrence	Yes	No	Yes	No	Yes	No	No	Yes
Thyroid size	Normal	Normal	Slightly enlarged	Slightly enlarged	Normal	Normal	Hypoplastic	Hypoplastic
Scintigraphy	^{123}iodine	^{123}iodine	^{123}iodine	^{123}iodine	^{123}iodine	Tc		Technetium negative
Uptake	23% (8–30)	Normal	Normal	Normal	Normal			
Echography							Hypoplastic gland	Hypoplastic gland
TSH mU/L	80–103 mU/L	129 mU/L Afterward: 26–115	44	34	99	92	89	>130 530
T4 nmol/L[b]	118–167	22	20	12.9	12	10	78	15.4
FT4 pmol/L		Normal	Normal	Normal	Normal			4.8
Tg μg/L								66 84 (normal range for age 2–106)

[a] Abbreviations: EHT, euthyroid hyperthyrotropinemia; CH, congenital hypothyroidism; TSH, thyrotropin; (F)T4, (free) thyroxin; Tg, thyroglobulin.
[b] T4 values in most original publications in μg/dL (T4 in μg/dL → nmol/L = ×12.87).

Table 5
Reports on Familial Cases with TSH Resistance with Absence of TSH Receptor Mutations[a]

	Takeshita et al.			Ahlbom et al.	Xie et al.		
Reference	94			114	95		
Diagnosis	3 Patients with CH			23 cases of CH with familial occurrence or consanguineous parents	3 Families with ETH		
Patient/family	Patient 1	Patient 2	Patient 3		Family 1	Family 2	Family 3
Consanguinity	Yes	Yes	?	In 13 families	Common ancestors	No	No
Familial occurrence	No	Yes	Yes	In 16 families	Yes	Yes	No
Inheritance	AR	AR	?	AR (AD in some families possible)	AD	AD	?
Type of study	Direct sequencing TSHR cDNA			Linkage with markers flanking the TSH receptor	Direct sequencing of TSH receptor cDNA and promoter Haplotyping with intragenic markers DGGE G$_s\alpha$ in one family		
Comment	Patients 1 previously reported in ref. 92 Patient 2 in ref. 95			Markers distance 2.3 and 1.2 cM from TSH receptor locus No intragenic markers No linkage assuming genetic homogeneity and heterogeneity	No linkage to TSH receptor locus with intragenic markers in families 1 and 2		

[a] Abbreviations: EHT, euthyroid hyperthyrotropinemia; CH, congenital hypothyroidism; AR, autosomal-recessive; AD, autosomal-dominant; DGGE, denaturing gradient gel electrophoresis.

of 18 mo. The thyroid gland was normal in location and size. He was treated with dessicated thyroid, which was stopped 1 mo before he was re-examined at age 17 yr. These studies demonstrated very low T3, T4, and thyroglobulin levels, and a TSH of 385 mU/L by RIA and 400 mU/L using a bioassay. Radioiodine uptake increased after infusion of dibutyryl-cAMP, but not after administration of exogenous TSH. A biopsy demonstrated follicles of variable size, which were largely devoid of colloid. Lymphocytic infiltration was absent. The basal plasma membrane showed thickening on electron microscopy, and the endoplasmic reticulum was distended. Binding of ^{125}iodine to thyroid membrane preparations was identical in comparison to normal control tissue. cAMP accumulation could be induced by sodium fluoride, but not by TSH, suggesting unresponsiveness to TSH owing to a molecular defect located upstream of adenylyl cyclase, possibly in the TSH receptor itself. The morphological aspects of the basal plasma membrane could theoretically point to an abnormality in membrane composition, although it is rather unlikely that such a defect would be restricted to thyrocytes.

Aarseth et al. studied a 53-yr-old man with unresponsiveness to endogenous and exogenous TSH *(91)*. He was diagnosed with hypothyroidism at the age of 16 mo. His parents were not consanguineous, but two of his five siblings were also treated for hypothyroidism of unknown origin. The thyroid was of normal size. The basal radioiodine uptake was normal, but did not increase after administration of exogenous TSH. In contrast, infusion of dibutyryl-cAMP resulted in an increased uptake and stimulated the release of protein bound ^{131}iodine. The other endocrine axes were found to be normal, and pseudohypoparathyroidism was excluded. His thyroid gland was studied histologically after he died from chronic lymphocytic leukemia and showed findings similar to these reported previously by Stanbury et al. *(86,90)*. The follicles were small, lined by a cuboid or flat epithelium, and mostly contained no colloid. In many areas, there was no obvious formation of follicles. Attempts to demonstrate absence or presence of the TSH receptor by an immunohistochemical approach were not successful, and electron microscopy did not show thickened basal membranes of the follicles as described by Codaccioni et al. *(90)*.

Familial unresponsivness to TSH in a brother and a sister was reported by Takamatsu *(92)*. The parents were first cousins. Congenital hypothyroidism was diagnosed at age 3 in the girl and at age 5 in the boy. The girl also had a situs inversus totalis. The mother had subclinical hypothyroidism with a minimally elevated TSH of 4.8 mU/L (upper norm 4). There were no signs suggestive of pseudohypoparathyroidism Ia. The thyroid glands were located normally and showed decreased uptake of ^{123}iodine. The elevated TSH increased further in response to TRH, but the T3 levels remained low. The thyroglobulin levels were undetectable. TSH was shown to have normal bioreactivity based on the amount of cAMP generated in FRTL-5 cell cultures. In a later study by Takeshita et al., DNA analysis was performed in the affected girl and a TSH resistant patient studied by Maesaka et al., but found to be negative for mutations in the coding region of the TSH receptor gene (*see below*) *(89,94)*.

TSH Resistance Owing to Mutations in the TSH Receptor Gene

THE *HYT/HYT* MOUSE

Loss-of-function mutations in the TSH receptor gene as a cause of TSH resistance were first discovered in the hypothyroid *hyt/hyt* mouse *(46,47)*. The phenotype of this mouse with autosomal-recessive congenital hypothyroidism, retarded growth, mild anemia, hearing loss, and infertility was initially described by Beamer et al. in 1981 *(96)*.

The hyt/hyt mouse has a 5- to 10-fold reduction in serum T4, a 16-fold decrease in T3, and a 100-fold elevation of TSH *(97)*. Before unraveling the genetic defect, the *hyt/hyt* mouse was carefully characterized in morphologic, behavioral, and biochemical studies *(97–100)*. The hypoplastic thyroids of these mutant mice are located in the proper position. Histologically, the thyroid follicular cells are developed, but incompletely differentiated, and the epithelial cells are not organized into structures recognizable as follicles (Fig. 4) *(98)*. The thyroid tissue of homozygous mice is unresponsive to TSH in vivo and in vitro, but activators of $G_s\alpha$ induce a normal cAMP response, suggesting a defect in the TSH receptor *(100)*. In addition, the defect could be linked to mouse chromosome 12, which contains the locus of the mouse TSH receptor gene. TSH receptor mRNA expression levels were found to be normal *(100)*. Direct sequencing of the mouse TSH receptor cDNA ultimately led to the identification of a transition of a single base pair at nucleotide 1666 (C→T), resulting in the substitution of the highly conserved proline556 in the fourth transmembrane domain by leucine (Fig. 3, Table 2) *(46,47)*. The mutation eliminates TSH binding, and thus normal receptor function, although the membrane localization of the receptor appears to be preserved. It is noteworthy that an identical mutation of proline to leucine at the corresponding location in rhodopsin leads to an autosomal-dominant form of retinitis pigmentosa *(101)*.

Consistent with the clinical observation in humans that the developing auditory system is particularly vulnerable to a hypothyroid metabolic state *(102)*, the hearing thresholds of the *hyt/hyt* mouse for auditory-evoked brainstem potentials are increased by 40–45 dB *(103)*. The macroscopic structure of the cochlea was found to be normal, but scanning electron microscopy demonstrated morphologic abnormalities of the stereociliae of both the inner and outer hair cell system *(103)*.

EUTHYROID HYPERTHYROTROPINEMIA CAUSED BY TSH RECEPTOR MUTATIONS

The first human case with TSH resistance owing to a documented defect in the TSH receptor gene was reported by Sunthornthepvarakul et al. in 1995 (Table 4) *(28)*. The three sisters of this kindred were euthyroid and had normal thyroid hormones but high TSH levels, a constellation referred to as euthyroid hyperthyrotropinemia. The parents were unrelated and from distinct ethnic extraction. The proposita was the second girl and was found to have a TSH of 103 mU/L on neonatal screening. The gland was normal on a radioiodine scan. The TSH elevation persisted, thus excluding transient neonatal hyperthyrotropinemia *(104)*. Subsequently, a similar constellation with a TSH of 80 mU/L and a T4 of 126 nmol/L (normal 77–167) was found in her 4-yr-old sister, and later in the third girl born to this family. Both parents showed discretely elevated TSH levels of 3.9 and 4.3 mU/L (0.4–3.6). None of the family members had clinical or biochemical signs of hypothyroidism and the children developed normally. Additional tests performed in the oldest girl excluded the presence of autoantibodies directed against the TSH receptor, pseudohypoparathyroidism, abnormalities in gonadotropin, or glycoprotein α-subunit concentrations. The minimal elevation of TSH found in the parents, together with the fact that all three siblings were affected, suggested an inherited, autosomal-recessive genetic defect.

Molecular genetic analyses ruled out that the disorder was caused by TSH with reduced biologic activity, since the three siblings had a normal TSHβ-subunit gene, but they were all found to be compound heterozygotes for mutations in the TSH receptor gene *(28,105)*. The patients inherited missense mutations from the father (aspara-

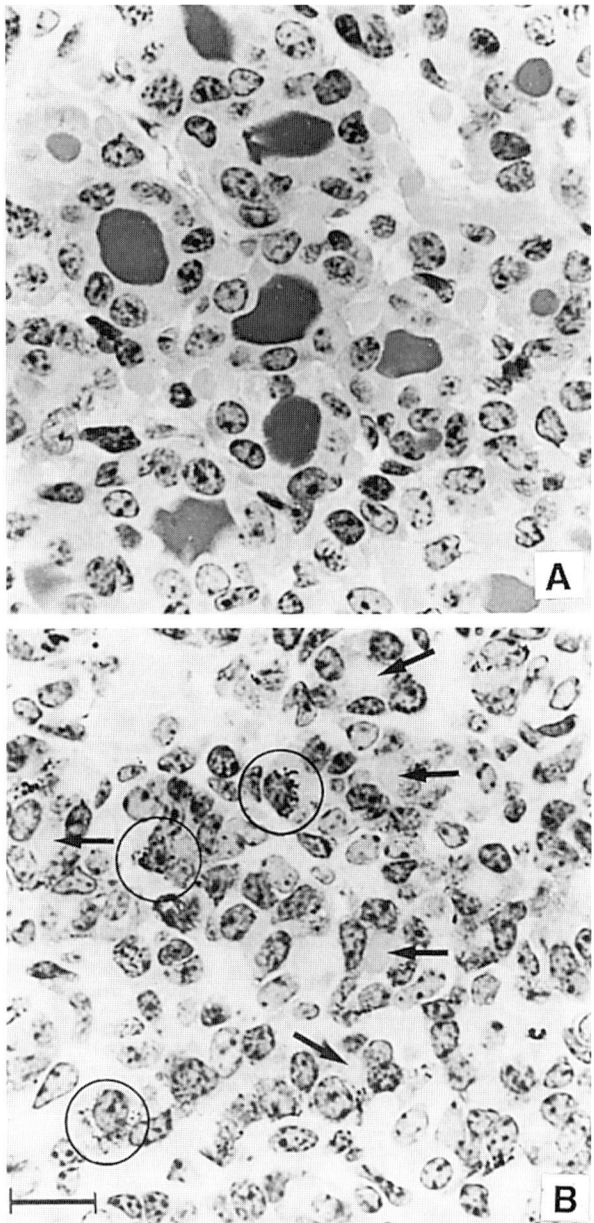

Fig. 4. Histological section of fetal thyroid tissue obtained at day 18 of gestation. (**A**) wild-type mouse, (**B**) *hyt/hyt* mouse. The *hyt/hyt* mouse shows poorly developed follicles with small lumina fille with lightly stained, PAS-positive colloid (arrows). Some follicular cells show granular deposits of PAS stain (circles) that are not present in sections from wild-type mice. Bar = 15 μm × 660. (From ref. *98*. Copyright Wiley & Sons, with permission from publisher and authors).

gine167isoleucine) and the mother (proline162alanine), both affecting the extracellular TSH binding domain of the receptor (Fig. 3, Table 2). Both amino acids are part of a putative α-helix formed by the fifth leucine-rich repeat motif and are thought to influence interaction of the receptor with the ligand *(61,62,105)*. After introducing the mutations in the TSH receptor cDNA of mammalian expression vectors, their functional conse-

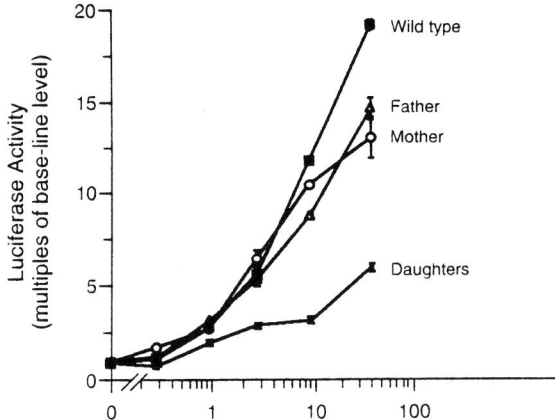

Fig. 5. TSH receptors were cotransfected with a cAMP-dependent luciferase reporter gene into COS-7 cells. The wild-type was transfected alone, or in combination with the paternal mutant (isoleucine167asparagine) and the maternal mutant (proline162alanine) to simulate the heterozygous state of the parents. The compound heterozyous situation in the daughters was mimicked by coexpression of equal amounts of the mutant TSH receptors. TSH stimulation is almost normal in the conditions simulating the heterozygous state of the parents. In the compound heterozygous situation, about 20 times more TSH is needed to achieve a similar cAMP-dependent luciferase activity. (From ref. *28.* Copyright Massachusetts Medical Society 1995, with permission from publisher and authors).

quences were tested in transfected cells. The mutant TSH receptor inherited from the father had almost no biologic activity, and that inherited from the mother displayed reduced activity. In cells transfected with both the paternal and maternal mutants, simulating the compound heterozygous situation found in the three sisters, almost 20 times higher levels of TSH were required for a similar degree of stimulation as for the wild-type receptor (Fig. 5). These findings are consistent with the clinical observation of 20-fold elevated TSH levels. It remains unclear why these patients have a chronically elevated TSH despite normal thyroid hormone levels. The current hypotheses include subclinical hypothyroidism, altered frequency of TSH secretion, or a resetting of the threshold for T3-dependent negative regulation of TSH gene expression *(28).*

Euthyroid hyperthyrotropinemia owing to mutations in the TSH receptor gene was also reported by de Roux et al. in four unrelated French families (Table 4) *(29).* All index patients were discovered through neonatal screening and had elevated plasma TSH concentrations, whereas T3 and T4 concentrations were normal. These children had unimpaired physical and intellectual development. The size of the thyroid gland was either normal or slightly increased. In three families, the affected individuals were compound heterozygotes for mutations in the receptor, and in one consanguineous relation the offspring were homozygous (Tables 2 and 4, Fig. 3). In the inbred kindred, two of the children have additional genetically determined malformations (Down's syndrome, Pierre Robin syndrome).

Among the loss-of-function mutations detected by de Roux et al. *(29),* some resulted in premature stop codons (glutamine234stop, tryptophan546stop), thus resulting in the synthesis of a nonfunctional truncated receptor. The proline162alanine substitution found in both alleles of the offspring of the related parents is identical to the mutation in the maternal allele in the patients reported by Sunthornthepvarakul et al. *(28,105),* and it was also found in a patient studied by Asteria et al. *(106).* In the latter case, no defect was found in the coding region of the other allele prompting the speculation that it may contain a mutation in the promoter region *(106)*. The cysteine41serine mutation was previously created by site-directed mutagenesis and found to abrogate TSH binding *(107)*. In order to define the mechanism by which the newly discovered mutants result in inactivation, they were transiently expressed in COS-7 cells. Cell-surface expression of the mutant receptor was determined by immunofluorescence along with their ability to bind TSH and stimulate cAMP accumulation. All studied mutants (cysteine390tryptophan; aspartic acid410 asparagine; phenylalanine525leucine) were found to be inserted into the cell membrane. The substitution of cysteine390 by tryptophan abolished high affinity hormone binding, further supporting the idea that this residue, which is conserved within the family of glycoprotein hormone receptors, is important for receptor structure by formation of a disulfide bond with cysteine 301 *(42,108)*. Replacement of aspartic acid 410 by asparagine abolished the TSH-induced increase of cAMP despite normal binding of TSH by the mutant. This observation suggests that certain residues in the extracellular domain may also influence signal transduction. Mutation of phenylalanine525 to leucine markedly impaired adenylyl cyclase activation, supporting a role for the second intracellular loop in receptor coupling *(109,110)*.

A further patient with euthyroid hyperthyrotropinemia identified on neonatal screening was reported by Clifton-Bligh et al. *(30)*. At age 8 wk, his thyroid function tests showed a serum-free thyroxine of 10 pmol/L (12–28) with a TSH of 92 mU/L. The serum levels of the glycoprotein hormone α-subunit were elevated (2.5 ng/mL, normal 0.24-1.05), but the molar ratio of the α-subunit to TSH was within the normal range. Clinically, there were no signs of hypothyroidism in this child. The TSH could be suppressed into the normal range with 60 μg levothyroxine. Under treatment with levothyroxine, the patient became irritable, and it was therefore stopped at 2 yr of age. The patient continued to develop normally without treatment. A thyroid scan showed a gland of normal size and location. His thyroid tests at 3 yr of age showed an increased TSH of 134 mU/L, and a free T4 of 12 pmol/L. His TSH was shown to be biologically active on cells expressing recombinant TSH receptors. Direct sequencing confirmed that the TSHβ-subunit gene was normal. Analysis of the TSH receptor gene revealed that the patient is a compound heterozygote for mutations in the TSH receptor gene. The maternal allele harbored a substitution of arginine109 by glutamine, whereas the paternal allele was found to contain a premature termination codon at tryptophan 546 located in the fourth transmembrane segment (Fig. 3, Table 2). Functionally, the mutants were tested in a cotransfection system with a cAMP-responsive luciferase construct. The arginine109glutamine mutation demonstrated a markedly impaired signal transduction in response to TSH in comparison to the wild type. The tryptophan546stop mutant did not stimulate reporter gene activity above basal levels. The heterozygous state found in the parents was mimicked in vitro by cotransfecting wild-type and mutant cDNAs. The resulting dose–response curves did not differ from that seen with wild-type alone, indicating that these mutants did not exert a dominant-negative effect and that one intact TSH receptor allele is sufficient to lead to normal function. Binding of ^{125}iodine-labeled TSH to membranes of COS cells transiently transfected with the cDNA encoding the arginine109glutamine mutant displayed a 40% reduction of the B_{max} and a raised EC_{50} (20 vs 7 mU). Binding on whole cells showed very poor binding to the arginine109glutamine mutant, suggesting impaired membrane insertion of the mutant. Based on the structural model for the extracellular domain of the TSH receptor *(61)*, arginine109glutamine is likely to project into solution and to be involved in TSH binding. As expected, the tryptophan546stop mutant did not show any significant specific binding of TSH.

Since chronic elevation of TSH secretion in untreated primary congenital hypothyroidism may lead to pituitary autonomy *(111)*, the authors addressed the question of thyrotrope autonomy in this patient. An MRI scan performed at the age of 2 yr showed two small areas of hypoattenuation in the pituitary after gadolinium administration. The significance of this finding is currently uncertain. In a functional test, administration of exogenous T3 led to a prompt suppression of TSH into the normal range.

CONGENITAL HYPOTHYROIDISM DUE TO TSH RECEPTOR MUTATIONS

Biebermann et al. screened patients with congenital hypothyroidism for mutations in the TSH receptor gene by single-strand conformation polymorphism (SSCP) analysis. In their initial report on 48 children, they found one girl with an abnormal migration pattern in the SSCP screening *(112)*. She was an offspring of non-consanguineous parents. She had congenital hypothyroidism with TSH levels of 82 mU/L, a free T4 of 8 pmol/L (18–33), and a T3 of 1.4 nmol/L (1.8–3.3) *(80,113)*. The hypothyroid state was thus not as severe as in patients suffering from athyreosis or complete hormone synthesis defects, but more similar to patients with ectopic thyroid tissue or partial dyshormogenesis. In contrast to the patients with euthyroid hyperthyrotropinemia, her thyroid gland had a reduced volume on ultrasonography. Her mother was found at one instance to have a moderately elevated TSH (5.5; normal <4) and a borderline T4 of 10.5 pmol/L (11–25), findings that could, however, not be confirmed on subsequent testing.

Sequence analysis demonstrated that both parents were heterozygous carriers of a defective allele, and that the affected child inherited the two defective copies. The mother had an 18-bp deletion between nucleotides 1217 and 1234 together with an insertion of 4 bp. This resulted in a frameshift and a premature stop codon after amino acid 419 (420×) in the extracellular domain of the receptor. The paternal allele displayed the same transversion of T to G at nucleotide 1170 found by Clifton-Bligh et al. resulting in a substitution of cysteine390 by tryptophan (Tables 2 and 4, Fig. 3) *(30)*. As can be expected, the severely truncated mutant maternal allele was non-functional. Consistent with the findings of de Roux et al. *(29)*, the tryptophan390cysteine mutant showed decreased binding of ^{125}iodine-labeled bovine TSH, and higher doses of TSH were needed to induce an equivalent cAMP response in comparison to the wild-type receptor. Strikingly, and in contrast to the previously reported patient with the cysteine390tryptophan mutation and euthyroid hyperthyrotropinemia, the patient studied by Biebermann et al. was overtly hypothyroid *(30,113)*. Since the second allele has been shown to be nonfunctional in both these patients *(29,80)*, additional factors modulating thyroid function have to be implied.

Using a monoclonal antibody (MAb) directed against the extracellular domain of the human TSH receptor, the cysteine390tryptophan mutant was found to be inserted into the membrane. The 420X mutant was transcribed and translated, but the cells contained large amounts of immunoreactive cytosolic aggregates, but no expression in the plasma membrane (Fig. 6), suggesting retention of the truncated protein in the endoplasmic reticulum.

These authors speculated that mutations in the TSH receptor gene may be the molecular cause of congenital hypothyroidism in a substantial number of patients *(113)*. As discussed in their subsequent study *(113a)*, this may however be a relatively rare cause of congenital hypothyroidism, because ectopic thyroid glands and athyreosis are much more frequent than hypoplastic glands, and the autosomal-recessive mode of inheritance

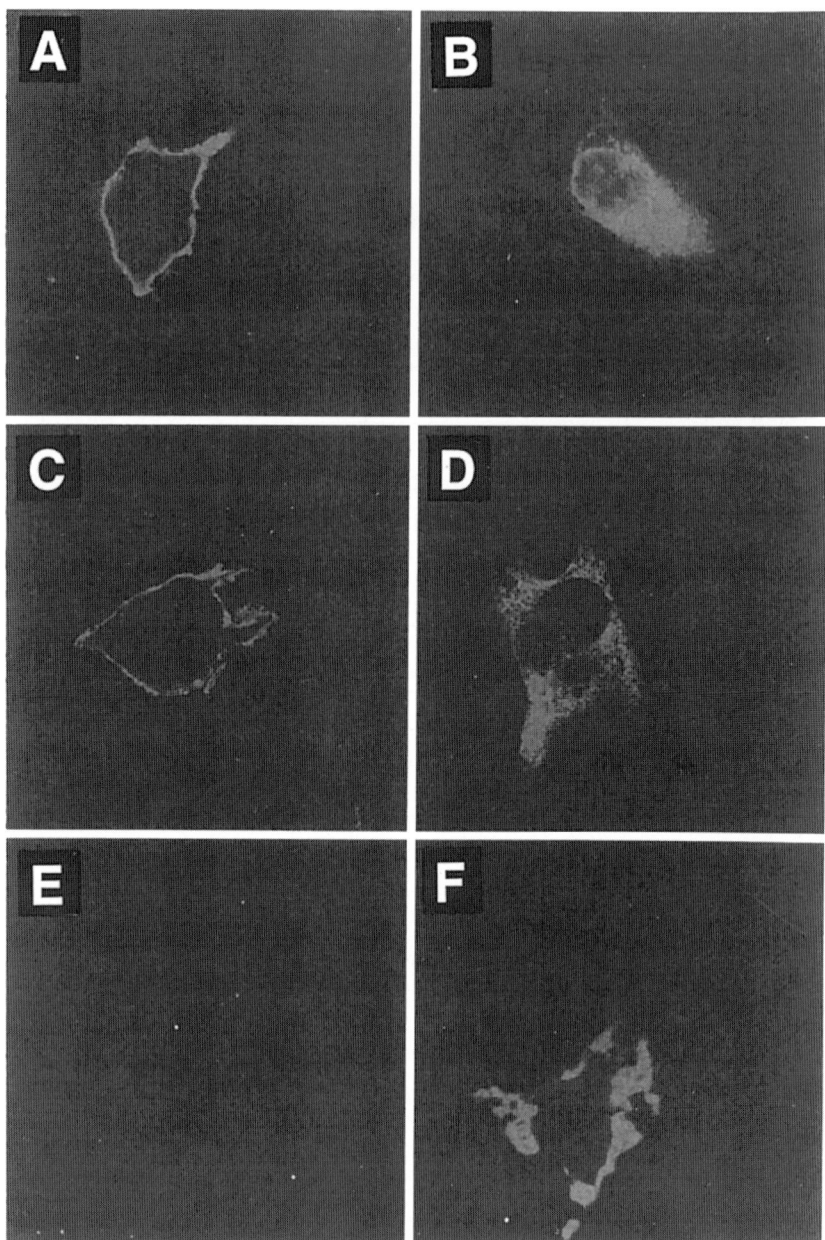

Fig. 6. Immunofluorescence with an antibody directed against the extracellular domain of the TSH receptor was performed in COS-7 cells transfected with the wild-type TSH receptor (**A,B**), the cysteine390tryptophan mutant (**C,D**), and the truncated TSH receptor 420X (**E,F**). (**A,C,E**) nonpermabilized cells. (**B,D,F**) permeabilized cells. Both the wild-type and the cysteine390tryptophan mutant are detectable in the cytosol and on the cell surface. In cells transfected with the 420X mutant, immunopositivity is present within the cells, but not on the cell membrane. (Reproduced from ref. *113b*. Copyright Endocrine Society 1997, with permission from publisher and authors).

is inconsistent with the sporadic occurrence of the majority of cases with congenital hypothyroidism *(80)*. This conclusion is supported by a study of familial or consanguineous cases of congenital hypothyroidism, which could not demonstrate linkage to the TSH receptor locus *(114)*. The frequency of individuals carrying two affected alleles with partial or complete inactivation of the TSH receptor remain to be defined. It has been estimated to occur in about 1:24,000 neonates *(105)*.

Abramowicz et al. identified a boy with congenital hypothyroidism on neonatal screening whose older sister was previously diagnosed with the same disorder (Fig. 7A, Table 4) *(93)*. The parents of the siblings were related. ^{99}Technetium scintigraphy showed no detectable thyroid tissue, prompting the diagnosis of "thyroid agenesis." In contrast to patients with real athyreosis, both patients had thyroglobulin levels in the high-normal range (Fig. 7A), and ultrasonography of the neck revealed the presence of correctly located hypoplastic thyroid glands. Direct DNA sequencing of the TSH receptor revealed the presence of a point mutation in exon 10 of the receptor resulting in the substitution of a highly conserved residue in the fourth transmembrane domain, alanine 553, by threonine. This segment also contains the inactivating mutation of the *hyt/hyt* mouse, and the tryptophan546stop mutant observed in two patients with euthyroid hyperthyrotropinemia *(29,30)*. Consistent with the pedigree analysis, which suggested an autosomal-recessive mode of inheritance, the mutation was homozygous in the affected siblings and heterozygous in both parents and the two unaffected brothers. Functional analysis of the mutated receptor in transfected cells demonstrated an extremely low level of expression of the mutated receptor at the cell surface, despite normal intracellular synthesis (Fig. 7B). The mutated receptor was able to bind TSH and stimulate cAMP accumulation in transfected cells, but appeared to have lost the ability to activate the phospholipase C pathway.

Although the reasons for the relatively high thyroglobulin levels are currently unclear, Abramowicz et al. made the interesting suggestion that measuring the thyroglobulin levels in newborns with congenital hypothyroidism may be helpful in delineating the etiology of these disorders *(93)*. A similar case was indeed reported by Gagné et al., who observed a neonate with congenital hypothyroidism, thyroid hypoplasia, and normal plasma thyroglobulin levels *(115)*. Molecular analysis of the TSH receptor gene revealed that the patient was a compound heterozygote for a transversion of G to C in the splice donor site of intron 6, the other a deletion of two nucleotides in exon 10 leading to a premature stop in the third extracellular loop. The mother harbored the splice site mutation in one allele, and although the father was not available for the studies, an autosomal-recessive inheritance is likely. A maternal great aunt of the patient also had congenital hypothyroidism, but no mutations were found in the TSH receptor gene in her case *(115)*.

Resistance to TSH Without Mutations in the TSH Receptor

Takeshita et al. analyzed the nucleotide sequence of the entire coding region of the TSH receptor gene in three patients with primary congenital hypothyroidism owing to TSH unresponsiveness (Table 5) *(94)*. Two of these patients have been reported in previous clinical studies and were the offspring of consanguineous parents (patients 1 and 2) *(89,92)*, and two of the patients had siblings with congenital hypothyroidism (patients 2 and 3), suggesting the possibility of an autosomal-recessive mode of inheritance in all three families. All three subjects had markedly elevated TSH levels (170, 125, and 522 mU/L) and low peripheral hormone levels. Autoantibodies were absent. Normal bioactivity of TSH was documented in two patients using a bioassay with FRTL-5

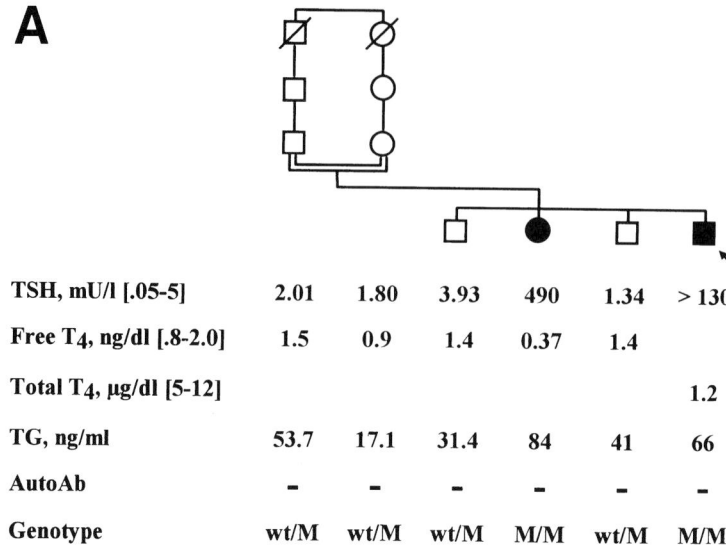

Fig. 7. (A) Pedigree and thyroid function tests of a family with familial congenital hypothyroidism (solid symbols) caused by a mutation in the TSH receptor (alanine553threonine). The parents are first cousins, and the propositus is indicated by an arrow. The thyroid gland was not detectable on physical examination and ^{99}technetium scanning. Blood values reflect the levels at the time of diagnosis and were determined in the absence of thyroid hormone. Note the relatively high thyroglobulin levels (Normal range: for newborns 2–106 ng/mL; <10 yr 2–65 ng/mL; adults 2–35 ng/mL). (Reproduced from ref. 93, with copyright permission from the American Society for Clinical Investigation and the authors).

thyroid cells (patients 2 and 3). In patient 1, resistance to TSH was diagnosed by absent increases in the radioiodine uptake, T3 and T4 after administration of exogenous TSH (89). The response to TRH was increased in all three individuals, but it was not followed by a rise of thyroid hormones. The glands were normally located as assessed by scintigraphy and/or ultrasound, but two of them were hypoplastic. The TSH receptor was sequenced directly by an RT-PCR approach using RNA extracted from leukocytes and did not show abnormalities. One patient was found to be heterozygous for a known polymorphism (tyrosine601histidine).

Linkage studies using genetic markers flanking the TSH receptor locus by a distance of 2.3 (D14S287) and 1.2 (D14S616) centiMorgans were performed in 23 kindreds with congenital hypothyroidism by Ahlbom et al. (114). In 13 of these families, the parents were related, and in 16, several offspring had congenital hypothyroidism. Ten of these families were from Sweden, 11 from Pakistan, and 1 from Syria and Egypt. These analyses did not support linkage to the TSH receptor locus, and these authors concluded that inactivating mutations in the TSH receptor gene account probably only for a minority of cases with congenital hypothyroidism.

In an analysis of four unrelated children with congenital hypothyroidism and normally located hypoplastic gland, we were unable to find defects in the coding region of the TSH receptor (Nogueira CR, Nguyen LQ, Coelho-Neto JR, Arseven OK, Jameson JL, Kopp P, Medeiros-Neto GA, unpublished results), and a similar observation was mentioned by de Roux et al. in their report on four children with euthyroid hyperthyrotropinemia (29).

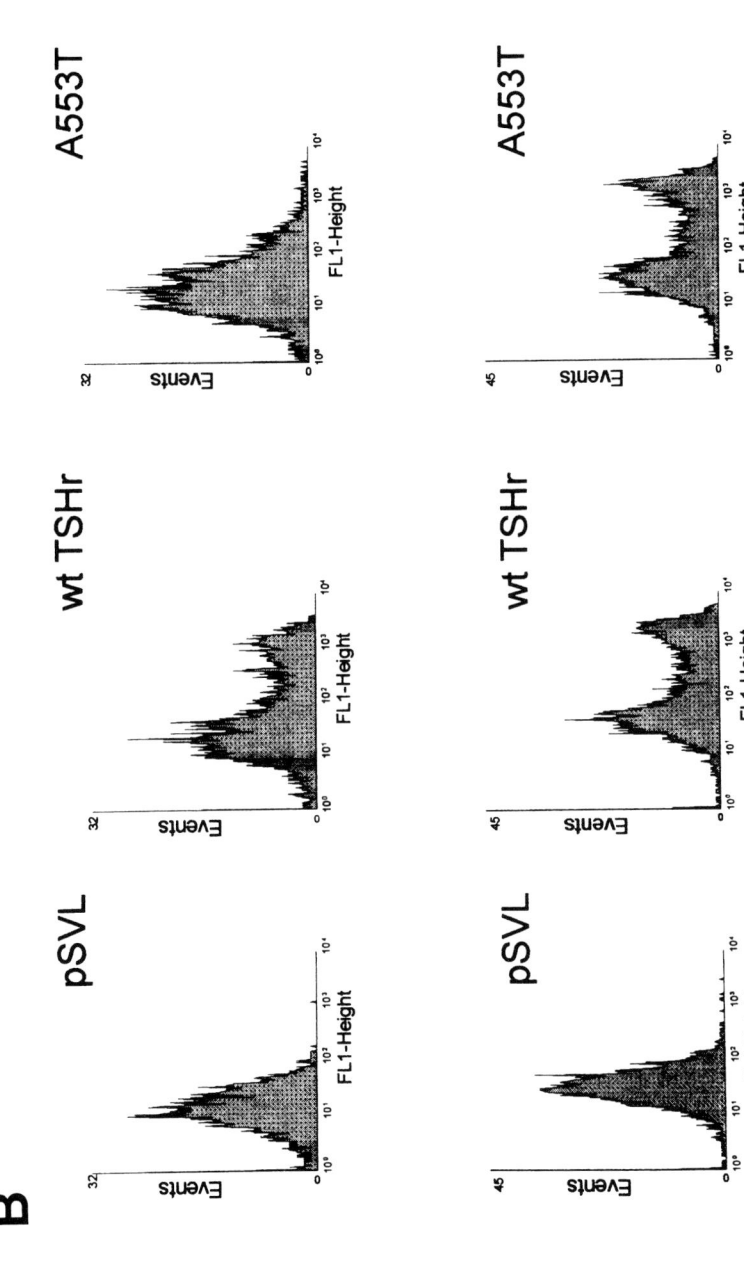

Fig. 7. (*Continued*). (**B**) Expression analysis of the alanine553threonine by flow immunocytometry. Fluorescence intensity was expressed in arbitrary units, as a function of cell numbers (10,000 propidium iodide-gated cells). (**A**) Nonpermeabilized cells assayed after transfection with pSVL vector (pSVL), wild-type TSH receptor (wt TSHR), or the alanine553threonine mutant (A553T). (**B**) Saponin-permeabilized cells with identical transfections as under (A). The wild-type receptor is expressed at the cell surface, but the alanine553threonine mutant is not inserted into the cell membrane (A), although it is detectable in the permeabilized cells (B). (Reproduced from ref. 93 by copyright permission from the American Society for Clinical Investigation and the authors).

A clinical study of a family with unresponsiveness to TSH in three generations studied by Mimouni et al. suggested that the disorder was transmitted in an autosomal-dominant fashion *(116)*. A similar observation was made in two of three families with euthyroid hyperthyrotropinemia studied by Xie et al. *(95)*. These authors presented a detailed study on three unrelated families with resistance to TSH without abnormalities in the coding region or promoter (−534 to +189 bp) of the TSH receptor gene *(95)*. The thyroid glands were normal in size and location. Interestingly, one family was found to have slightly elevated thyroglobulin levels, a constellation reminiscent of the findings in the patients with congenital hypothyroidism reported by Abramowicz et al. *(93)* (Table 4). The TSH receptor gene was excluded as a candidate gene for the defect in two families by haplotype analysis using the currently known intragenic markers *(117,118)*. In a bioassay using Chinese hamster ovary cells expressing the TSH receptor, the TSH was shown to be bioactive, and a defect in the TSHβ-subunit gene was further excluded by direct sequencing. The $G_s\alpha$ gene was analyzed in one family by denaturing gradient gel electrophoresis (DGGE), but did not reveal abnormal migration patterns.

The study by Xie et al. underscores that TSH resistance is genetically heterogeneous, since other molecular defects than TSH receptor mutations or defects in $G_s\alpha$ result in an identical phenotype *(95,119)*. Obviously, the list of candidate genes is long and includes G-proteins, G-protein receptor kinases, adenylyl cyclases, protein kinases, or transcription factors *(119)*. Unless the responsible gene is thyroid-specific, such a defect is expected to result in manifestations in other tissues. This is the case in PHP Ia, also referred to as Albright's hereditary osteodystrophy (AHO), where the affected subjects show resistance to a variety of hormones that couple to G-protein-coupled receptors, including TSH and PTH *(31)* (*see* Chapter 3). Other postreceptor defects resulting in resistance to TSH remain to be defined. In this context, it is noteworthy that no mutations were found in the thyroid transcription factor 1 (TTF1) gene in patients with thyroid dysgenesis *(120,121)*. Severe defects in TTF1 are expected to result in a more dramatic phenotype as illustrated by the murine knockout model *(122)*. This transcription factor is involved in the development of the forebrain, the thyroid, and the lung, and its targeted disruption was lethal *(122)*.

ACTIVATING MUTATIONS IN THE TSH RECEPTOR

The detection of inactivating mutations in the TSH receptor has been preceded by the discovery of gain-of-function mutations in several forms of hyperthyroidism. They are discussed briefly here in order to illustrate the broad spectrum of phenotypes caused by molecular alterations in the same gene.

Somatic TSH Receptor Mutations in Toxic Adenomas

Since chronic stimulation of the cAMP cascade results in enhanced proliferation and function of thyroid follicular cells, it was predicted that any molecular alteration leading to constitutive activation of the cAMP pathway would result in clonal autonomous growth and function of thyroid follicular cells and ultimately in a toxic adenoma *(70)*. Such somatic mutations were first discovered in the stimulatory $G_s\alpha$ subunit (*gsp* mutations) in toxic adenomas, as well as in nonfunctioning adenomas, and differentiated thyroid carcinomas *(123–125)*. These mutations, most commonly found in amino acids arginine 201 and glutamine 227, impair the hydrolysis of GTP to GDP, resulting in persistent activation of adenylyl cyclase. The same $G_s\alpha$ mutations are found in 35–40% of somatotroph tumors in acromegalic patients *(126)*. Mosaicism for $G_s\alpha$ mutations that occur early in development cause the McCune Albright syndrome and may affect multiple tissues, including the thyroid *(127)*.

Somatic mutations in the TSH receptor were first discovered by Parma et al. in toxic adenomas *(85)*. The first identified mutations were clustered in the third intracellular loop and the sixth transmembrane domain of the receptor, but a wide variety of activating somatic mutations have been found in subsequent studies (Fig. 3, Table 2) *(35,85, 128–137)*. In contrast to activating mutations in many other GPCRs, there is a striking diversity in the affected residues, which are scattered over almost the entire transmembrane domain as well as the carboxy-terminal region of the extracellular domain. All these mutations increase basal cAMP levels, but only few amino acid substitutions activate the phospholipase C cascade. The prevalence of these mutations in toxic adenomas varies between 2.5 and 80% *(35,131)*, a finding that might, in part, be influenced by differences in iodine intake, sampling technique, and methodological approaches *(138,139)*.

Since gain-of-function mutations are by definition dominant, one mutated allele is sufficient to result in disease. Gain-of-function mutations in the transmembrane domain or the intracellular loops are thought to modify the relative positions of the helices, thus mimicking the conformational changes induced by binding of ligand. Alternatively, they could alter the structure of domains that inhibit receptor coupling to G-proteins in the absence of TSH *(129,140)*. Mutations in the extracellular domain have been proposed to result in the relief of a negative constraint present in the unliganded carboxy-terminal part of the extracellular domain *(35,134,135,141,142)*.

Germline Mutations in Familial Nonautoimmune Hyperthyroidism

Autosomal-dominant familial hyperthyroidism without evidence of an autoimmune etiology was first recognized by Thomas et al. *(143)*. The typical signs associated with autoimmune hyperthyroidism, exophthalmos, myxedema, stimulating autoantibodies, and lymphocytic infiltration of the thyroid gland are absent in this form of hyperthyroidism. Because all thyroid follicular cells display an increased growth rate, these patients have a diffuse goiter. Duprez et al. *(140)* elucidated the molecular basis of inherited nonautoimmune hyperthyroidism by detecting activating germline mutations in the TSH receptor in the family reported by Thomas et al., an observation that has subsequently been reported in other families *(144–146)*. The onset of hyperthyroidism may vary in carriers of the same mutation in a given kindred. This suggests that other factors, for example, genetic background and/or iodine intake, can modulate the phenotypic expression *(143,147,148)*.

Sporadic Germline Mutations in Congenital Hyperthyroidism

Congenital hyperthyroidism is usually caused by transplacental passage of stimulating TSH receptor autoantibodies, and most cases of neonatal hyperthyroidism resolve within the first 3–7 mo as the maternal antibodies are cleared from the circulation *(149–151)*. Autoimmune neonatal hyperthyroidism is rare and occurs in <2% of infants born to a mother with a history of Graves' disease *(152)*, a condition with an estimated incidence of about 2 of every 1000 pregnancies *(143)*.

Constitutively activating *de novo* mutations of the TSH receptor have been found in a few patients with sporadic congenital nonautoimmune hyperthyroidism *(145, 153–157)*. Congenital hyperthyroidism owing to a toxic adenoma harboring a somatic TSH receptor mutation was recently reported as another unusual variant of congenital hyperthyroidism *(135)*. These rare cases with nonautoimmune congenital hyperthyroidism owing to TSH receptor mutations must be differentiated from the much more frequent and transient autoimmune form of hyperthyroidism, because most of these patients have pronounced hyperthyroidism requiring a more aggressive therapeutic approach (surgery, ablative radiotherapy). It is noteworthy that several of these children with severe neonatal hyperthyroidism seem to have mild mental retardation *(153,156,157)*, suggesting that high levels of thyroid hormone may impair brain development *(158)*.

TSH Receptor Mutations in Thyroid Carcinomas

In well-differentiated thyroid cancers, mutations in the $G_s\alpha$ and the TSH receptor genes seem to be rare and have only been reported occasionally *(125,159–162)*. Although constitutive activation of the cAMP pathway does result in enhanced growth, it is insufficient to result in malignant transformation of thyrocytes. This suggest that other or additional genetic and epigenetic alterations are required for the development of thyroid malignancies *(163–165)*.

TSH RESISTANCE: SUMMARY AND SIGNIFICANCE

Reduced or absent thyroid responsiveness to biologically active TSH caused by loss-of-function mutations in the TSH receptor results in a spectrum of clinical phenotypes. These range from euthyroid hyperthyrotropinemia, a newly defined entity, to overt congenital hypothyroidsm with thyroid hypoplasia. The size of the thyroid gland has been reported to be normal, slightly increased or decreased in the patients with euthyroid hyperthyrotropinemia *(28–30)*, and found to be hypoplastic in the hypothyroid patients *(80,93)*.

The *hyt/hyt* mouse and patients with TSH-resistant congenital hypothyroidism with correctly located hypoplastic glands confirm that development and migration of the thyroid is independent of TSH stimulation *(80,93,98)*. This is consistent with the observation that the genes for thyroperoxidase, thyroglobulin, and the TSH receptor are only expressed once the gland has reached its pretracheal location *(122)*. In mice, the TSH receptor is found to be expressed on day 14 of embryonic development, a time when the thyroid anlage has just arrived in the position of the anterior neck and follicle differentiation is starting *(55)*. Although TSH and its receptor are thus not crucial for early events of thyroid development, they seem to be of importance to achieve complete differentiation, growth, and function of thyroid follicular cells. The largely normal findings in heterozygous carriers of defective alleles indicated that one normal TSH-receptor copy is sufficient for normal function.

The finding of high thyroglobulin levels in hypothyroid patients with TSH resistance is currently not understood. It suggests that thyroglobulin measurements, combined with ultrasound of the neck, may be helpful in distinguishing congenital athyreosis from thyroid hypoplasia, an entity that may eventually escape detection by scintigraphy *(93)*. The few patients with congenital TSH-resistant hypothyroidism caused by receptor mutations are not as severely hypothyroid as patients with athyreosis or complete hormone synthesis defects, and resemble more the phenotype typically encountered in congenital hypothyroidism owing to partial dyshormogenesis or heterotopic thyroid tissue.

Among the patients with documented loss-of-function mutations in the TSH receptor, there is allelic heterogeneity (Tables 2 and 4, Fig. 3). Some mutations have, however, been found in several patients. As illustrated by differences in patients with the same loss-of-function mutation, additional genetic or environmental factors probably modulate phenotypic expression *(29,80)*. It is also apparent that the same phenotype may result from other postreceptor defects, indicating genetic heterogeneity *(95,114,116)*.

The ongoing molecular analysis of patients with TSH resistance will undoubtedly continue to contribute to our understanding of the pathophysiology of the thyroid and congenital hypothyroidism. Altogether, hormone resistance may prove to be a more common cause of human endocrinopathy then anticipated, and possibly explain physiologic variants.

ACKNOWLEDGMENTS

The author wishes to express his thanks to J. L. Jameson for his support, and L. Q. Nguyen and W. Lowe for critically reading the manuscript. This work was supported in part by a New Investigator Award from the Howard Hughes Medical Institute.

REFERENCES

1. Albright F, Burnett CH, Smith PH. Pseudohypoparathyroidism: an example of "Seabright-Bantam syndrome." Endocrinology 1942;30:922–932.
2. Refetoff S, Weiss RE, Usala SJ. The syndromes of resistance to thyroid hormone. Endocr Rev 1993; 14:348–399.
3. Kopp P, Kitajima K, Jameson JL. Syndrome of resistance to thyroid hormone: insights into thyroid hormone action. Proc Soc Exp Biol Med 1996;211:49–61.
4. Alm J, Hagenfeldt L, Larsson A, Lundberg K. Incidence of congenital hypothyroidism: retrospective study of neonatal laboratory screening versus clinical symptoms as indicators leading to diagnosis. Br Med J 1984;289:1171–1175.
5. Toublanc JE. Comparison of epidemiological data on congenital hypothyroidism in Europe with those of other parts in the world. Horm Res 1992;138:230–235.
6. Delane F, Czernichow P. Thyroid hormones, biochemistry and physiology. In: Bertrand J, Rappaport R, Sizonenko P. eds. Pediatric Endocrinology. William and Wilkins, Baltimore, 1993, pp. 242–251.
7. New England Congenital Hypothyroidism Collaborative. Effects of neonatal screening for hypothyroidism: prevention of mental retardation by treatment before clinical manifestation. Lancet 1981; 2:1095–1098.
8. LaFranchi S. Congenital hypothyroidism: a newborn screening success story? Endocrinologist 1994;4: 477–486.
9. Klein RZ Mitchell ML. Neonatal screening for hypothyroidism. In: Braverman LE, Utiger RD, eds. The Thyroid. Lippincott-Raven, Philadelphia, 1996, pp. 984–988.
10. Fisher DA, Dussault JH, Foley TPJ, Klein A, LaFranchi S, Larsen P, et al. Screening for congenital hypothyroidism: results of screening of one million North American infants. J Pediatr 1979;94: 700–705.
11. Kaplan EL, Shukla M, Hara H, Ito K. Developmental abnormalities of the thyroid. In: De Groot, LJ, ed. Endocrinology. W.B. Saunders, Philadelphia, 1994, pp. 893–899.
12. Yamada M, Saga Y, Shibusawa N, Hirato J, Murakami M, Iwasaki T, et al. Tertiary hypothyroidism and hyperglycemia in mice with targeted disruption of the thyrotropin-releasing hormone gene. Proc Natl Acad Sci USA 1997;94:10,862–10,867.
13. Illig R, Krawczynska H, Torresani T, Prader A. Elevated plasma TSH and hypothyroidism in children with hypothalamic hypopituitarism. J Clin Endocrinol Metab 1975;41:722–728.
14. Faglia G, Bitensky L, Pinchera A, Ferrari C, Paracchi A, Beck-Peccoz P, et al. Thyrotropin secretion in patients with central hypothyroidism: evidence for reduced biological activity of immunoreactive thyrotropin. J Clin Endocrinol Metab 1979;48:989–998.
15. Beck-Peccoz P, Amr S, Menezes-Ferreira MM, Faglia G, Weintraub BD. Decreased receptor binding of biologically inactive thyrotropin in central hypothyroidism. Effect of treatment with thyrotropin-releasing hormone. N Engl J Med 1985;312:1085–1090.
16. Collu R, Tang J, Castagné J, Lagacé G, Masson N, Huot C, et al. A novel mechanism for isolated central hypothyroidism: inactivating mutations in the thyrotropin-releasing hormone receptor gene. J Clin Endocrinol Metab 1997;82:1361–1365.

17. Samuels MH, Ridgway EC. Central hypothyroidism. Endocrinol Metab Clin 1992;21:903–919.
18. Nogueira CR, Leite CC, Chedid EPT, Liberman B, Pimentel-Filho FR, Kopp P, et al. Autosomal recessive deficiency of combined pituitary hormones (except ACTH) in a consanguineous Brazilian kindred. J Endocrinol Invest 1997;20:629–633.
19. Wu W, Cogan JD, Pfäffle RW, Dasen JS, Frisch H, O'Connell SM, et al. Mutations in PROP1 cause familial combined pituitary hormone deficiency. Nature Genet 1998;18:147–149.
20. Woods KA, Weber A, Clark AJ. The molecular pathology of pituitary hormone deficiency and resistance. Baillieres Clin Endocrinol Metab 1995;9:453–487.
21. Li S, Crenshaw EBD, Rawson EJ, Simmons DM, Swanson LW, Rosenfeld MG. Dwarf locus mutants lacking three pituitary cell types result from mutations in the POU-domain gene pit-1. Nature 1990;347:528–533.
22. Radovick S, Nations M, Du Y, Berg LA, Weintraub BD, Wondisford FE. A mutation in the POU-homeodomain of Pit-1 responsible for combined pituitary hormone deficiency. Science 1992;257:1115–1118.
23. Haugen B., Ridgway CE. Transcription factor Pit-1 and its clinical implications: from bench to bedside. Endocrinologist 1995;5:132–139.
24. Hayashizaki Y, Hiraoka Y, Tatsumi Y, Hashimoto T, Furuyama J, Miyai K, et al. Deoxyribonucleic acid analysis of five families with familial inherited thyroid stimulating hormone deficiency. J Clin Endocrinol Metab 1990;71:792–796.
25. Dacou-Voutetakis G, Feltquate DM, Drakopoulou M, Kourides IA, Dracopoli NC. Familial hypothyroidism caused by a nonsense mutation in the thyroid-stimulating hormone subunit gene. Am J Hum Genet 1990;46:988–993.
26. Mori R, Sawai T, Kwoshita E, Baba T, Matsumoto T, Yoshimoto M, et al. Rapid detection of a point mutation in thyroid-stimulating hormone β-subunit gene causing isolated thyroid-stimulating hormone deficiency. Jpn J Hum Genet 1991;36:313–316.
27. Medeiros-Neto G, Heodotou DT, Rajan S, Kommareddi S, de Lacerda L, Sandrini R, et al. A circulating, biologically inactive thyrotropin caused by a mutation in the beta subunit gene. J Clin Invest 1996;97:1250–1256.
28. Sunthornthepvarakul T, Gottschalk ME, Hayashi Y, Refetoff S. Resistance to thyrotropin caused by mutations in the thyrotropin-receptor gene. N Engl J Med 1995;332:155–160.
29. De Roux N, Misrahi M, Brauner R, Houang M, Carel JC, Granier M, et al. Four families with loss of function mutations of the thyrotropin receptor. J Clin Endocrinol Metab 1996;81:4229–4235.
30. Clifton-Bligh RJ, Gregory JW, Ludgate M, John R, Persani L, Asteria C, et al. Two novel mutations in the thyrotropin (TSH) receptor gene in a child with resistance to TSH. J Clin Endocrinol Metab 1997;82:1094–1100.
31. Spiegel, AM Weinstein, LS Shenker A. Abnormalities in G protein coupled-receptors signal transduction pathways in human disease. J Clin Invest 1993;92:1119–1125.
32. Spiegel AM. Mutations in G proteins and G protein-coupled receptors in endocrine disease. J Clin Endocrinol Metab 1996;81:2434–2442.
33. DeGroot LJ. Congenital defects in thyroid hormone formation and action. In; DeGroot, LJ, ed. Endocrinology. W.B. Saunders, Philadelphia, 1994, pp. 871–892.
34. Rees Smith B, McLachlan SM, Furmaniak J. Autoantibodies to the thyrotropin receptor. Endocr Rev 1988;9:106–121.
35. Parma J, van Sande J, Swillens S, Tonacchera M, Dumont J, Vassart G. Somatic mutations causing constitutive activity of the thyrotropin receptor are the major cause of hyperfunctioning thyroid adenomas: identification of additional mutations activating both the cyclic adenosine 3',5'-monophosphate and inositol phosphate-Ca2+ cascades. Mol Endocrinol 1995;9:725–733.
36. Tonacchera M, van Sande J, Parma J, Duprez L, Cetani F, Costagliola S, et al. TSH receptor and disease. Clin Endocrinol 1996;44:621–633.
37. Paschke R, Ludgate M. The thyrotropin receptor in thyroid diseases. N Engl J Med 1997;337:1675–1681.
38. Nagayama Y, Rapoport B. The thyrotropin receptor 25 years after its discovery: new insights after its molecular cloning. Mol Endocrinol 1992;92:145–156.
39. Vassart G, Dumont JE. The thyrotropin receptor and the regulation of thyrocyte function and growth. Endocr Rev 1992:13:596–611.
40. Kohn LD, Shimura H, Shimura Y, Hidaka A, Giuliani C, Napolitano G, et al. The thyrotropin receptor. Vit Horm 1995;50:287–384.
41. Parmentier M, Libert F, Maenhaut C, Lefort A, Gerard C, Perret J, et al. Molecular cloning of the thyrotropin receptor. Science 1989;246:1620–1622.

42. Nagayama Y, Kaufman KD, Seto P, Rapoport B. Molecular cloning, sequence and functional expression of the cDNA for the human thyrotropin receptor. Biochem Biophys Res Commun 1989;165: 1184–1190.
43. Libert F, Lefort A, Gerard C, Parmentier M, Perret J, Ludgate M, et al. Cloning, sequencing and expression of the human thyrotropin (TSH) receptor: evidence for binding of autoantibodies. Biochem Biophys Res Commun 1989;165:1250–1255.
44. Misrahi M, Loosfelt H, Atger M, Sar S, Guiochon-Mantel A, Milgrom E. Cloning, sequencing and expression of human TSH receptor. Biochem Biophys Res Commun 1990;166:394–403.
45. Akamizu T, Ikuyama S, Saji M, Kosugi S, Kozak C, McBride OW, et al. Cloning, chromosomal assignment, and regulation of the rat thyrotropin receptor: expression of the gene is regulated by thyrotropin, agents that increase cAMP levels, and thyroid autoantibodies. Proc Natl Acad Sci USA 1990;87:5677–5681.
46. Stein SA, Oates EL, Hall CR, Grumbles RM, Fernandez LM, Taylor NA, et al. Identification of a point mutation in the thyrotropin receptor of the hyt/hyt hypothyroid mouse. Mol Endocrinol 1994;8: 129–138.
47. Gu WX, Du GG, Kopp P, Rentoumis A, Albanese C, Kohn LD, et al. The thyrotropin (TSH) receptor transmembrane domain mutation (Pro556-Leu) in the hypothyroid hyt/hyt mouse results in plasma membrane targeting but defective TSH binding. Endocrinology 1995;136:3146–3153.
48. Silversides DW, Houde A, Ethier JF, Lussier JG. Bovine thyrotropin receptor cDNA is characterized by full-length and truncated transcripts. J Mol Endocrinol 1997;18:101–112.
49. Bockmann J, Winter C, Wittkowski W, Kreutz MR, Bockers TM. Cloning and expression of a brain-derived TSH receptor. Biochem Biophys Res Commun 1997;238:173–178.
50. Nguyen LQ, Jameson JL, Stein BS, Kopp P. Molecular cloning of the cat thyrotropin (TSH) receptor: evidence against an autoimmune etiology of feline hyperthyroidism. 80th Meeting of the Endocrine Society, New Orleans, 1988, abstract P3-110.
51. Rousseau-Merck MF, Misrahi M, Loosfelt H, Atger M, Milgrom E, Berger R. Assignment of the human thyroid stimulating hormone receptor (TSHR) gene to chromosome 14q31. Genomics 1990;8: 233–236.
52. Libert F, Passage E, Lefort A, Vassart G, Mattei MG. Localization of human thyrotropin receptor gene to chromosome region 14q31 by in situ hybridization. Cytogenet Cell Genet 1990;54:82,83.
53. Gross B, Misrahi M, Sar S, Milgrom E. Composite structure of the human thyrotropin receptor gene. Biochem Biophys Res Commun 1991;177:679–687.
54. Vassart G, Parma J, van Sande J, Dumont JE. The thyrotropin receptor and the regulation of thyrocyte function and growth: update 1994. In: Braverman LE, Refetoff S, eds. Endocrine Review Monographs, 1994, pp. 77–80.
55. Damante G, DiLauro R. Thyroid-specific gene expression. Biochim Biophys Acta 1994;1218: 255–266.
56. Endo T, Ohno M, Kotani S, Gunji K, Onaya T. TSH receptor in non-thyroid tissues. Biochem Biophys Res Commun 1993;190:774–779.
57. Heufelder AE, Bahn RS. Evidence for the presence of a functional TSH-receptor in retroocular fibroblasts from patients with Graves' ophthalmopathy. Exp Clin Endocrinol 1992;100,:62–67.
58. Heufelder AE. Involvement of the orbital fibroblast and TSH receptor in the pathogenesis of Graves' ophthalmopathy. Thyroid 1995;5:331–40.
59. Paschke R, Vassart G, Ludgate M. Current evidence for and against the TSH receptor being the common antigen in Graves' disease and thyroid associated ophthalmopathy. Clin Endocrinol (Oxford) 1995;42:565–569.
60. Russo D, Chazenbalk G, Nagayama Y, Wadsworth HL, Rapoport B. Site-directed mutagenesis of the human thyrotropin receptor: role of asparagine-linked oligosacharides in the expression of a functional receptor. Mol Endocrinol 1991;5:29–33.
61. Kajava AV, Vassart G, Wodak SJ. Modeling of the three-dimensional structure of proteins with the typical leucine-rich repeats. Structure 1995;3:867–877.
62. Van Sande J, Parma J, Tonacchera M, Swillens S, Dumont J, Vassart G. Somatic and germ-line mutations of the TSH receptor gene in thyroid disease. J Clin Endocrinol Metab 1995;9: 2577–2585.
63. Furmaniak J, Hashim FA, Buckland PR, Petersen VB, Beever K, Howells RD, et al. Photoaffinity labeling of the TSH receptor on FRTL5 cells. FEBS Lett 1987;215:16–322.
64. Loosfelt H, Pichon C, Jolivet A, Misrahi M, Caillou B, Jamous M, et al. Two-subunit structure of the human thyrotropin receptor. Proc Natl Acad Sci USA 1992;89:3765–3769.

65. Chazenbalk GD, Tanaka K, Nagayama Y, Kakinuma A, Jaume JC, McLachlan SM, et al. Evidence that the thyrotropin receptor ectodomain contains not one, but two, cleavage sites. J Clin Endocrinol Metab 1997;138:2893–2899.
66. Misrahi, M, Milgrom, E. Cleavage and shedding of the TSH receptor. Eur J Endocrinol 1997;137:599–602.
67. Nagayama Y, Tanaka K, Haras T, Namba H, Yamashita S, Taniyama K, et al. Involvement of G protein-coupled receptor kinase 5 in homologous desensitization of the thyrotropin receptor. J Biol Chem 1996;271:10,143–10,148.
68. Iacovelli L, Franchetti R, Masini M, De Blasi A. GRK2 and β-arrestin1 as negative regulators of thyrotropin receptor-stimulated response. Mol Endocrinol 1996;10:1138–1146.
69. Lalli E, Sassone-Corsi P. Thyroid-stimulating hormone (TSH)-directed induction of the CREM gene in the thyroid gland participates in the long-term desensitization of the TSH receptor. Proc Natl Acad Sci USA 1995;92:9633–9637.
70. Dumont JE, Lamy F, Roger P, Maenhaut C. Physiological and pathological regulation of thyroid cell proliferation and differentiation by thyrotropin and other factors. Physiol Rev 1992;72:667–697.
71. Laurent E, Mockel J, van Sande J, Graff I, Dumont JE. Dual activation by thyrotropin of the phospholipase C and cAMP cascades in human thyroid. Mol Cell Endocrinol 1987;52:273–278.
72. Van Sande J, Raspe E, Perret J, Lejeune C, Maenhaut C, Vassart G, et al. Thyrotropin activates both the cyclic AMP and the PIP2 cascades in CHO cells expressing the human cDNA of TSH receptor. Mol Cell Endocrinol 1990;74:R1–R6.
73. Raspé E, Laurent E, Andry G, Dumont JE. ATP, bradykinin, TRH and TSH activate the Ca2+-phosphatidylinositol cascade of human thyrocytes in primary culture. Mol Cell Endocrinol 1991;81:175–183.
74. Allgeier A, Offermans S, van Sande J, Spicher K, Schultz G, Dumont JE. The human thyrotropin receptor activates G-protein Gs and Gq11. J Biol Chem 1994;269:13,733–13,735.
75. Laugwitz KL, Allgeier A, Offermanns S, Spicher K, van Sande J, Dumont JE, et al. The human thyrotropin receptor: a heptahelical receptor capable of stimulating members of all four G protein families. Proc Natl Acad Sci USA 1996;93:116–120.
76. Ledent C, Dumont JE, Vassart G, Parmentier M. Thyroid expression of an A2 adenosine receptor transgene induces thyroid hyperplasia and hyperthyroidism. EMBO J 1992;11:537–542.
77. Michiels FM, Caillou B, Talbot M, Dessarps-Freichey F, Maunoury MT, Schlumberger M, et al. Oncogenic potential of guanine nucleotide stimulatory factor alpha subunit in thyroid glands of transgenic mice. Proc Natl Acad Sci USA 1994;91:10,488–10,492.
78. Zeiger MA, Saji M, Gusev Y, Westra WH, Takiyama Y, Dooley WC, et al. Thyroid-specific expression of cholera toxin A1 subunit causes thyroid hyperplasia and hyperthyroidism in transgenic mice. Endocrinology 1997;138:3133–3140.
79. Shenker AG. protein-coupled receptor structure and funtion: the impact of disease-causing mutations. Baillière's Clin Endocrinol Metab 1995;9:427–451.
80. Biebermann H, Schöneber T, Krude H, Schultz G, Gudermann T, Grüters A. Mutations of the human thyrotropin receptor gene causing thyroid hypoplasia and persistent congenital hypothyroidism. J Clin Endocrinol Metab 1997;82:3471–3480.
81. Bond RA, Leff P, Johnson TD, Milano CA, Rockman HA, McMinn TR, et al. Physiological effects of inverse agonists in transgenic mice with myocardial overexpression of the beta 2-adrenoreceptor. Nature 1995;374:272–276.
82. Kjelsberg MA, Cotecchia S, Ostrowski J, Caron MG, Lefkowitz RJ. Constitutive activation of the alpha 1B-adrenergic receptor by all amino acid substitutions at a single site. Evidence for a region which constrains receptor activation. J Biol Chem 1992;267:1430.
83. Robinson PR, Cohen GB, Zhukovsky EA, Oprian DD. Constitutively active mutants of rhodopsin. Neuron 1992;9:719–725.
84. Shenker A, Laue L, Kosugi S, Merendino JJJ, Minegishi T, Cutler GBJ. A constitutively activating mutation of the luteinizing hormone receptor in familial male precocious puberty. Nature 1993;365:652–654.
85. Parma J, Duprez L, van Sande J, Cochaux P, Gervy C, Mockel J, et al. Somatic mutations in the thyrotropin receptor gene cause hyperfunctioning thyroid adenomas. Nature 1993;356:649–651.
86. Stanbury JB, Rocmans P, Buhler UK, Ochi Y. Congenital hypothyroidism with impaired thyroid response to thyrotropin. N Engl J Med 1968;279:1132–1136.

87. Job JC, Canlorbe P, Thomassin N, Vassal J. L'hypothyroidie infantile à début précoce avec glande en place, fixation faible de radioiode et défaut de réponse à la thyréostimuline. Ann Endocrinol (Paris) 1969;30:696–701.
88. Medeiros-Neto GA, Knobel M, Bronstein MD, Simonetti J, Filho FF, Mattar E. Impaired cyclic-AMP response to thyrotrophin in congenital hypothyroidism with thyroglobulin deficiency. Acta Endocrinol 1979;92:62–72.
89. Maesaka H, Takahashi K, Yokoya S, Tokuhiro E, Suwa S. Two cases of congenital hypothyroidism with TSH unresponsiveness. Horumon to Rinsho 1979;27:948–953.
90. Codaccioni JL, Carayon P, Michel-Bechet M, Foucault F, Lefort G, Pierron H. Congenital hypothyroidism associated with thyrotropin unresponsiveness and thyroid cell membrane alterations. J Clin Endocrinol Metab 1980;50:932–937.
91. Aarseth HP, Haug E, Raknerud N, Frey HM. TSH unresponsiveness, a case report. Acta Endocrinol 1983;102:358–366.
92. Takamatsu J, Nishikawa M, Horimoto M, Ohsawa N. Familial unresponsiveness to thyrotropin by autosomal recessive inheritance. J Clin Endocrinol Metab 1993;77:1569–1573.
93. Abramowicz MJ, Duprez L, Parma J, Vassart G, Heinrichs C. Familial congenital hypotyroidism due to inactivating mutation of the thyrotropin receptor causing profound hypoplasia of the thyroid gland. J Clin Invest 1997;99:3018–3024.
94. Takeshita A, Nagayama Y, Yamashita S, Takamatsu J, Ohsawa N, Maesaka H, et al. Sequence analysis of the thyrotropin (TSH) receptor gene in congenital primary hypothyroidism associated with TSH unresponsiveness. Thyroid 1994;4:255–259.
95. Xie J, Pannain S, Pohlenz J, Weiss RE, Moltz K, Morlot M, et al. Resistance to thyrotropin (TSH) in three families is not associated with mutations in the TSH receptor or TSH. J Clin Endocrinol Metab 1997;82:3933–3940.
96. Beamer WG, Eicher EM, Maltais LJ, Southard JL. Inherited primary hypothyroidism in mice. Science 1981;212:61–62.
97. Stein SA, Shanklin DR, Kruilich L, Roth MG, Chubb CM, Adams PM. Evaluation and characterization of the hyt/hyt hypothyroid mouse. II. Abnormabilities of TSH and the thyroid gland. Neuroendocrinology 1989;49:509–519.
98. Beamer WD, Creswell LA. Defective thyroid ontogenesis in fetal hypothyroid (*hyt/hyt*) mice. Anat Rec 1982;202:387–393.
99. Adams P M, Stein S A, Palnitkar M, Anthony A, Gerrity L. Evaluation and characterization of the hyt/hyt hypothyroid mouse. I. Somatic and behavioral studies. Neuroendocrinology 1989;49:138–143.
100. Stein SA, Zakarija M, McKenzie JM, Shanklin DR, Palnitkar MB, Adams PM. The site of the molecular defect in the thyroid gland of the hyt/hyt mouse: abnormabilities in the TSHR receptor-G protein-adenylyl cyclase complex. Thyroid 1991;1:257–265.
101. Dryja TP, Hahn LB, Crowley GS, McGee TL, Berson EL. Mutation spectrum of the rhodopsin gene among patients with autosomal dominant retinitis pigmentosa. Proc Natl Acad Sci USA 1991;88: 9370–9374.
102. Meyerhoff WL. Hypothyroidism and the ear: electrophysiological, morphological and chemical considerations. Laryngoscope 1979;89:1–25.
103. O'Malley BW, Li D, Turner DS. Hearing loss and cochlear abnormabilities in the congenital hypothyroid (hyt/hyt) mouse. Hearing Res 1995;88:181–189.
104. Brown RS, Bellisario RL, Botero D, Fournier L, Abrams CA, Cowger ML, et al. Incidence of transient congenital hypothyroidism due to maternal thyrotropin receptor-blocking antibodies in over one million babies. J Clin Endocrinol Metab 1996;81:1147–1151.
105. Refetoff S, Sunthornthepvarakul T, Gottschalk M, Hayashi Y. Resistance to thyrotropin and other abnormabilities of the thyrotropin receptor. Rec Prog Hormone Res 1996;51:97–122.
106. Asteria C, Persani L, Romoli R, Beck-Peccoz P. Resistance to thyrotropin action resulting from inactivating mutation of thyrotropin receptor (TSH-R) gene. J Endocrinol Invest 1996;19(Suppl. 6) 26 (Abtract).
107. Wadsworth HL, Russo D, Nagayama Y, Chazenbalk GD, Rapoport B. Studies on the role of amino acids 38-45 in the expression of a functional thyrotropin receptor. Mol Endocrinol 1990;6:394–398.
108. Kosugi S, Mori T. TSH receptor and LH receptor. Endocr J 1995;42:587–606.
109. Chazenbalk GD, Nagayama Y, Russo D, Wadsworth HL, Rapoport B. Functional analysis of the cytoplasmic domains of the human thyrotropin receptor by site-directed mutagenesis. J Biol Chem 1990;265:20,970–20,975.

110. Kosugi S, Kohn LD, Akamizu T, Mori T. The middle portion in the second cytoplasmic loop of the thyrotropin receptor plays a crucial role in adenylate cyclase activation. Mol Endocrinol 1994;8:498–509.
111. Medeiros-Neto GA, Kourides IA, Almeida F, Gomes E, Cavaliere H, Ingbar SH. Enlargement of the sella turcica in some patients with longstanding untreated endemic cretinism. Serum TSH, alpha, TSH-beta, and prolactin responses to TRH. J Endocrinol Invest 1981;4:303–307.
112. Biebermann H, Krude H, Thiede C, Kotulla P, Grüters A. Sporadic congenital hypothyroidism due to compound heterozygosity for two mutations of the coding sequence of the thyrotropin receptor gene. 78th Meeting of the Endocrine Society, San Franciso. 1996. Abstract P2–954.
113. Biebermann H, Grüters A, Schöneberg T, Gudermann T. Congenital hypothyroidism caused by mutations in the thyrotropin-receptor gene. N Engl J Med 1997;336:1390–1391.
113a. Biebermann H, Schöneberg T, Krude H, Schultz G, Gudermann T, Grüters A. Mutations of the human thyrotropin receptor gene causing thyroid hypoplasia and persistent congenital hypothyroidism. J Clin Endocrinol Metab 1997;82:3471–3480.
114. Ahlbom BD, Yaqoob M, Larsson A, Illicki A, Annerén G, Wadelius C. Genetic and linkage analysis of familial congenital hypothyroidism: exclusion of linkage to the TSH receptor. Hum Genet 1997;99: 186–190.
115. Gagné N, Parma J, Deal C, Vassart G, van Vliet G. Apparent congenital athyreosis due to compound heterozygosity for inactivating mutations in the thyrotropin receptor (TSH-R) gene. 79th Annual Meeting of the Endocrine Society, Minneapolis, 1997, Abstract OR30-2.
116. Mimouni M, Mimouni-Bloch A, Schachter J, Shohat M. Familial hypothyroidism with autosomal dominant inheritance. Arch Dis Child 1996;75:245–246.
117. Sunthornthepvarakul T, Hayashi Y, Refetoff S. Polymorphism of a variant human thyrotropin receptor (hTSHR) gene. Thyroid 1994;4:147–149.
118. De Roux N, Misrahi M, Chatelain N, Gross B, Milgrom E. Microsatellites and PCR primers for genetic studies and genomic sequencing of the human TSH receptor gene. Mol Cell Endocrinol 1996;117: 253–256.
119. Levine MA, Ringel MD. Editorial: Resistance to TSH in patients with normal TSH receptors—where do we turn when "Sutton's Law" proves false? J Clin Endocrinol Metab 1997;82:3930–3932.
120. Perna MG, Civitareale D, De Filippis V, Sacco M, Cisternino C, Tassi V. Absence of mutations in the gene encoding thyroid transcription factor-1 (TTF-1) in patients with thyroid dysgenesis. Thyroid 1997;7:377–381.
121. Lapi P, Macchia PE, Chiovato L, Biffali E, Moschini L, Larizza D, et al. Mutations in the gene encoding thyroid transcription factor-1 (TTF-1) are not a frequent cause of congenital hypothyroidism (CH) with thyroid dysgenesis. Thyroid 1997;7: 383–387.
122. Kimura S, Hara Y, Pineau T, Fernandez-Salguero P, Fox CH, Ward JM, et al. The T/ebp null mouse: thyroid-specific enhancer-binding protein is essential for the organogenesis of the thyroid, lung, ventral forebrain, and pituitary. Genes Dev 1996;10:60–69.
123. Lyons J, Landis, CA, Harsh G, Vallar L, Grünewald K, Feichtinger H, et al. Two G protein oncogenes in human endocrine tumors. Science 1990;249:655–659.
124. O'Sullivan C, Barton CM, Staddon SL, Brown CL, Lemoine NR. Activating point mutations of the gsp oncogene in human thyroid adenomas. Molec Carcinogen 1991;4:345–349.
125. Suarez HG, du Villard JA, Caillou B, Schlumberger M, Parmentier C, Monier R. Gsp mutations in human thyroid tumors. Oncogene 1991;6:677–679.
126. Spada A, Vallar L, Faglia G. G protein oncogenes in pituitary tumors. Trends Endocrinol Metab 1992; 3:355–360.
127. Weinstein LS, Shenker A, Gejman PV, Merino MJ, Friedman E, Spiegel AM. Activating mutations of the stimulatory G protein in the McCune-Albright syndrome. N Engl J Med 1991;325:1688–1695.
128. Porcellini A, Ciullo I, Laviola L, Amabile G, Fenzi G, Avvedimento VE. Novel mutations of thyrotropin receptor gene in thyroid hyperfunctioning adenomas. J Clin Endocrinol Metab 1994;79:657–661.
129. Paschke R, Tonacchera M, van Sande J, Parma J, Vassart G. Identification and functional characterization of two new somatic mutations causing constitutive activation of the thyrotropin receptor in hyperfunctioning autonomous adenomas of the thyroid. J Clin Endocrinol Metab 1994;79:1785–1789.
130. Russo D, Arturi F, Wicker R, Chazenbalk GD, Schlumberger M, DuVillard JA, et al. Genetic alterations in thyroid hyperfunctioning adenomas. J Clin Endocrinol Metab 1995;80:1347–51.
131. Takeshita A, Nagayama Y, Yokoyama N, Ishikawa N, Ito K, Yamashita S, et al. Rarity of oncogenic mutations in the thyrotropin receptor of autonomously functioning thyroid adenomas. J Clin Endocrinol Metab 1995;80:2607–2611.

132. Russo D, Arturi F, Suarez HG, Schlumberger M, Du Villard JA, Crocetti U, et al. Thyrotropin receptor gene alterations in thyroid hyperfunctioning adenomas. J Clin Endocrinol Metab 1996;81:548–51.
133. Parma J, Duprez L, van Sande J, Hermans J, Rocmans P, van Vliet G, et al. Diversity and prevalence of somatic mutations in the thyrotropin receptor and Gsα genes as a cause of toxic thyroid adenomas. J Clin Endocrinol Metab 1997;82:2695–2701.
134. Duprez L, Parma J, Costagliola S, Hermans J, van Sande J, Dumont JE, Vassart, G. Constitutive activation of the TSH receptor by spontaneous mutations affecting the N-terminal extracellular domain. FEBS Letters 1997;409:469–474.
135. Kopp P, Muirhead S, Jourdain N, Gu WX, Jameson JL, Rodd C. Congenital hyperthyroidism caused by a solitary toxic adenoma harboring a novel somatic mutation (serine281→isoleucine) in the extracellular domain of the thyrotropin receptor. J Clin Invest 1997;100:1634–1639.
136. Führer D, Holzapfel HP, Wonerow P, Scherbaum WA, Pascke R. Somatic mutations in the thyrotropin receptor gene and not in the Gsa protein gene in 31 toxic thyroid nodules. J Clin Endocrinol Metab 1997;82:3885–3891.
137. Holzapfel HP, Führer D, Wonwerow P, Weinland G, Scherbaum WA, Paschke R. Identification of constitutively activating somatic thyrotropin receptor mutations in a subset of toxic multinodular goiters. J Clin Endocrinol Metab 1997;82:4229–4233.
138. Derwahl M. TSH receptor and Gs-a gene mutations in the pathogenesis of toxic thyroid adenomas - a note of caution. J Clin Endocrinol Metab 1996;81:2783–2785.
139. Tonacchera M, Cetani F, Parma J, van Sande J, Vassart G, Dumont J. Oncogenic mutations in thyroid adenoma: methodological criteria. Eur J Endocrinol 1996;135:444–446.
140. Duprez L, Parma J, van Sande J, Allgeier A, Leclère J, Schvartz C, et al. Germline mutations in the thyrotropin receptor gene cause non-autoimmune autosomal dominant hyperthyroidism. Nature Genet 1994;7:396–401.
141. Zhang ML, Sugawa H, Kosugi S, Mori T. Constitutive activation of the thyrotropin receptor by deletion of a portion of the extracellular domain. Biochem Biophys Res Commun 1995;211:205–209.
142. van Sande J, Massart C, Costagliola S, Allgeier A, Cetani F, Vassart G, et al. Specific activation of the thyrotropin receptor by trypsin. Mol Cell Endocrinol 1996;119:161–168.
143. Thomas JL, Leclère J, Hartemann P, Duheille J, Orgiazzi J, Petersen M, et al. Familial hyperthyroidism without evidence of autoimmunity. Acta Endocrinol 1982;100:512–518.
144. Tonacchera M, van Sande J, Cetani F, Swillens S, Schvartz C, Winizewski P, et al. Functional characteristics of three new germline mutations of the thyrotropin receptor gene causing autosomal dominant toxic thyroid hyperplasia. J Clin Endocrinol Metab 1996;81:547–554.
145. Schwab KO, Söhlemann P, Gerlich M, Broecker M, Petrykowski W, Holzapfel HP, et al. Mutations of the TSH receptor as cause of congenital hyperthyroidism. Exp Clin Endocrinol Diabetes 1996; 104(Suppl 4), 124–128.
146. Führer D, Wonerow P, Willgerodt H, Paschke R. Identification of a new thyrotropin receptor germline mutation (Leu629Phe) in a family with neonatal onset of autosomal dominant nonautoimmune hyperthyroidism. J Clin Endocrinol Metab 1997;82:4234–4238.
147. Horton GL. Hyperthyroidie héréditaire par hyperactivité diffuse non-autoimmune de la thyroide avec autonomie de fonction et de croissance. In: Hyperthyroidie héréditaire par hyperactivité diffuse non-autoimmune de la thyroide avec autonomie de fonction et de croissance. Thesis, University of Lausanne, Switzerland, 1987.
148. Leclère J, Béné MC, Aubert V, Klein M, Pascal-Vigneron V, Weryha G, et al. Clinical consequences of activating germline mutations of TSH receptor, the concept of toxic hyperplasia. Horm Res 1997; 47:158–162.
149. Zakarija M, McKenzie JM. Pregnancy-associated changes in the thyroid-stimulating antibody of Graves' disease and the relationship to neonatal hyperthyroidism. J Clin Endocrinol Metab 1983;57: 1036–1040.
150. Zakarija M, McKenzie JM, Hoffmann WH. Prediction and therapy of intrauterine and late-onset neonatal hyperthyroidism. J Clin Endocrinol Metab 1986;62:368–371.
151. Fort P, Lifshitz F, Pugliese M, Klein I. Neonatal thyroid disease: differential expression in three successive offsprings. J Clin Endocrinol Metab 1988;66:645–647.
152. Ramsay I, Kaur S, Krassas G. Thyrotoxicosis in pregnancy: results of treatment by antithyroid drugs combined with T4. Clin Endocrinol (Oxford) 1983;18:73–85.
153. Kopp P, van Sande J, Parma J, Duprez L, Gerber H, Joss E et al. Congenital hyperthyroidism caused by a mutation in the thyrotropin-receptor gene. N Engl J Med 1995;332:150–154.

154. De Roux N, Polak M, Couet J, Legher J, Czernichow P, Milgrom E, et al. A neomutation of the thyroid-stimulating hormone receptor in a severe neonatal hyperthyroidism. J Clin Endocrinol Metab 1996; 81:2023–2026.
155. Esapa CT, Betts P, Kendall-Taylor P, Harris PE. A novel TSH receptor mutation in an infant with thyrotoxicosis. J Endocrinol Invest 1996;19:71 (Abstract).
156. Kopp P, Roe T, Jameson JL. Congenital non-autoimmune hyperthyroidism in a non-identical twin caused by a sporadic germline mutation in the thyrotropin receptor gene. Thyroid 1997;7:765–770.
157. Holzapfel HP, Wonwerow P, von Petrykowski W, Henschen M, Scherbaum WA, et al. Sporadic congenital hyperthyroidism due to a spontaneous germline mutation in the thyrotropin receptor gene. J Clin Endocrinol Metab 1997;82:3879–3884.
158. Hollingsworth DR. Neonatal hyperthyroidism. In: Delange F, Fisher DA, Malvaux P, eds. Pediatric Thyroidology. Karger, Basel, 1985, pp. 210–222.
159. Matsuo K, Friedman E, Gejman PV, Fagin JA. The thyrotropin receptor (TSH-R) is not an oncogene for thyroid tumors: structural studies of the TSH-R and the alpha-subunit of Gs in human thyroid neoplasms. J Clin Endocrinol Metab 1993;76:1446–1451.
160. Russo D, Arturi F, Schlumberger M, Caillou B, Monier R, Filetti S, et al. Activating mutations of the TSH receptor in differentiated thyroid carcinomas. Oncogene 1996;11:1907–1911.
161. Spambalg D, Sharifi N, Elisei F, Gross J. L, Medeiros-Neto G, Fagin JA. Structural studies on the thyrotropin receptor and $G_s\alpha$ in human thyroid cancers: low prevalence of mutations predicts infrequent involvement in malignant transformation. J Clin Endocrinol Metab 1996;81:3898–3901.
162. Russo D, Tumino S, Arturi F, Vigneri P, Grasso G, Pontecorvi A, et al. Detection of an activating mutation of the thyrotropin receptor in a case of an autonomously hyperfunctioning thyroid insular carcinoma. J Clin Endocrinol Metab 1997;82:735–738.
163. Fagin JA. Molecular defects in thyroid gland neoplasia. J Clin Endocrinol Metab 1992;75:1398–1400.
164. Wynford-Thomas D. Molecular genetics of thyroid cancer. Trends Endocrinol Metab 1993;4:224–232.
165. Farid NR, Shi Y, Zou M. Molecular basis of thyroid cancer. Endocr Rev 1994;15:202–232.

7 Thyroid Hormone Resistance

*V. Krishna K. Chatterjee, MD, BMBCh, FRCP,
Roderick J. Clifton-Bligh, BSc (Med),
and Mark Gurnell, BSc (Hons), MBBS MRCP*

CONTENTS

INTRODUCTION
CLINICAL FEATURES
MOLECULAR GENETICS
PROPERTIES OF MUTANT RECEPTORS
PATHOGENESIS OF VARIABLE TISSUE RESISTANCE
ANIMAL MODELS OF RESISTANCE
MANAGEMENT
FUTURE DIRECTIONS
HORMONE RESISTANCE
ACKNOWLEDGMENTS
REFERENCES

INTRODUCTION

Thyroid hormones (thyroxine T4, triiodothyronine T3) regulate many cellular functions in virtually every type of tissue. The diverse responses to thyroid hormone include regulation of growth, the metabolic rate, muscular activity, and myocardial contractility. Thyroid hormone is also required for developmental processes, such as functional differentiation of the central nervous system and metamorphosis in amphibia *(1)*. The synthesis of thyroid hormones is controlled by hypothalamic thyrotropin-releasing hormone (TRH) and pituitary thyroid-stimulating hormone (TSH), and in turn, T4 and T3 regulate TRH and TSH production as part of a negative feedback loop.

Many of these effects of thyroid hormone on physiological processes are mediated by changes in expression of specific target genes in different tissues. Thus, the feedback effects of thyroid hormones on TSH production are mediated by inhibition of hypothalamic TRH and pituitary TSHα- and β-subunit gene expression. Target genes that are induced by thyroid hormone include malic enzyme and sex hormone binding globulin (SHBG) in the liver, myosin heavy chain and sodium-calcium ATPase in myocardium, myelin basic protein in brain, and sodium-potassium ATPase in skeletal muscle. The regulation of these genes by thyroid hormone is now known to be mediated by a protein

From: *Contemporary Endocrinology: Hormone Resistance Syndromes*
Edited by: J. L. Jameson © Humana Press Inc., Totowa, NJ

that is a member of the steroid/nuclear receptor superfamily of ligand-inducible transcription factors (2). In keeping with other members of the superfamily, the thyroid hormone receptor (TR) is organized into distinct functional domains. A central DNA binding domain (DBD) containing two "zinc finger" motifs mediates receptor binding to specific regulatory DNA sequences or thyroid response elements (TREs), usually located in the promoter regions of target genes. Many TREs consist of a tandem or direct repeat arrangement of the hexanucleotide sequence AGGTCA. Other recognized TRE configurations include palindromic and everted repeat arrangments of this motif, and some negatively regulated promoters (TRH, TSHα, TSHβ) also contain single hexameric binding sites. Although TR can bind these sequences as a monomer, it interacts preferentially with a heterodimer partner—the retinoid X receptor (RXR). In addition, homodimeric receptor–DNA interactions have also been observed particularly with everted repeat TREs (3). It is now recognized that in the absence of ligand, many positively regulated promoters are repressed or "silenced" by unliganded receptor. Hormone binding to the carboxy-terminal domain of TR results in relief of repression followed by ligand-dependent transactivation (Fig. 1). Recently, specific cofactors, which may mediate silencing and transcription activation functions, have been isolated: a family of corepressor proteins (nuclear receptor corepressor, N-CoR; silencing mediator for RAR/TR, SMRT) interact with unliganded TR, but dissociate following T3 binding (4); conversely, a number of putative coactivators (steroid receptor coactivator 1 [SRC-1], CREB binding protein [CBP], CBP/cointegrator-associated protein [p/CIP]) that are recruited by TR and other nuclear receptors in a hormone-dependent manner have also been identified (5). Most recently, some of these cofactors have been shown to possess intrinsic enzymatic activity. Thus SRC-1, p/CIP, CBP, and a CBP-associated factor (p/CAF) exhibit histone acetylase activity (6); conversely, the corepressors recruit a factor (histone deacetylase, HDAC), which can deacetylate histones (7). One model of receptor action that follows from this proposes that such enzymatic modification of core histones within nucleosomes modulates the accessibility or binding of basal transcription factors to DNA, thereby regulating levels of target gene transcription (8) (Fig. 1).

In humans, two highly homologous thyroid hormone receptors, denoted TRα and TRβ, are encoded by separate genes on chromosomes 17 and 3, respectively. Alternate splicing of each gene generates two major receptor isoforms, TRα1 and TRβ1, which are widely expressed, but with differing tissue distributions. In the rat, mouse, and human, a third isoform (TRβ2) is also produced by alternate splicing of the β gene. This receptor variant is most highly expressed in rodent pituitary and hypothalamus (9), and the human counterpart is thought to have a similar tissue distribution. Following the cloning of thyroid hormone receptors, resistance to thyroid hormone (RTH), was shown to be tightly linked to the TRβ gene locus in a single family (10). This prompted analysis of the TRβ gene in other cases, and a large number of receptor mutations have since been associated with this disorder (11). In this chapter, the clinical features and molecular genetics of RTH and current concepts relating to the pathogenesis of the disorder and implications for optimal treatment of these cases will be reviewed.

CLINICAL FEATURES

The syndrome of resistance to thyroid hormone is an uncommon disorder characterized by reduced responsiveness of target tissues to circulating thyroid hormones. The biochemical hallmark of RTH reflects resistance to thyroid hormone action in the hypo-

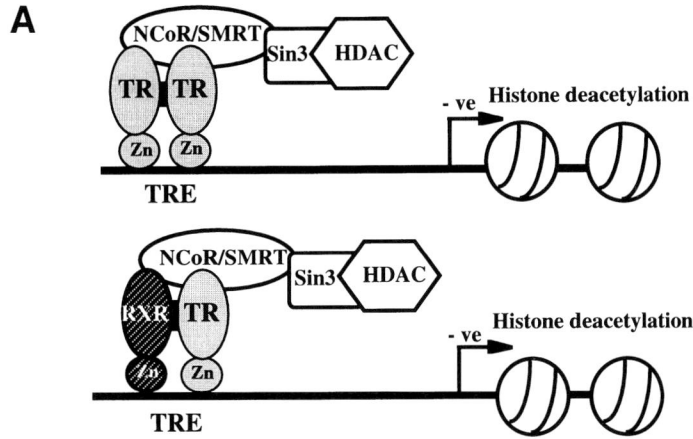

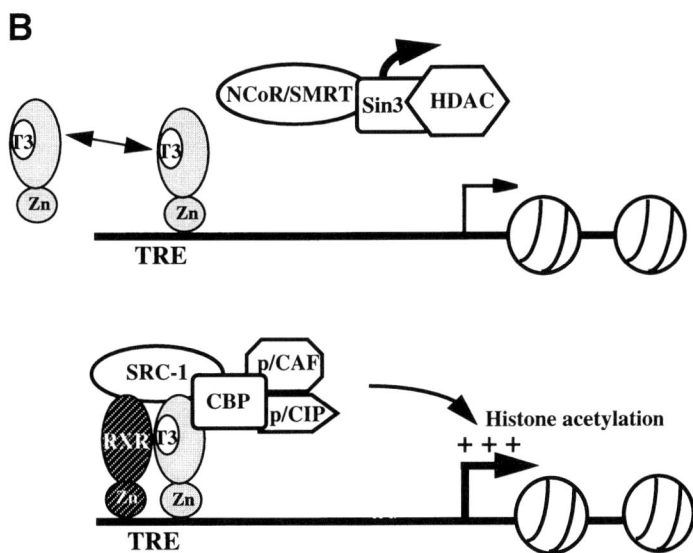

Fig. 1. A schematic outline of the mechanism of action of TR on positively regulated target genes. (**A**) Repression of basal transcription. In the absence of ligand, TR binds to target gene response elements as either a homodimer (upper panel) or heterodimer with RXR (lower panel). Basal gene transcription is inhibited by recruitment of a complex containing corepressor (NCoR or SMRT), which associates with histone deacetylase (HDAC) via an intermediary protein (Sin3). The deacetylation of core histones and possibly of general transcription factors results in transcriptional repression. (**B**) Hormone-induced changes. TR homodimers dissociate following T3 binding, whereas the heterodimer–DNA complex is stable. The corepressor complex is also released, enabling recruitment of coactivators (SRC-1, CBP, p/CIP, p/CAF). The intrinsic histone acetylase activities of the latter enables remodeling of chromatin enhancing transcription.

thalamic–pituitary–thyroid axis, with inappropriate TSH secretion driving T4 and T3 production, to establish a new equilibrium with high serum levels of thyroid hormones together with a nonsuppressed TSH. These biochemical features may also be observed in other clinical contexts (Table 1). Resistance to thyroid hormone was first described in 1967 in two siblings who were clinically euthyroid despite high circulating thyroid

Table 1
Causes of Hyperthyroxinaemia and Normal TSH Levels

Raised serum binding proteins
Familial dysalbuminaemic hyperthyroxinaemia
Anti-iodothyronine antibodies
Anti-TSH antibodies
Nonthyroidal illness
Acute psychiatric disorders
Neonatal period
Drugs (e.g., amiodarone, heparin)
Thyroxine replacement therapy
TSH-secreting pituitary tumor
Resistance to thyroid hormone

hormone levels, and exhibited a number of other abnormalities, including deaf-mutism, stippled femoral epiphyses with delayed bone maturation, and short stature, as well as dysmorphic facies, winging of the scapulae, and pectus carinatum *(12)*. It is now clear that some of these features are unique to this kindred in which the disorder was recessively inherited. The majority of RTH cases that have been described since then are dominantly inherited with highly variable clinical features. Many patients with RTH are either asymptomatic, or have nonspecific symptoms and may be noted to have a goiter, prompting thyroid function tests, which suggest the diagnosis. Attempts to treat the biochemical abnormality with surgery or radioiodine are often unsuccessful, with recrudescence of the goiter and thyroid dysfunction *(13)*. In these individuals, classified as exhibiting generalized resistance (GRTH), the high thyroid hormone levels are thought to compensate for ubiquitous tissue resistance, resulting in a euthyroid state. In contrast, a number of individuals with the same biochemical abnormalities exhibit clinical features of thyrotoxicosis; in adults, these can include weight loss, tremor, palpitations, insomnia, and heat intolerance; in children, failure to thrive, accelerated growth, and hyperkinetic behavior have also been noted. When this clinical entity was first described, patients were thought to have "selective" pituitary resistance to thyroid hormone action (PRTH) with preservation of normal hormonal responses in peripheral tissues *(14)*. Hypothyroid features, such as growth retardation and a delayed bone age in children or hypercholesterolaemia in adults, have also been observed in RTH and may coexist with thyrotoxic symptoms in the same individual *(15)*.

However, a comparison of the clinical and biochemical characteristics of individuals classified as either GRTH or PRTH indicates that there is a wide overlap between these entities. For example, there are no differences in age, sex ratio, frequency of goiter, or levels of free T4, free T3, or TSH between patients with the two types of disorder. Significantly, features, such as tachycardia, hyperkinetic behavior, and emotional disturbance, have been documented in individuals with GRTH *(16)*. Conversely, serum SHBG—a hepatic index of thyroid hormone action—is normal in patients with PRTH, suggesting that target tissue resistance is not solely confined to the pituitary–thyroid axis in this group *(17)*. Temporal variation in clinical symptoms is another factor that can confound the assessment and classification of these patients. Thus, in two cases of RTH, affected individuals exhibited thyrotoxic symptoms and signs that varied spontaneously over several years *(16)*. Overall, these observations indicate that although all patients

Table 2
Indices of Thyroid Hormone Action Used in RTH

Pituitary: TSH
General: Basal metabolic rate
Hepatic: SHBG, ferritin, cholesterol
Muscle: Creatine kinase, ankle jerk relaxation time
Cardiac: Sleeping pulse rate, systolic time interval, diastolic isovolumetric relaxation time
Bone: Height, bone age, bone density, osteocalcin, alkaline phosphatase, pyridinium crosslinks
Hematological: Soluble interleukin-2 receptor
Lung: Angiotensin-converting enzyme

with RTH exhibit abnormal thyroid function, the clinical presentation varies widely both between and within individuals. Nevertheless, the absence or presence of overt thyrotoxic features allows patients to be classified as either GRTH or PRTH, and this clinical distinction will probably remain useful as a guide to the most appropriate form of treatment (see below).

In addition to clinical features, the measurement of indices of thyroid hormone action is of use in evaluating the differing responses of various target organs and tissues to elevated circulating thyroid hormones (Table 2). Although these measurements are most useful in assessing the effects of marked thyroid hormone excess states, such as overt hyperthyroidism, they may be less discriminatory in individuals with borderline thyroid dysfunction or with some exceptions (e.g., bone age, systolic time intervals, BMR, CPK) in hypothyroidism. In order to improve the sensitivity and specificity of these parameters, it has been suggested that RTH patients be assessed following the administration of graded supraphysiological doses of T3 (50, 100, and 200 µg/d, each given for a period of 3 d) with comparison of any change in indices to baseline values and responses in normal subjects (18).

The clinical phenotype of RTH has been further refined by a prospective study of 104 cases from 42 families at the National Institutes of Health (19). A palpable goiter was present in 65% of individuals (especially adult females), and interestingly, fewer children with RTH born to affected mothers exhibited thyroid enlargement (35%) compared to offspring of unaffected mothers (87%). Approximately one-third of children had low body weight. Childhood short stature (height <5th centile) was noted in 18% and delayed bone age (>2 SD) in 29%, comparing favorably with figures from another recent study (20), but final adult height was not affected. Significant hearing loss—a novel clinical feature—was documented in 21% of RTH cases: in the majority, audiometry indicated a conductive defect, probably related to an increased incidence of recurrent ear infections in childhood; abnormal otoacoustic emissions, suggestive of cochlear dysfunction, were also documented in those with hearing deficit (21). Finally, the bioactivity of circulating TSH has been shown to be significantly enhanced in RTH, perhaps accounting for the goiter and markedly elevated serum thyroid hormones, despite the normal immunoreactive TSH levels observed in some cases (22).

Two recent studies have documented neuropsychological abnormalities in a large number of patients with RTH. First, a history of attention-deficit hyperactivity disorder

Table 3
Recognized Features of RTH

Elevated serum free thyroid hormones
Normal TSH with enhanced bioactivity
Goiter
Growth retardation, short stature
Low body mass index in childhood
ADHD, low IQ
Tachycardia, atrial fibrillation, heart failure
Hearing loss
Ear, nose, and throat infections
Osteopenia

(ADHD) in childhood was elicited more frequently in patients with RTH compared to their unaffected relatives (23). A second study showed that both children and adults with RTH exhibited problems with language development, manifested by poor reading skills and problems with articulation (24). However, in a subsequent analysis of one family, RTH cosegregated with lower IQ rather than ADHD (25). A direct comparison of individuals with ADHD and RTH vs ADHD alone indicates an association with lower nonverbal intelligence and academic achievement in the former group (26). On the other hand, two different surveys of unselected ADHD cases using thyroid function test failed to detect any cases of RTH, suggesting that screening for this disorder in this cohort is not likely to be beneficial (27,28). Although magnetic resonance imaging shows that anomalies of the Sylvian fissure or Heschl's gyri are more frequent in RTH, these features do not correlate with ADHD (29). The major clinical features that are recognized in association with RTH are summarized in Table 3.

MOLECULAR GENETICS

In the majority of cases, RTH is familial and dominantly inherited. Consonant with this mode of inheritance, many groups have reported that affected individuals are heterozygous for mutations in the TRβ-gene (11,30,31). Approximately 10% of cases are sporadic and associated with *de novo* heterozygous receptor mutations. Over 70 different defects, including point mutations, in-frame deletions, and frame-shift insertions, have been documented to date, all of which localize to the hormone binding domain of the receptor (Fig. 2). Consequently, the ability of in vitro synthesized mutant proteins to bind T3 is moderately or markedly reduced, and their ability to activate or repress target gene expression is impaired (32,33). In contrast, in the first RTH family described, with the recessively inherited form of the disorder, the two affected siblings were found to be homozygous for a complete deletion of both alleles of the TRβ receptor gene (34). Importantly, the obligate heterozygotes in this family, harboring a deletion of one TRβ allele, were completely normal with no evidence of thyroid dysfunction. This suggested that mere deficiency of functional β receptor, as a consequence of the single deleted TRβ allele, was insufficient to generate the resistance phenotype. Accordingly, we and others put forward the hypothesis that the mutant receptors in dominantly inherited RTH were not simply functionally impaired, but also capable of inhibiting wild-type receptor action. Indeed, in vitro experiments indicate that when coexpressed, the mutant proteins

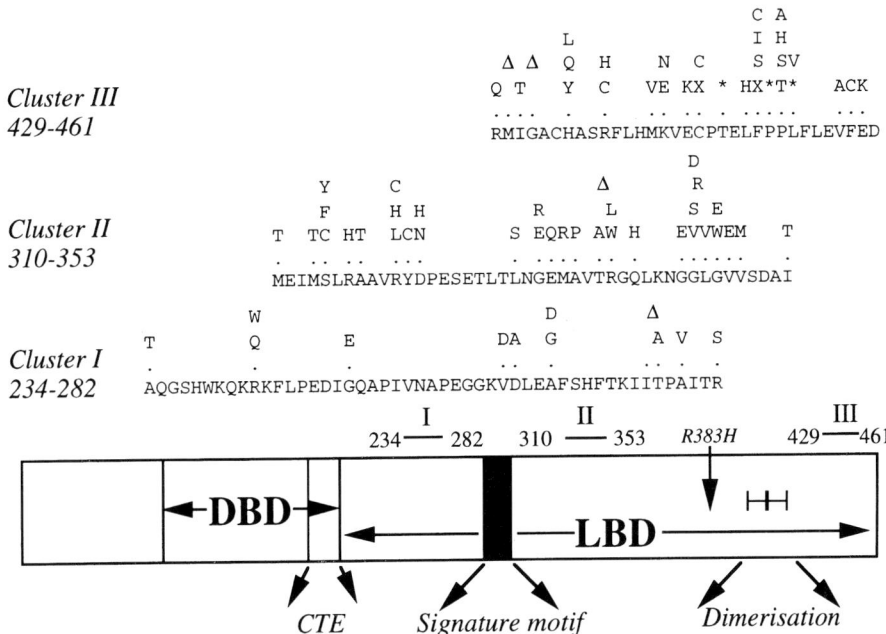

Fig. 2. A schematic representation of the domains of TRβ, showing that with one exception (R383H), RTH mutations localize to three clusters within the LBD. The receptor defects include missense mutations, in-frame codon deletions (Δ), and premature termination codons (*), but frameshift mutations have been omitted for clarity. Mutations include published information together with our unpublished data, and their nomenclature conforms with the consensus statement from the First Workshop on RTH *(31a)*. No RTH mutations have been identified in the zinc finger region or an α-helical CTE that represent the DBD. Similarly, regions in the LBD that are important for maintaining its conformation (signature motif) or ability to dimerize with RXR (dimerization) are devoid of natural mutations. This correlates with the observation that DNA binding and dimerization are critical for the dominant negative activity of RTH mutants.

are able to inhibit the function of their wild-type counterparts in a "dominant negative" manner *(35,36)*. Further clinical and genetic evidence to support this hypothesis is provided by a single case in which severe resistance was associated with marked developmental delay and growth retardation *(37)*. This individual was homozygous for a mutation in both alleles of the TRβ gene and the extreme phenotype presumably reflects the inhibitory effects of two dominant negative mutant β receptors *(38)*.

Based on the supposition that PRTH was associated with selective pituitary resistance, it had been hypothesized that this disorder might be associated with defects in the pituitary type II 5'-deiodinase enzyme or the TRβ2 receptor isoform *(39)*. However, two reports have documented TRβ mutations in PRTH *(40,41)*, and we have extended these observations and identified further receptor mutations in a number of other PRTH cases *(31)*. Some of the mutations we have observed in individuals with PRTH have also been reported in GRTH cases from unrelated kindreds. Furthermore, we have found that even within a single family, the same receptor mutation can be associated with abnormal thyroid function and thyrotoxic features consistent with PRTH in some individuals, but similar biochemical abnormalities and a lack of symptoms indicative of GRTH in other members. Overall, these findings indicate that GRTH and PRTH represent differing phenotypic manifestations of a single genetic entity.

In a small number of cases, clear-cut biochemical evidence of RTH is not associated with a mutation in the coding region of TRβ1. Possible explanations in such individuals include a somatic TRβ1 mutation whose expression is limited in order to be undetectable in peripheral blood leukocyte DNA, or a mutation in TRβ2, the pituitary-expressed receptor isoform. Finally, the possibility of novel, nonreceptor mechanisms by which thyroid hormone action could be disrupted to produce RTH should also be considered. This latter hypothesis is further supported by a recent report of one family in which clear biochemical evidence of RTH appeared not to be associated with mutations in TRβ2 or TRβ1, or linked to the TRα gene locus *(42)*. Far-Western blotting showed aberrant binding of TRβ to a unique 84-kDa protein from patient, but not control fibroblast nuclear extracts, but the identity of this putative receptor-associated cofactor remains to be elucidated. However, it is known that patients with Rubinstein-Taybi syndrome, a disorder associated with defects in the nuclear receptor coactivator CBP, exhibit a number of somatic abnormalities (broad thumbs, mental retardation, short stature), yet have normal circulating free T4 and TSH levels *(43)*.

PROPERTIES OF MUTANT RECEPTORS

In X-linked androgen insensitivity syndrome, deleterious mutations have been described throughout the androgen receptor, whereas recessively inherited vitamin D-resistant rickets is characterized by mutations mainly in the DBD of the vitamin D receptor. In contrast, all the mutations described thus far in RTH, localize to the hormone binding domain. The majority occur in two areas (αα 310–353; αα 429–461), and the recent identification of additional novel mutations *(44–47)* has delineated a third cluster (αα 234–282) (Fig. 2). A number of functional studies of mutant receptors indicate that although they are transcriptionally impaired, their ability to bind DNA, form heterodimers with RXR, and exert a dominant negative effect on positively and negatively regulated target genes is preserved *(32,33,48)*. Conversely, it has been shown that the introduction of additional artificial mutations that abolish DNA binding or heterodimer formation abrogates the dominant negative activity of mutant receptors *(49,50)*. These observations suggest that receptor mutants with impaired transcriptional function, but normal DNA binding and dimerization properties, retain dominant negative potential leading to resistance to hormone action. Conversely, we speculate that heterozygous mutations in the DBD or dimerization domain of the receptor elude discovery, because they lack dominant negative activity and are therefore clinically and biochemically silent.

Structural studies of the DBDs of TR and RXR bound to a TRE *(51)* and the ligand binding domain (LBD) of TRα *(52)* have provided further insights into the clustered distribution of RTH mutations. As expected from their impaired ligand binding properties, most mutations are located around the hormone binding cavity *(52)*. In addition, the amino-terminal part of the TR LBD has been shown uniquely to contain an α helix that constitutes a C-terminal extension (CTE) of the DBD *(51)*. The LBD also contains a "signature motif," which is highly conserved amongst nuclear receptors and implicated in maintaining the structural integrity of this domain *(53)* and its ability to heterodimerize with RXR *(54)*. With their involvement in DNA binding or dimerization, these two regions are devoid of naturally occuring mutations and may represent the boundaries of the first mutation cluster (Fig. 2). Likewise, the virtually mutation-free region between αα 353 and 429 contains a region (helix 10 and 11) implicated in dimerization *(52)*.

However, such structure–function correlations might not be the sole determinants of the clustered distribution of receptor mutations in RTH. It has also been observed that mutations are nonuniformly distributed within the three major clusters, such that some codon changes (e.g., R243W, R338W, R438H) are particularly frequent *(55)*. These mutations represent transitions in CpG dinucleotides that are known to be frequent sites of point mutation in several other genes, suggesting that the concurrence of CpG dinucleotides within a cluster leads to the overrepresentation of certain codon changes in RTH.

Most recently, a number of natural receptor mutants have been shown to interact aberrantly with transcriptional cofactors. For example, a subset of RTH mutations are associated with markedly abnormal thyroid function in vivo and altered transcriptional function in vitro, despite little impairment in ligand binding. For a natural mutation involving a conserved hydrophobic residue (Leu 454 Val) in the carboxy-terminal transactivation domain of TR, such properties were explicable on the basis of attenuated mutant receptor interaction with putative transcriptional coactivators *(56)*.

Previous studies had also established that the ability of RTH mutant receptors to repress or "silence" basal gene transcription was likely to be an important factor contributing to their dominant negative potency. Non-T3 binding mutants exhibited constitutive silencing function, particularly when bound to DNA as homodimers, which could not be relieved by ligand *(57,58)* (Fig. 3B). Conversely, RTH mutants with impaired homodimerization properties were weaker dominant negative inhibitors *(59)*. With the recent identification of corepressors, Yoh et al. have extended these observations, showing that some RTH mutants either bind corepressor more avidly when unliganded or fail to dissociate from corepressor on T3 binding *(60)*. Furthermore, artificial mutations that abolish corepressor binding abrogate the dominant negative activity of natural receptor mutants, suggesting that this is a further functional attribute retained by mutant receptors in RTH *(60)*.

In contrast to silencing of basal transcription observed with positively regulated genes, unliganded TR has been shown to enhance the basal activity of negatively regulated (TRH, TSHβ) promoters, with transrepression on the addition of T3 *(61)*. Most recently, coexpressed corepressors have been shown to augment such unliganded activation, and an artificial mutant lacking corepressor binding is impaired for this function *(62)*. In this context, it is interesting to note that we have recently identified an unusual natural mutant (R383H) *(63)*, which when synthesized artificially had not been expected to be associated with RTH given its mildly reduced hormone binding *(64)*. However, we have shown that R383H exhibits delayed corepressor release and is transcriptionally impaired, particularly with negatively regulated promoters *(63)*. Given the pivotal role of negatively regulated target genes in the pathogenesis of RTH, aberrant corepressor recruitment or release may well prove to be the critical receptor abnormality in this disorder (Fig. 3A).

PATHOGENESIS OF VARIABLE TISSUE RESISTANCE

Genetic and functional evidence suggests that the ability to exert a dominant negative effect within the hypothalamic–pituitary–thyroid axis is a key property of RTH mutant receptors, which generates the characteristically abnormal thyroid function tests that lead to the identification of the disorder. Indeed, one study indicates that for a subset of RTH mutants, there is a correlation between their functional impairment in vitro and the degree of central pituitary resistance as measured by the elevation in serum free T4 in

A POSITIVELY REGULATED GENES

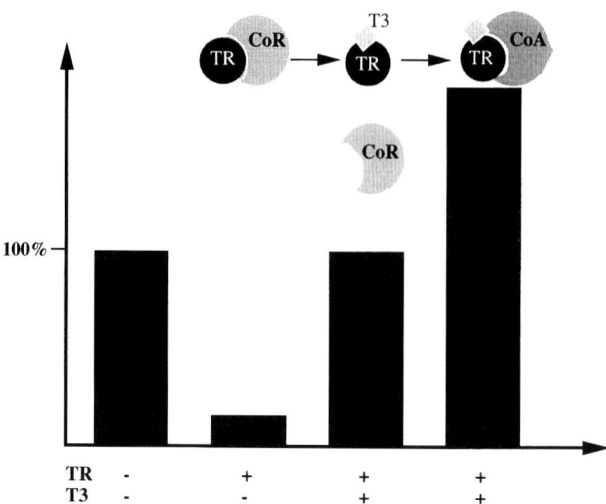

NEGATIVELY REGULATED GENES

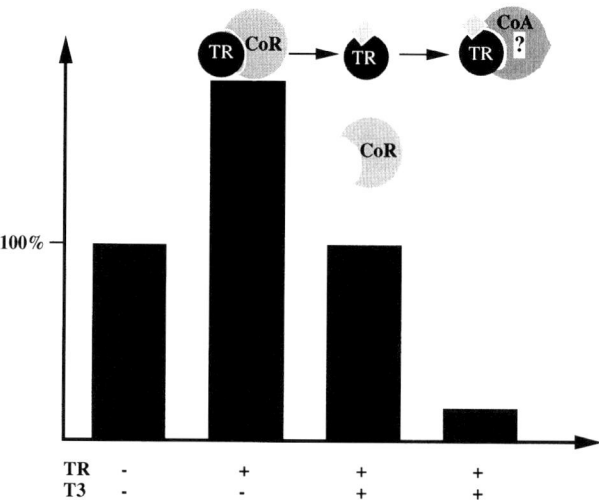

Fig. 3. (A) Transcriptional cofactors may exert opposite effects in the context of positively vs negatively regulated promoters. On a positive TRE, unliganded TR recruits corepressor to inhibit basal transcription. Corepressor dissociates following hormone binding, leading to derepression. Coactivator recruitment mediates transcriptional activation.

On negative TREs, the unliganded receptor–corepressor complex enhances basal transcription. T3-dependent corepressor release reverses this. Recruitment of other cofactors (coactivators/related proteins) mediates ligand-dependent transrepression.

Three types of functional impairment have been described with RTH mutants: the majority exhibit reduced ligand binding to account for impaired corepressor release or coactivator recruitment; we have recently described a mutant with impaired coactivator recruitment, yet relatively normal T3 binding *(56)*; a third subset of mutants exhibit disproportionately impaired corepressor release relative to their altered hormone binding affinities *(60)*. Since impaired receptor function on negatively regulated promoters (TRH, TSHα, TSHβ) is required to generate the characteristic biochemical phenotype of RTH, we speculate that delayed corepressor release on negative TREs may be the minimal receptor abnormality required to mediate this disorder.

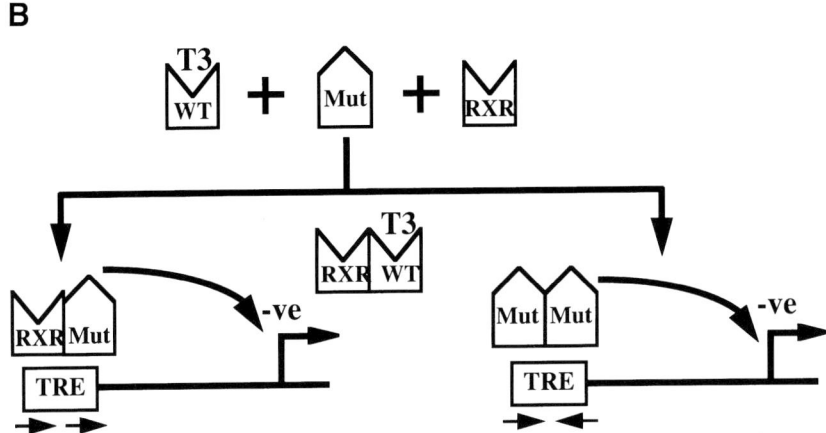

Fig. 3. *(Continued)* **(B)** Possible mechanisms for dominant negative inhibition of wild-type receptor action by RTH mutants. Direct repeat TREs found in some target genes (left) interact with TR-RXR heterodimers. Here, functionally impaired mutant receptor–RXR complexes compete with transcriptionally competent wild-type TR-RXR heterodimers. Everted repeat TREs (right) may be particularly susceptible to dominant negative inhibition by mutant TRs, since these response elements can also bind receptor homodimers. Here, mutant TR homodimers with impaired T3 binding properties fail to dissociate in the presence of ligand, preventing wild-type TR–RXR interaction. The mutant homodimer also represses transcription by maintaining interaction with the corepressor complex.

vivo *(65)*. On this background, the diverse clinical phenotypes may be owing to variable degrees of peripheral resistance in different patients, as well as variable resistance in different tissues within a single individual. A number of factors may contribute to such variable tissue resistance.

One contributory element may be the differing tissue distributions of receptor isoforms. The hypothalamus/pituitary and liver express predominantly TRβ2 and TRβ1 receptors, respectively *(9,66)*, whereas TRα1 is the major species detected in myocardium *(67)*. Therefore, mutations in the TRβ gene are likely to be associated with pituitary and liver resistance, as exemplified by normal SHBG and nonsuppressed TSH levels, whereas the tachycardia and cardiac abnormalities often seen in RTH may represent retention of myocardial sensitivity to thyroid hormone action mediated by a normal α-receptor.

Another factor that may regulate the degree of tissue resistance is the relative expression of mutant vs wild-type alleles of the TRβ gene. Although one study has suggested that both alleles are equally expressed *(68)*, another showed marked differences in the relative levels of wild-type and mutant receptor messenger RNA in skin fibroblasts from two RTH cases *(69)*. In one of these individuals, a temporal variation in expression of the mutant allele in fibroblasts appeared to correlate with the degree of skeletal tissue resistance.

Attempts to correlate the phenotype of RTH with the nature of the underlying TRβ mutation have been confounded by three factors: first, the imprecision of clinical criteria used to define GRTH and PRTH; second, the apparent spontaneous temporal variation in thyrotoxic features in some RTH cases *(16)*; third, the relatively small number of individuals with any given mutation that have been identified thus far. Nevertheless, some interesting correlations are emerging from the published literature. The first patient reported to have PRTH *(40)* was found to harbor an R338W receptor mutation, and the same phenotype has been described in the majority of individuals with this codon substitution *(31,41)*. Interestingly, when tested in vitro, this mutant exhibits dominant negative activity with the negatively regulated pituitary TSHα-subunit gene promoter, but is a relatively poor inhibitor of wild-type receptor action in other TRE contexts *(33,70)*. Furthermore, when introduced into other RTH mutant backgrounds, this mutation weakens their dominant negative potency on positively regulated reporter genes *(71)*. An R429Q mutant, variably associated with either PRTH or GRTH, is also selectively impaired when tested with negatively regulated (TRH, TSH) gene promoters *(72)*, and it has been observed that the dominant negative potential of other mutant receptors can differ depending on the nature and configuration of TREs *(33,73,74)*, suggesting that target gene promoter context is a further variable which may influence resistance to thyroid hormone action.

Since the TRβ2 receptor isoform is highly expressed in negatively regulated target tissues (pituitary, hypothalamus), RTH mutants have also been analyzed in this receptor context. TRβ2 has been shown to form homodimers more readily on some response elements *(75)*, which might enhance its dominant negative potency and its amino-terminal domain does contain a ligand-independent transcription activation function *(76)*. Safer et al. have studied the properties of some RTH mutants (R338W or L, V349M, R429Q, I431T) that may be associated with predominant pituitary resistance (PRTH) *(77)*. They found that such mutants exert a greater dominant negative effect in a TRβ2 than TRβ1 context, and correlated this with the observation that the β2 receptor is less dependent on RXR for its transcriptional function.

Finally, nonreceptor mutation-related factors may also affect the clinical and biochemical phenotype. For example, a deleterious R316H mutation was associated with normal thyroid function in some members of one kindred *(78)*, but in an unrelated family from our series, the same mutation was associated with abnormal thyroid function *(31)*, suggesting that other variables in the pituitary–thyroid axis can modulate mutant receptor action. Indirect evidence in favor of this notion is also provided by the observation that the unaffected first-degree relatives in a kindred with RTH had normal, but above average total T4 levels compared to unrelated controls *(79)*.

ANIMAL MODELS OF RESISTANCE

Many of the features in the recessively inherited cases of RTH associated with a deletion encompassing the TRβ gene have been recapitulated by an animal model involving targeted disruption of the mouse TRβ locus *(80)*. Homozygous TRβ knockout mice had elevated serum thyroid hormones and an inappropriately elevated TSH analogous to RTH. Also, importantly heterozygous animals were biochemically normal, corroborating findings in their human counterparts. The homozygous mice also exhibited profound sensorineural deafness without obvious cochlear malformation, indicating that the deaf-mutism in recessive human RTH is also related to a defect in TRβ, rather than deletion of a contiguous gene *(81)*. Together with the hearing abnormalities found in patients with dominantly inherited RTH, these findings underscore the importance of TRβ in auditory development and function. Whether this role is mediated by TRβ2 or TRβ1 awaits isoform-specific knockout models.

A recombinant adenovirus containing an RTH mutant TRβ has been used to infect hepatocytes both in vitro and in vivo. In this somatic gene-transfer model, hypothyroid mice exhibited attenuated T3-dependent responses of biochemical markers in liver *(82)*. An animal model with more generalized hormone resistance is represented by transgenic mice with ubiquitous mutant TRβ expression directed by a β-actin promoter *(83)*. Although thyroid function was only mildly abnormal, these mice did display decreased weight and hyperactivity, which are recognized features of the human syndrome. Most recently, RTH mutants have been targeted transgenically to the pituitary using a tissue-specific promoter *(84)*. Interestingly, these mice also showed only mildly deranged thyroid function, suggesting that additional dominant negative activity of mutant receptors on the hypothalamic TRH gene might contribute to the pathogenesis of the biochemical phenotype.

The introduction of homologous point mutations into TRα1 generates mutant receptor proteins that are also capable of exerting dominant negative effects in vitro *(74)*, raising the possibility that naturally occuring defects in the TRα gene might also be associated with an RTH phenotype. However, the abnormal pituitary function in TRβ knockout mice, without compensation by TRα, suggests a minor role for the α isoform in this tissue. This may provide one explanation for the failure to identify mutations in human TRα1 in RTH. This notion is further supported by the observation that mice with a selective knockout of TRα1 exhibit low or normal serum thyroid hormones with decreased heart rate and body temperature—a phenotype quite dissimilar to RTH *(85)*. Interestingly, disruption of the entire murine TRα locus (including TRα1 and TRα2) results in a severe neonatal hypothyroid phenotype, which is lethal if untreated *(86)*.

MANAGEMENT

One of the most important reasons for recognizing RTH is that its management differs from that of other common forms of thyroid dysfunction. In addition, the distinction between GRTH and PRTH on the basis of clinical criteria remains useful, since the management of the two states also differs. In most individuals with GRTH, the receptor defect is compensated by high circulating thyroid hormone levels, leading to a euthyroid state not associated with abnormalities other than a goiter. Certain circumstances, such as hypercholesterolaemia in adults or growth retardation in young children, may warrant the administration of supraphysiological doses of L-T4 to overcome the high degree of resistance in certain tissues. Although successful in some cases *(11)*, such therapy needs careful monitoring of a number of other indices of peripheral thyroid hormone action to avoid the adverse cardiac effects or excess catabolism associated with thyroxine overtreatment. Misdiagnosis of RTH, followed by inappropriate thyroid ablation, invariably renders the RTH patient hypothyroid and is another context in which thyroxine replacement in supraphysiologic dosage is indicated.

In contrast, a general reduction in thyroid hormone levels may be of benefit in the management of PRTH patients with thyrotoxic symptoms. However, the administration of conventional antithyroid drugs usually causes a further rise in serum TSH levels with consequent thyroid enlargment, and may also be associated with a theoretical risk of inducing autonomous pituitary TSH-secreting neoplasms. Accordingly, agents that inhibit pituitary TSH secretion, yet are devoid of peripheral thyromimetic effects, are used to reduce thyroid hormone levels. The most widely used example is the thyroid hormone analog 3,3,5 triiodothyroacetic acid (TRIAC), which has been shown to be beneficial in both childhood and adult cases *(87–90)*. This compound has a number of interesting properties that make it an attractive therapeutic option in RTH: first, it exerts predominantly pituitary and hepatic thyromimetic effects in vivo *(91)*—target tissues that are relatively refractory to thyroid hormones in RTH; second, it exhibits a higher affinity for TRβ than TRα in vitro *(92)*. A dose of 1.4–2.8 mg has been used, and a recent study suggested that twice daily administration might be optimal in inhibiting TSH secretion *(93)*. However, TRIAC treatment did not result in a clinical response in one instance *(94)*. Dextro thyroxine (D-T4) is another useful agent that has been shown to be effective in some cases *(95–97)*. If these compounds fail, the dopaminergic agent bromocriptine *(98)* or the somatostatin analog octreotide *(99)* may be administered. However, past experience indicates that TSH secretion escapes from the inhibitory effects of bromocriptine *(87,96)* as well as octreotide *(100)*. In view of the spontaneous variation in thyrotoxic symptoms in RTH, we recommend periodic cessation of all therapy and re-evaluation of the clinical status of the patient. Thyroid ablation could be used in rare circumstances, such as severe thyrotoxic cardiac failure in PRTH, but is probably best avoided otherwise, since it is irreversible and may worsen the compensated euthyroidism in some tissues with potentially harmful consequences.

The treatment of PRTH in childhood also requires careful monitoring to ensure that any reduction in thyroid hormone levels is not associated with growth retardation or adverse neurological sequelae. Indeed, control of cardiac and sympathomimetic manifestations with β-blockade may be the safest course in this context. Finally, a recent study showed that L-T3 therapy improved hyperactivity in nine children with ADHD and RTH, including three individuals who were unresponsive to methylphenidate *(101)*.

FUTURE DIRECTIONS

From a genetic standpoint, RTH has been found to exhibit considerable molecular heterogeneity, with an expanding repertoire of TRβ mutations. Recent studies have continued to provide insights into structure–function relationships in the receptor, particularly its interaction with transcription intermediary proteins—the corepressors and coactivators. It is anticipated that future studies of RTH mutants will help delineate novel receptor regions implicated in cofactor recruitment and, in particular, clarify their role in negative regulation of target gene transcription.

The clinical phenotype of RTH is proving to be complex, with the increasing recognition of features, such as ADHD, hearing deficit, osteopenia, and cardiac dysfunction, refuting the notion that RTH is a compensated euthyroid state. The challenge for the future will be to devise rational therapies for such complications—perhaps analogs with highly selective TRβ agonist or TRα antagonist activity. Although both GRTH and PRTH are associated with TRβ mutations, there are preliminary indications that some codon changes might predispose to predominant pituitary resistance. A more complete understanding of the factors that contribute to pituitary resistance is hampered by the lack of pituitary thyrotrope or hypothalamic cell lines in which to assay mutant receptor function with negatively regulated gene promoters in a eutopic context.

The pathogenesis of variable peripheral tissue resistance in RTH is also ill-understood. The differing tissue distributions of TRα1 vs TRβ1, and the variable dominant negative activity of mutant receptors in different cell type and promoter contexts are likely variables, but the ability to explore these in a clinical context is limited. The future availability of an animal model, involving the introduction of a dominant negative RTH mutation into the mouse TRβ gene, will greatly facilitate analyses of variable tissue resistance and also provide an environment to test novel treatments.

ACKNOWLEDGMENTS

V. K. K. C. and M. G. are supported by the Wellcome Trust (U.K) and R. C.-B. is a Commonwealth Scholar. We are also greatly indebted to many physicians for referrals of RTH cases without which our studies would not be possible.

REFERENCES

1. Chatterjee VKK, Tata JR. Thyroid hormone receptors and their role in development. Cancer Surveys 1992;14:147–167.
2. Mangelsdorf DJ, Evans RM. The RXR heterodimers and orphan receptors. Cell 1995;83:841–850.
3. Chin WW. Nuclear thyroid hormone receptors. In: Parker MG, ed. Nuclear Hormone Receptors. London, Academic, 1991, pp. 79–102.
4. Horlein AJ, Heinzel T, Rosenfeld MG. Gene regulation by thyroid hormone receptors. Curr Opinion Endocrinol Diabetes 1996;3:412–416.
5. Horwitz KB, Jackson TA, Bain DL, Richer JK, Takimoto GS, Tung L. Nuclear receptor coactivators and corepressors. Mol Endocrinol 1996;10:1167–1177.
6. Montminy M. Transcriptional activation: Something new to hang your HAT on. Nature 1997;387:654–655.
7. Heinzel T, Lavinsky RM, Mullen T-M, Soderstrom M, Laherty CD, Torchia J, et al. A complex containing N-CoR, mSin3 and histone deacetylase mediates transcriptional repression. Nature 1997;387:43–48.
8. Wolffe A. Transcriptional control: Sinful repression. Nature 1997;387:16–17.
9. Lazar MA. Thyroid hormone receptors: multiple forms, multiple possibilities. Endocr Rev 1993;14:184–193.
10. Usala SJ, Bale AE, Gesundheit N, Weinberger C, Lash RW, Wondisford FE, et al. Tight linkage between the syndrome of generalized thyroid hormone resistance and the human c-erb A β gene. Mol Endocrinol 1988;2:1217–1220.
11. Refetoff S, Weiss RE, Usala SJ. The syndromes of resistance to thyroid hormone. Endocr Rev 1993;14:348–399.
12. Refetoff S, De Wind LT, De Groot LJ. Familial syndrome combining deaf-mutism, stippled epiphyses, goiter and abnormally high PBI: possible target organ refractoriness to thyroid hormone. J Clin Endocrinol Metab 1967;27:279–294.
13. Refetoff S, Salazar A, Smith TJ, Scherberg NH. The consequences of inappropriate treatment due to failure to recognise the syndrome of pituitary and peripheral tissue resistance to thyroid hormone. Metabolism 1983;32: 822–834.
14. Gershengorn MC, Weintraub BD. Thyrotropin-induced hyperthyroidism caused by selective pituitary resistance to thyroid hormone. A new syndrome of inappropriate secretion of TSH. J Clin Invest 1975;56:633–642.
15. Magner JA, Petrick P, Menezes-Ferreira M, Weintraub BD. Familial generalized resistance of thyroid hormones: a report of three kindreds and correlation of patterns of affected tissues with the binding of [^{125}I]triiodothyronine to fibroblast nuclei. J Endocrinol Invest 1986;9:459–469.
16. Beck-Peccoz P, Chatterjee VKK. The variable clinical phenotype in thyroid hormone resistance syndrome. Thyroid 1994;4:225–232.
17. Beck-Peccoz P, Roncoroni R, Mariotti S, Medri G, Marcocci C, Brabant G, et al. Sex hormone-binding globulin measurement in patients with inappropriate secretion of thyrotropin (IST): evidence against selective pituitary thyroid hormone resistance in nonneoplastic IST. J Clin Endocrinol Metab 1990;71:19–25.
18. Sarne DH, Refetoff S, Rosenfield RL, Farriaux JP. Sex hormone-binding globulin in the diagnosis of peripheral tissue resistance to thyroid hormone: the value of changes after short term triiodothyronine administration. J Clin Endocrinol Metab 1988;66:740–746.
19. Brucker-Davis F, Skarulis MC, Grace MB, Benichou J, Hauser P, Wiggs E, et al. Genetic and clinical features of 42 kindreds with resistance to thyroid hormone. Ann Intern Med 1995;123:572–583.
20. Weiss RE, Refetoff S. Effect of thyroid hormone on growth: lessons from the syndrome of resistance to thyroid hormone. Endocrinol Metab Clin North Am 1996;25:719–730.

21. Brucker-Davis F, Skarulis MC, Pikus A, Ishizawar D, Mastroianni M-A, Koby M, et al. Prevalence and mechanisms of hearing loss in patients with resistance to thyroid hormone. J Clin Endocrinol Metab 1996;81:2768–2772.
22. Persani L, Asteria C, Tonacchera M, Vitti P, Chatterjee VKK, Beck-Peccoz P. Evidence for the secretion of thyrotropin with enhanced bioactivity in syndromes of thyroid hormone resistance. J Clin Endocrinol Metab 1994;78:1034–1039.
23. Hauser P, Zametkin AJ, Martinez P, Vitiello B, Matochik JA, Mixson AJ, et al. Attention deficit-hyperactivity disorder in people with generalized resistance to thyroid hormone. N Engl J Med 1993; 328:997–1001.
24. Mixson AJ, Parrilla R, Ransom SC, Wiggs EA, McClaskey JH, Hauser P, et al. Correlation of language abnormalities with localization of mutations in the β-thyroid hormone receptor in 13 kindreds with generalized resistance to thyroid hormone: identification of four new mutations. J Clin Endocrinol Metab 1992;75:1039–1045.
25. Weiss RE, Stein MA, Duck SC, Chyna B, Phillips W, O'Brien T, et al. Low intelligence but not attention deficit hyperactivity disorder is associated with resistance to thyroid hormone caused by mutation R316H in the thyroid hormone receptor β gene. J Clin Endocrinol Metab 1994;78:1525–1528.
26. Stein MA, Weiss RE, Refetoff S. Neurocognitive characteristics of individuals with resistance to thyroid hormone: comparisons with individuals with attention-deficit hyperactivity disorder. J Dev Behav Pediatr 1995;16:406–411.
27. Weiss RE, Stein MA, Trommer B, Refetoff S. Attention-deficit hyperactivity disorder and thyroid function. J Paediatr 1993;123:539–545.
28. Valentine J, Rossi E, O'Leary P, Parry TS, Kurinczuk JJ, Sly P. Thyroid function in a population of children with attention deficit hyperactivity disorder. J Paediatr Child Health 1997;33:117–120.
29. Leonard CM, Martinez P, Weintraub BD, Hauser P. Magnetic resonance imaging of cerebral anomalies in subjects with resistance to thyroid hormone. Am J Med Genet 1995;60:238–243.
30. Parrilla R, Mixson AJ, McPherson JA, McClaskey JH, Weintraub BD. Characterization of seven novel mutations of the c-erbAβ gene in unrelated kindreds with generalized thyroid hormone resistance: evidence for two "hot spot" regions of the ligand binding domain. J Clin Invest 1991;88:2123–2130.
31. Adams M, Matthews C, Collingwood TN, Tone Y, Beck-Peccoz P, Chatterjee VKK. Genetic analysis of 29 kindreds with generalized and pituitary resistance to thyroid hormone. J Clin Invest 1994;94: 506–515.
31a. Beck-Peccoz P, Chatterjee VKK, Chin WW, DeGroot LJ, Jameson JL, Nakamura H, et al. Nomenclature of thyroid hormone receptor β gene mutations in resistance to thyroid hormone: consensus statement from the First Workshop on Thyroid Hormone Resistance, July 10–11, 1993, Cambridge, United Kingdom. J Clin Endocrinol Metab 1994;78:990–993.
32. Meier CA, Dickstein BM, Ashizawa K, McClaskey JH, Muchmore P, Ransom SC, et al. Variable transcriptional activity and ligand binding of mutant β1 3,5,3'-triiodothyronine receptors from four families with generalised resistance to thyroid hormone. Mol Endocrinol 1992;6:248–258.
33. Collingwood TN, Adams M, Tone Y, Chatterjee VKK. Spectrum of transcriptional, dimerization and dominant negative properties of twenty different mutant thyroid hormone β-receptors in thyroid hormone resistance syndrome. Mol Endocrinol 1994;8:1262–1277.
34. Takeda K, Sakurai A, De Groot LJ, Refetoff S. Recessive inheritance of thyroid hormone resistance caused by complete deletion of the protein-coding region of the thyroid hormone receptor-β gene. J Clin Endocrinol Metab 1992;74:49–55.
35. Sakurai A, Miyamoto T, Refetoff S, DeGroot LJ. Dominant negative transcriptional regulation by a mutant thyroid hormone receptor β in a family with generalised resistance to thyroid hormone. Mol Endocrinol 1990;4: 1988–1994.
36. Chatterjee VKK, Nagaya T, Madison LD, Datta S, Rentoumis A, Jameson JL. Thyroid hormone resistance syndrome: inhibition of normal receptor function by mutant thyroid hormone receptors. J Clin Invest 1991;87:1977–1984.
37. Ono S, Schwartz ID, Mueller OT, Root AW, Usala SJ, Bercu BB. Homozygosity for a dominant negative thyroid hormone receptor gene responsible for generalized thyroid hormone resistance. J Clin Endocrinol Metab 1991;73:990–994.
38. Usala SJ, Menke JB, Watson TL, Wondisford FE, Weintraub BD, Brard J, et al. A homozygous deletion in the c-erbAβ thyroid hormone receptor gene in a patient with generalised thyroid hormone resistance: isolation and characterization of the mutant receptor. Mol Endocrinol 1991;5: 327–335
39. Franklyn JA. Syndromes of thyroid hormone resistance. Clin Endocrinol (Oxford) 1991;34: 237–245.

37. Hone J, Accili D, Psiachou H, Zadeh JA, Mitton S, Wertheimer E, et al. Homozygosity for a null allele of the insulin receptor gene in a patient with leprechaunism. Hum Mutat 1995;6:17–22.
38. Kadowaki T, Kadowaki H, Taylor SI. A nonsense mutation causing decreased levels of insulin receptor mRNA: detection by a simplified technique for direct sequencing of genomic DNA amplified by the polymerase chain reaction. Proc Natl Acad Sci USA 1990;87:658–62.
39. Imano E, Kadowaki H, Kadowaki T, Iwama N, Watarai T, Kawamori R, et al. Two patients with insulin resistance due to decreased levels of insulin-receptor mRNA. Diabetes 1991;40:548–557.
40. Kadowaki T, Kadowaki H, Accili D, Taylor SI. Substitution of lysine for asparagine at position 15 in the alpha-subunit of the human insulin receptor. A mutation that impairs transport of receptors to the cell surface and decreases the affinity of insulin binding. J Biol Chem 1990;265:19,143–19,150.
41. Accili D, Mosthaf L, Ullrich A, Taylor SI. A mutation in the extracellular domain of the insulin receptor impairs the ability of insulin to stimulate receptor autophosphorylation. J Biol Chem 1991;66:434–439.
42. Taouis M, Levy-Toledano R, Roach P, Taylor SI, Gorden P. Structural basis by which a recessive mutation in the alpha-subunit of the insulin receptor affects insulin binding. J Biol Chem 1994;269: 14,912–14,918.
43. Taouis M, Levy-Toledano R, Roach P, Taylor SI, Gorden P. Rescue and activation of binding-deficient insulin receptor: evidence for intermolecular transphosphorylation. J Biol Chem 1994;269:27,762–27,766.
44. Yoshimasa Y, Seino S, Whittaker J, Kakehi T, Kosaki A, Kuzuya H, et al. Insulin-resistant diabetes due to a point mutation that prevents insulin proreceptor processing. Science 1988;240:784–787.
45. Yoshimasa Y, Paul JI, Whittaker J, Steiner DF. Effects of amino acid replacements within the tetrabasic cleavage site on the processing of the human insulin receptor precursor expressed in Chinese hamster ovary cells. J Biol Chem 1990;265:17,230–17,237.
46. Treadway JL, Morrison BD, Soos MA, Siddle K, Olefsky J, Ullrich A, et al. Transdominant inhibition of tyrosine kinase activity in mutant insulin/insulin-like growth factor I hybrid receptors. Proc Natl Acad Sci USA 1991;88:214–218.
47. Frattali AL, Treadway JL, Pessin JE. Transmembrane signaling by the human insulin receptor kinase. Relationship between intramolecular beta subunit trans- and cis-autophosphorylation and substrate kinase activation. J Biol Chem 1992;267:19,521–19,528.
48. Levy-Toledano R, Caro LHP, Accili D, Taylor SI. Investigations of the mechanism of the dominant negative effect of mutations in the tyrosine kinase domain of the insulin receptor. EMBO J 1994;13: 835–842.
49. Taylor SI, Roth J, Blizzard RM, Elders MJ. Qualitative abnormalities in insulin binding in a patient with extreme insulin resistance: Decreased sensitivity to alterations in temperature and pH. Proc Natl Acad Sci USA 1981;78:7157–7161.
50. Kadowaki T, Bevins CL, Cama A, Ojamaa K, Marcus Samuels B, Kadowaki H, et al. Two mutant alleles of the insulin receptor gene in a patient with extreme insulin resistance. Science 1988;240:787–790.
51. Kadowaki H, Kadowaki T, Cama A, Marcus Samuels B, Rovira A, Bevins CL, et al. Mutagenesis of lysine 460 in the human insulin receptor. Effects upon receptor recycling and cooperative interactions among binding sites. J Biol Chem 1990;265:21,285–21,296.
52. Cama A, Sierra ML, Kadowaki T, Kadowaki H, Quon MJ, Rüdiger HW, et al. Two mutant alleles of the insulin receptor gene in a family with a genetic form of insulin resistance: a 10 base pair deletion in exon 1 and a mutation substituting serine for asparagine-462. Hum Genet 1995;95:174–182.
53. Sun XJ, Wang LM, Zhang Y, Yenush L, Myers MGJ, Glasheen E, et al. Role of IRS-2 in insulin and cytokine signalling. Nature 1995;377:173–177.
54. Lavan BE, Lane WS, Lienhard GE. The 60-kDa phosphotyrosine protein in insulin-treated adipocytes is a new member of the insulin receptor substrate family. J Biol Chem 1997;272:11,439–11,443.
55. Lavan BE, Fantin VR, Chang ET, Lane WS, Keller SR, Lienhard GE. A novel 160-kDa phosphotyrosine protein in insulin-treated embryonic kidney cells is a new member of the insulin receptor substrate family. J Biol Chem 1997;272:21,403–21,407.
56. Sciacchitano S, Taylor SI. Cloning, tissue expression, chromosomal localization of the mouse IRS-3 gene. Endocrinology 1997;138:4931–4940.
57. White MF, Livingston JN, Backer JM, Lauris V, Dull TJ, Ullrich A, et al. Mutation of the insulin receptor at tyrosine 960 inhibits signal transmission but does not affect its tyrosine kinase activity. Cell 1988;54:641–649.
58. Tamemoto H, Kadowaki T, Tobe K, Yagi T, Sakura H, Hayakawa T, et al. Insulin resistance and growth retardation in mice lacking insulin receptor substrate-1. Nature 1994;372:182–186.
59. Araki E, Lipes MA, Patti ME, Bruning JC, Haag BI, Johnson RS, et al. Alternative pathway of insulin signaling in mice with targeted disruption of the IRS-1 gene. Nature 1994;372:186–191.

13. Cheatham B, Kahn CR. Insulin action and the insulin signaling network. Endocr Rev 1995;16:117–142.
14. Taylor SI, Cama A, Accili D, Barbetti F, Quon MJ, Sierra ML, et al. Mutations in the insulin receptor gene. Endocr Rev 1992;13:566–595.
15. Taylor SI, Wertheimer E, Accili D, Cama A, Hone J, Roach P, et al. Mutations in the insulin receptor gene: Update 1994. In: Underwood LE, ed. Endocrine Reviews Monographs. 2. The Endocrine Pancreas, Insulin Action, Diabetes. The Endocrine Society, Bethesda, 1994, pp. 58–65.
16. Hubbard SR, Wei L, Ellis L, Hendrickson WA. Crystal structure of the tyrosine kinase domain of the human insulin receptor. Nature 1994;372:746–754.
17. Hubbard SR. Crystal structure of the activated insulin receptor tyrosine kinase in complex with peptide substrate and ATP analog. EMBO J 1997;16:5572–5581.
18. Gustafson TA, He W, Craparo A, Schaub CD, O'Neill TJ. Phosphotyrosine-dependent interaction of SHC and insulin receptor substrate-1 with the NPEY motif of the insulin receptor via a novel non-SH2 domain. Mol Cell Biol 1995;15:2500–2508.
19. Kavanaugh MW, Turck CW, Williams LT. PTB domain binding to signaling proteins through a sequence motif containing phosphotyrosine. Science 1995;268:1177–1179.
20. Isakoff SJ, Yu Y-P, Su Y-C, Blaikie P, Yajnik V, Rose E, et al. Interaction between the phosphotyrosine binding domain of Shc and the insulin receptor is required for Shc phosphorylation by insulin in vivo. J Biol Chem 1996;271:3959–3962.
21. Levy-Toledano R, Taouis M, Blaettler DH, Gorden P, Taylor SI. Insulin-induced activation of phosphatidyl inositol 3-kinase: Demonstration that the p85 subunit binds directly to the COOH-terminus of the insulin receptor in intact cells. J Biol Chem 1994;269:31,178–31,182.
22. Sun XJ, Rothenberg P, Kahn CR, Backer JM, Araki E, Wilden PA, et al. Structure of the insulin receptor substrate IRS-1 defines a unique signal transduction protein. Nature 1991;352:73–77.
23. Taylor SI. Lilly Lecture: molecular mechanisms of insulin resistance. Lessons from patients with mutations in the insulin-receptor gene. Diabetes 1992;41:1473–1490.
24. Brown MS, Goldstein JL. A receptor-mediated pathway for cholesterol homeostasis. Science 1986;232:34–47.
25. Donohue WL, Uchida I. Leprechaunism: a euphuism for a rare familial disorder. J Pediatr 1954;45:505–519.
26. Kahn CR, Flier JS, Bar RS, Archer JA, Gorden P, Martin MM, et al. The syndromes of insulin resistance and acanthosis nigricans. Insulin-receptor disorders in man. N Engl J Med 1976;294:739–745.
27. Accili D, Frapier C, Mosthaf L, McKeon C, Elbein SC, Permutt MA, et al. A mutation in the insulin receptor gene that impairs transport of the receptor to the plasma membrane and causes insulin-resistant diabetes. EMBO J 1989;8:2509–2517.
28. Kadowaki T, Kadowaki H, Rechler MM, Serrano Rios M, Roth J, Gorden P, et al. Five mutant alleles of the insulin receptor gene in patients with genetic forms of insulin resistance. J Clin Invest 1990;86:254–264.
29. Cama A, Sierra ML, Ottini L, Kadowaki T, Gorden P, Imperato McGinley J, et al. A mutation in the tyrosine kinase domain of the insulin receptor associated with insulin resistance in an obese woman. J Clin Endocrinol Metab. 1991;73:894–901.
30. Cama A, Quon MJ, Sierra ML, Taylor SI. Substitution of isoleucine for methionine at position 1153 in the beta-subunit of the human insulin receptor. A mutation that impairs receptor tyrosine kinase activity, receptor endocytosis, insulin action. J Biol Chem 1992;267:8383–8389.
31. Cama A, Sierra ML, Quon MJ, Ottini L, Gorden P, Taylor SI. Substitution of glutamic acid for alanine-1135 in the putative "catalytic loop" of the tyrosine kinase domain of the human insulin receptor: A mutation that impairs proteolytic processing into subunits and inhibits receptor tyrosine kinase activity. J Biol Chem 1993;268:8060–8069.
32. Odawara M, Kadowaki T, Yamamoto R, Shibasaki Y, Tobe K, Accili D, et al. Human diatetes associated with a mutation in the tyrosine kinase domain of the insulin receptor. Science 1989;245:66–68.
33. Barnes ND, Palumbo PJ, Hayles AB, Folgar H. Insulin resistance, skin changes, and virilization: a recessively inherited syndrome possibly due to pineal gland dysfunction. Diabetologia 1974;10:285–289.
34. Rabson SM, Mendenhall EN. Familial hypertrophy of pineal body, hyperplasia of adrenal cortex and diabetes mellitus. Am J Clin Pathol 1956;26:283–290.
35. Roach P, Zick Y, Accili D, Taylor SI, Gorden P. A novel human insulin receptor gene mutation uniquely inhibits insulin binding without impairing post-translational processing. Diabetes 1994;43:1096–1102.
36. Wertheimer E, Lu SP, Backeljauw PF, Davenport ML, Taylor SI. Homozygous deletion of the human insulin receptor gene. Nature Genet 1993;5:71–73.

Because of the simple Mendelian inheritance, it is likely that positional cloning will enable the identification of the genes causing this family of syndromes.

CONCLUSION

Type 2 diabetes mellitus is characterized by two physiological defects: insulin deficiency and insulin resistance. The insulin deficiency is generally not as severe as observed in type 1 diabetes mellitus. Nevertheless, the combination of insulin deficiency with impaired insulin action is sufficient to cause diabetes and overt hyperglycemia. Although there is considerable evidence suggesting that genetic factors contribute importantly to determining the risk to develop type 2 diabetes, the pattern of inheritance is complex, suggesting polygenic inheritance. Thus far, genome-wide scans have not led to the identification of major genes that cause type 2 diabetes. On the other hand, there has been greater success in identifying genes that cause unusual variants of noninsulin-dependent diabetes that are transmitted with Mendelian inheritance. We have emphasized the role of mutations in the insulin-receptor gene that cause genetic syndromes of insulin-resistant diabetes mellitus (e.g., leprechaunism, Rabson-Mendenhall syndrome, and type A insulin resistance). In addition, mutations have been identified in several genes that contribute to the ability of the pancreatic β-cell to secrete insulin appropriately in response to the usual regulatory influences. For example, mutations in the genes encoding glucokinase or various transcription factors (i.e., HNF-1α, HNF-4α, and IPF-1) lead to impaired insulin secretion in various forms of maturity-onset diabetes of youth (MODY). In recent years, there has been rapid progress in our understanding of the molecular mechanisms involved in regulating insulin secretion and mediating insulin action. It is likely that understanding these aspects of basic science will lead to a better understanding of the causes of type 2 diabetes mellitus.

REFERENCES

1. Taylor SI. Diabetes mellitus. In: Scriver CR, Beaudet AL, Sly WS, Valle D, eds. The Metabolic and Molecular Bases of Inherited Disease. McGraw-Hill, New York, 1995, pp. 843–896.
2. Taylor SI, Accili D, Imai Y. Insulin resistance and insulin deficiency: which is the primary cause of NIDDM? Diabetes 1994;43:735–740.
3. Steiner DF, Tager HS, Chan SJ, Nanjo K, Sanke T, Rubenstein AH. Lessons learned from molecular biology of insulin-gene mutations. Diabetes Care 1990;13:600–609.
4. Matschinsky FM. Banting Lecture 1995. A lesson in metabolic regulation inspired by the glucokinase glucose sensor paradigm. Diabetes 1996;45:223–241.
5. Bell GI, Pilkis SJ, Weber IT, Polonsky KS. Glucokinase mutations, insulin secretion, and diabetes mellitus. Annu Rev Physiol 1996;58:171–186.
6. Permutt MA, Chiu KC, Tanizawa Y. Glucokinase and NIDDM. A candidate gene that paid off. Diabetes 1992;41:1367–1372.
7. Fajans SS, Bell GI, Bowden DW, Halter J, Polonsky KS. Maturity onset diabetes of the young (MODY). Diabet Med 1996;13(Suppl 6):S90–S95.
8. Froguel P, Zouali H, Vionnet N, Velho G, Vaxillaire M, Sun F, et al. Familial hyperglycemia due to mutations in glucokinase. Definition of a subtype of diabetes mellitus. N Engl J Med 1993;328:697–702.
9. Yamagata K, Furuta H, Oda N, Kaisaki PJ, Menzel S, Cox NJ, et al. Mutations in the hepatocyte nuclear factor-4alpha gene in maturity-onset diabetes of the young (MODY1). Nature 1996;384:458–460.
10. Yamagata K, Oda N, Kaisaki PJ, Menzel S, Furuta H, Vaxillaire M, et al. Mutations in the hepatocyte nuclear factor-1alpha gene in maturity-onset diabetes of the young (MODY3). Nature 1996;384:455–458.
11. Stoffers DA, Ferrer J, Clarke WL, Habener JF. Early-onset type-II diabetes mellitus (MODY4) linked to IPF1. Nature Genet 1997;17:138–139.
12. Quon MJ, Butte AJ, Taylor SI. Insulin signal transduction pathways. Trends Endocrinol Metab 1994;5:369–376.

Table 1
Selected Clinical Syndromes Associated with Lipoatrophy and Insulin Resistance

Genetic syndromes
  Seip-Berardinelli syndrome *(83,84)*
    Generalized lipoatrophy with onset at birth
    Autosomal-recessive pattern of inheritance
    Associated features: hepatomegaly (occasionally cirrhosis of liver),
      increased basal metabolic rate
  Kobberling-Dunnigan syndrome *(85,86)*
    Lipoatrophy of trunk and extremities, usually sparing head and neck;
      onset usually before the age of puberty
    Autosomal-dominant pattern of inheritance
Autoimmune syndromes
  Lawrence syndrome *(87)*
    Generalized lipoatrophy with onset generally during childhood
    Generally associated with overt autoimmune disease (e.g, dermatomyositis)
    Associated features: hepatomegaly (occasionally cirrhosis of liver),
      increased basal metabolic rate
  Barraquer-Simons syndrome *(88)*
    Lipoatrophy of face and upper body, usually sparing lower body;
      onset usually in adulthood
    Generally associated with overt autoimmune disease

biological actions of insulin. For example, considerable effort has been devoted to investigations of the genes encoding GLUT4 (the predominant glucose transporter in insulin-sensitive tissues, such as muscle and adipose tissue) and the enzymes related to glycogen metabolism (e.g., glycogen synthase). Complete discussion of this area of research is beyond the scope of this chapter. Suffice it to say that mutations in these genes do not appear to be a major cause contributing to the development of type 2 diabetes mellitus.

Other Syndromes Associated with Insulin Resistance. This chapter has emphasized syndromes resulting from mutations in the insulin receptor gene. Indeed, the insulin receptor gene is the only gene in which mutations have been unambiguously identified as the cause of insulin resistance in human disease. Although the molecular genetics of the common form of type 2 diabetes mellitus has also been discussed, the precise molecular causes of insulin resistance are not well understood in this complex polygenic disorder. Extensive discussion of other syndromes associated with severe insulin resistance is beyond the scope of this chapter. Nevertheless, it is important to mention the syndromes of lipodystrophy and lipoatrophic diabetes (Table 1). These are a collection of clinical syndromes that share several features in common: insulin resistance, acanthosis nigricans, hyperandrogenism (in female patients), and hypertriglyceridemia. In addition, the disorders are characterized by lipoatrophy, which may be either generalized or localized to specific regions of the body. The onset of lipoatrophy may be congenital or acquired later in life. Although several forms of the syndrome are believed to be genetic, there may be an autoimmune pathogenesis in some patients. The genetic forms of lipoatrophy appear to be monogenic. For example, the congenital generalized form of the syndrome (Berardinelli-Seip syndrome) appears to be transmitted according to an autosomal-recessive pattern of inheritance. In contrast, inheritance of the face-sparing congenital form (Dunnigan syndrome) is consistent with an autosomal-dominant pattern.

Interestingly, all four members of the IRS family (i.e., IRS-1, 2, 3, and 4) contain multiple potential tyrosine phosphorylation sites that are predicted to activate PI 3-kinase. In addition, it has been demonstrated directly that treatment of target cells with insulin leads to the association of activated PI 3-kinase with IRS-1, 2, and 3.

Activation of PI 3-kinase leads to phosphorylation of PI, a phospholipid that is located in membranes. Apparently, the products of the PI 3-kinase retain their membrane association where they bind to various intracellular proteins, possibly via pleckstrin homology domains. In any case, activation of PI 3-kinase leads to the activation of a cascade of protein kinases. The product of PI 3-kinase activates phosphoinositide-dependent kinase (PDK), which, in turn, phosphorylates protein kinase B (PKB, also known as Akt). PKB/Akt initiates a pathway that leads to the translocation of GLUT4 to the plasma membrane, one of the central metabolic actions of insulin *(76,77)*. In addition, PKB/Akt is known to phosphorylate glycogen synthase kinase- (GSK) 3, thereby leading to the inactivation of GSK3 and the activation of glycogen synthase, another of the important metabolic actions of insulin *(78)*. In addition to PKB/Akt, there are other protein kinases that appear to be downstream from PI 3-kinase, e.g., protein kinase C-zeta (PKC-ζ, another protein kinase that has been reported to trigger the translocation of GLUT4 to the plasma membrane) *(79)*.

Because of the importance of these pathways in mediating the metabolic actions of insulin, the genes encoding some of these proteins have been screened for genetic variation. For example, a polymorphism substituting Ile for Met^{326} was identified in the gene encoding the p85α regulatory subunit of PI 3-kinase *(80)*. Although the allelic frequency for the M326I substitution was normal (approx 0.15) in the diabetic population, homozygosity for this polymorphism (present in ≈2% of the population) was reported to be associated with a 40% decrease in the rate at which glucose was cleared from the plasma in an iv glucose tolerance test. In contrast, genetic variation was not identified in the genes encoding several protein kinases downstream from PI 3-kinase (e.g., protein kinase B-α and β; glycogen synthase kinase 3α and 3β) *(81)*.

ACTIVATION OF RAS AND MITOGEN-ACTIVATED PROTEIN (MAP) KINASE

GRB2 is an adapter protein that contains one SH2 and two SH3 domains *(12)*. When the SH2 domain of GRB2 binds to phosphotyrosine residues, this leads to the activation of m-Sos (mammalian homolog of Drosophila son-of-sevenless). m-Sos, a protein that binds to the SH3 domain of GRB2, is a guanine nucleotide exchange factor that accelerates the binding of GTP to Ras (and also the concomitant dissociation of GDP from Ras). When GTP is bound to Ras, this results in activation of Ras as well as activation of a cascade of protein kinases, including Raf, MEK, and MAP kinase. A large body of evidence has demonstrated that Ras and the MAP kinase pathway do not mediate the metabolic actions of insulin *(82)*. Nevertheless, this pathway may contribute importantly to the ability of insulin (and the closely related insulin-like growth factors) to promote cell growth and cell division.

PROTEINS FURTHER DOWNSTREAM IN THE INSULIN SIGNALING PATHWAY

This chapter has focused primarily on the proteins that function in the early steps of the insulin action pathway because this portion of the pathway, is common to most (if not all) of the metabolic actions of insulin. However, it is conceivable that there might also be mutations in proteins with specific functions that relate to a limited number of the subjects *(39,70)*. However, other studies have not confirmed the presence of this genetic association *(71,72)*. Although the cause of the contradictory data is not clear, it is possible that the differences may be explained in part by racial or ethnic differences among the populations that were studied. Furthermore, Clausen et al. have proposed that there is a complex interaction between obesity and genetic variation in the IRS-1 gene *(73)*. According to their study, the G972R substitution only predisposes to insulin resistance in obese individuals, not in normal weight individuals.

The epidemiologic investigation of the IRS-1 gene has been supplemented by a biochemical approach. Variant forms of the IRS-1 protein have been expressed by transfection of mutant cDNA in mammalian cells. This has made it possible to investigate the impact of amino acids substitutions on the biological function of IRS-1. Unfortunately, this area is also controversial. For example, Almind et al. *(74)* have reported that the G972R substitution caused a 30–50% decrease in the ability of IRS-1 to activate PI 3-kinase. In contrast, Imai et al. *(69)* did not detect any function of G972R-IRS-1. In any case, even if it is true that there is a 30–50% defect in the function of G972R-IRS-1, heterozygosity for this variant sequence would lead to a 15–25% decrease in the total activity of IRS-1. Would such a small defect would be physiologically significant? Animal studies have been designed in an effort to address this question. "Knockout" mice have been generated by targeted inactivation of the IRS-1 gene *(58,59)*. Homozygous knockout mice lack IRS-1, and exhibit marked growth retardation and mild insulin resistance. In contrast, heterozygous knockout mice (with a 50% defect in IRS-1) did not exhibit an obvious phenotype. Nevertheless, when IRS-1 knockout mice were bred with insulin receptor knockout mice *(75)*, it was possible to obtain mice that were heterozygous for null alleles at both loci. These double heterozygotes exhibited a marked degree of insulin resistance and had a signficant probability of developing overt diabetes mellitus. This study demonstrated that mild genetic defects at two loci (in this case the genes encoding IRS-1 and the insulin receptor) can exert additive (or possibly synergistic) effects to predispose animals to develop diabetes mellitus. Thus, this study adds credibility to the hypothesis that even a mild genetic defect in IRS-1 can contribute to the polygenic inheritance of type 2 diabetes mellitus when combined with high-risk alleles at other genetic loci.

Downstream Effectors in Insulin Action

Insulin is a pleiotropic hormone that elicits a large number of responses in a wide variety of target cells. Accordingly, there is a bewildering number of biochemical processes that must be elucidated to provide a complete description of the mechanism of insulin action. For example, it is apparent that IRS proteins interact with a large number of SH2 domain-containing proteins. We will focus on the two best-characterized pathways, i.e., the pathways involving PI 3-kinase and Ras.

THE PI 3-KINASE PATHWAY

PI 3-kinase consists of two distinct subunits: p85, an M_r 85,000 regulatory subunit that contains two SH2 domains, and p110, an M_r 110,000 catalytic subunit *(12)*. Both SH2 domains bind phosphotyrosine residues in the context of pTyr-X-X-Met or pTyr-Met-X-Met sequences. Binding of phosphotyrosine-containing proteins to the SH2 domains of p85 leads to activation of the catalytic activity of p110 *(22)*. Maximal activity of the PI 3-kinase is achieved when binding occurs simultaneously to both SH2 domains of p85.

p85 subunit of PI 3-kinase). In contrast, there are several distinctive tyrosine phosphorylation sites that are not shared by all of the IRS proteins (e.g., binding sites for growth factor receptor binding protein-2, GRB2).

With many other receptor tyrosine kinases, a closely related mechanism determines the specificity for phosphorylation of intracellular proteins. For example, the SH2 domains of phospholipase Cγ and Shc bind to phosphotyrosine residues in the receptor for platelet-derived growth factor (PDGF). The ability to bind to the phosphorylated PDGF receptor enables the receptor to phosphorylate these substrates. Although it is likely that there may be substrates that interact through their SH2 domains with the insulin receptor, this has not yet been reported.

Integral Membrane Proteins That Are Phosphorylated by the Insulin Receptor. There is a family of M_r 120,000 integral membrane glycoproteins that are phosphorylated by the insulin receptor. These proteins share several structural features in common: extracellular domains containing "immunoglobulin-like loops," a single-transmembrane domain, and short intracellular domains containing sites for tyrosine phosphorylation. It is likely that their location in the plasma membrane increases the "local concentration" of the substrate in the vicinity of the insulin receptor (another integral plasma membrane protein), thereby nonspecifically promoting a binding interaction between the two proteins. Of course, it is possible that there may also be sites of protein–protein interaction that specifically favor binding of the insulin receptor to this class of substrates.

pp120/HA4 was the first member of this family to be identified as a substrate for the insulin receptor *(63–65)*. Based on the sequence flanking the principal tyrosine phosphorylation site in pp120/HA4, Najjar et al. *(65)* predicted that the protein would bind to the SH2 domain of the SH2-containing phosphotyrosine phosphatases (SHP). Subsequently, two other laboratories identified two homologous glycoproteins (SHP substrate-1, SHPS-1 *[66]*) and signal-regulatory protein (SIRP) *(67)* that were phosphorylated by the insulin receptor and other tyrosine kinases. Furthermore, the phosphorylated proteins did indeed bind to SHP-1 and SHP-2, and served as substrates for these two phosphotyrosine phosphatases. Moreover, SIRP was demonstrated to exert an inhibitory effect on signaling through receptor tyrosine kinases.

MUTATIONS IN THE INSULIN RECEPTOR SUBSTRATE-1 (IRS-1) GENE

As discussed above (*see* Substrates Containing Phosphotyrosine Binding [PTB] Domains), IRS-1 is one of several intracellular proteins phosphorylated by the insulin receptor tyrosine kinase. After tyrosine residues in the molecule become phosphorylated, IRS-1 binds SH2 domains in several proteins, thereby triggering various signaling pathways that mediate many of the biological actions of insulin in the target cell. Because of its role in mediating insulin action, mutations in the IRS-1 gene have the potential to cause insulin resistance, thereby predisposing to the development of type 2 diabetes mellitus. Based on this hypothesis, several laboratories have screened for mutations in the IRS-1 gene *(68–72)*. Several variant forms of IRS-1 have been identified in patients with type 2 diabetes, but the pathological significance of these variant sequences is controversial. Some studies have reported an increase in the prevalence of variant forms of IRS-1 in patients with type 2 diabetes, thus suggesting a pathophysiologic role for the amino acid substitutions in IRS-1. For example, in some populations, the G972R substitution (the most common genetic variant in the IRS-1 gene) has been reported to be present in 10–20% of patients with type 2 diabetes mellitus, but only 5–10% of control

Class 5. Mutations that Retard Ligand Dissociation at Acidic pH and Accelerate Receptor Degradation

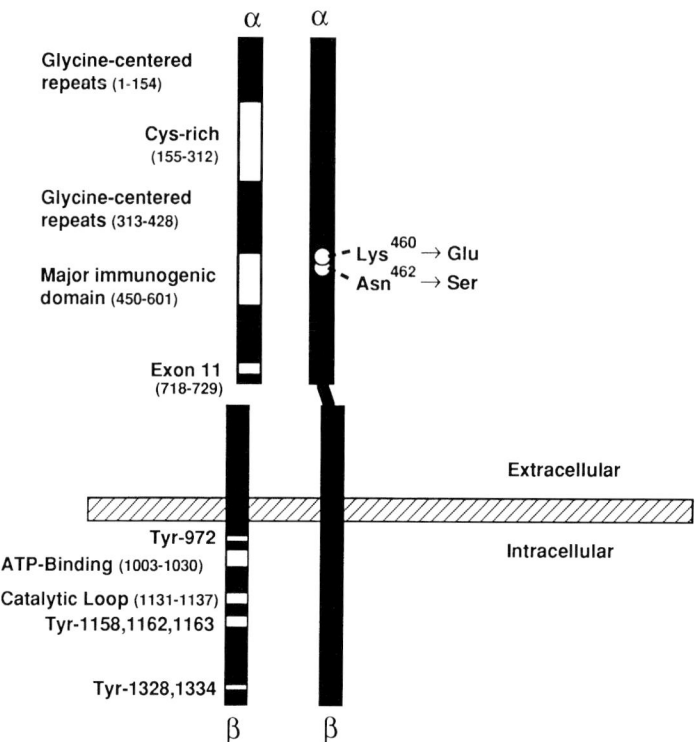

Fig. 7. Mutations that accelerate receptor degradation (class 5). Mutations that accelerate receptor degradation are illustrated as described in the legend to Fig. 3.

Substrates Containing Phosphotyrosine Binding (PTB) Domains. PTB domains were first identified in Shc, and subsequently identified in the "insulin receptor substrate" (IRS) family of proteins *(18–20,22,53–56)*. In both Shc and IRS proteins, the PTB domains bind to a motif containing phosphotyrosine (Asn-Pro-Xaa-pTyr). Substitution of Phe for Tyr^{972} in the Asn-Pro-Xaa-pTyr motif of the insulin receptor markedly impairs the ability of the receptor to phosphorylate IRS-1 *(57)*. This observation suggests that the ability of the PTB domain to bind to the phosphorylated receptor may be responsible for the fact that the insulin receptor can phosphorylate both Shc and IRS proteins. Several lines of evidence suggest that IRS proteins are important mediators of insulin action in vivo. For example, targeted inactivation of the IRS-1 by homologous recombination causes mice to become insulin-resistant *(58,59)*. Furthermore, overexpression of various IRS proteins mimics insulin action in adipose cells *(60,61)*. In addition, overexpression of a dominant negative mutant of IRS-3 inhibits the metabolic action of insulin in adipose cells *(62)*. There are at least four members of the family of IRS proteins *(22,53-56)*, of which at least three (IRS-1, 2, and 3) are expressed in adipose cells, a classical target cell for insulin action. Each of the IRS proteins contains multiple sites for tyrosine phosphorylation. All of the IRS proteins possess phosphotyrosine-containing motifs with the potential to bind SH2 domains of certain downstream effector molecules (most notably, the

Class 4. Mutations that Impair Receptor Tyrosine Kinase Activity

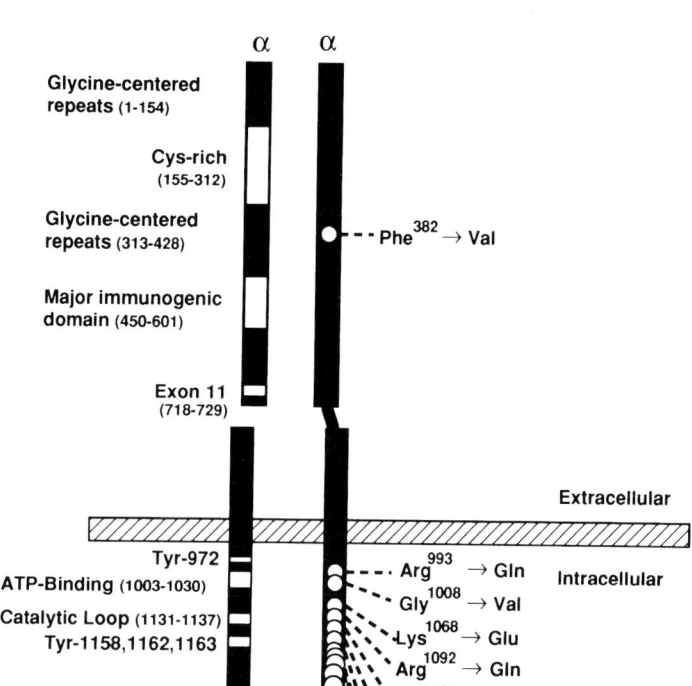

Fig. 6. Mutations that impair receptor tyrosine kinase activity (class 4). Mutations that impair receptor tyrosine kinase activity are illustrated as described in the legend to Fig. 3.

for degradation. The accelerated rate of receptor degradation leads to a decrease in the number of receptors on the cell surface, which in turn causes a decrease in the sensitivity of the dose-response curve for insulin action.

Substrates for Phosphorylation by the Insulin Receptor

ROLE OF TYROSINE PHOSPHORYLATION IN INSULIN ACTION

Activated insulin receptors phosphorylate many proteins within the cell. There appear to be at least two factors that determine which intracellular proteins are phosphorylated by the insulin receptor. First, it is necessary for the protein to bind to the receptor. Second, it is necessary that the protein must contain a tyrosine residue embedded in a sequence that conforms to the substrate specificity of the insulin receptor tyrosine kinase. We will discuss two classes of substrates for the insulin receptor, each of which illustrates a different mechanism to promote binding of the receptor to the substrate.

Class 3. Mutations that Decrease Affinity of Insulin Binding

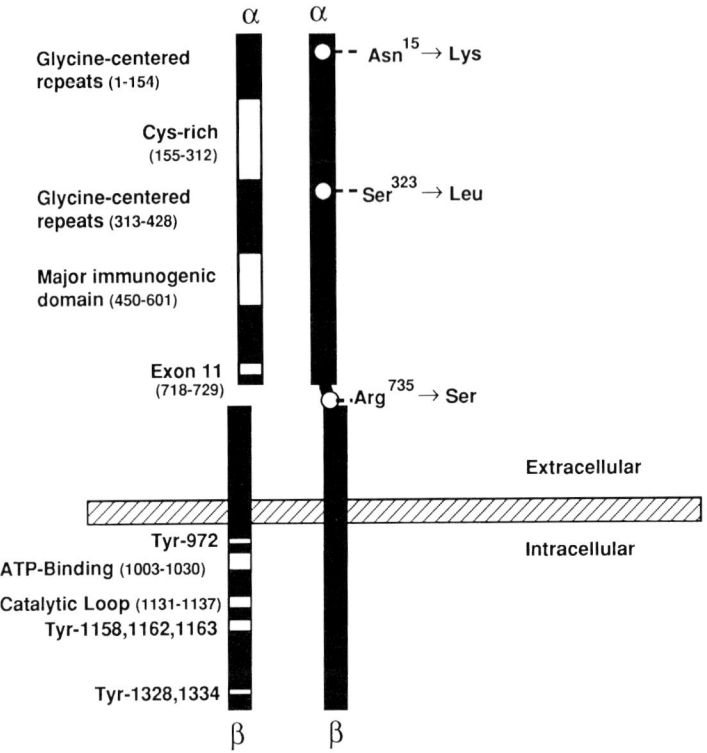

Fig. 5. Mutations that decrease the affinity of insulin binding (class 3). Mutations that decrease the affinity of insulin binding are illustrated as described in the legend to Fig. 3.

symmetrical mutant [$\alpha_2(\beta_{mut})_2$], in addition to a hybrid form ($\alpha_2\beta_{wt}\beta_{mut}$). The wild-type receptor [$\alpha_2(\beta_{wt})_2$] is active as a tyrosine kinase. However, the hybrid $\alpha_2\beta_{wt}\beta_{mut}$ and the mutant receptors [$\alpha_2(\beta_{mut})_2$] are both inactive as tyrosine kinase enzymes. Thus, a mutation in a single allele inactivates 75% of the receptors, thereby providing a plausible molecular explanation of the dominant pattern of inheritance observed with mutations in the tyrosine kinase domain *(46,47)*. However, this hypothesis remains controversial in that experimental efforts to demonstrate the formation of these hybrid receptors have not always been successful *(48)*.

Mutations That Accelerate Receptor Degradation (Class 5). Insulin binding triggers endocytosis of its receptor into endosomes. Internalized receptors are oriented with the insulin binding site located inside the endosome. The endosomal proton pump acidifies the lumen of the endosome (pH≈5.5), where the acidic pH promotes dissociation of insulin from its receptor. Two pathways are available to internalized receptors: recycling back to the plasma membrane for reutilization or degradation within the lysosome. At least two mutations (i.e., K460E and N462S) have been reported to impair the ability of acidic pH to dissociate insulin from the receptor (Fig. 7) *(49–52)*. In addition, these mutations appear to inhibit the pathway whereby receptors are recycled back to the plasma membrane. Consequently, receptors are preferentially targeted toward lysosomes

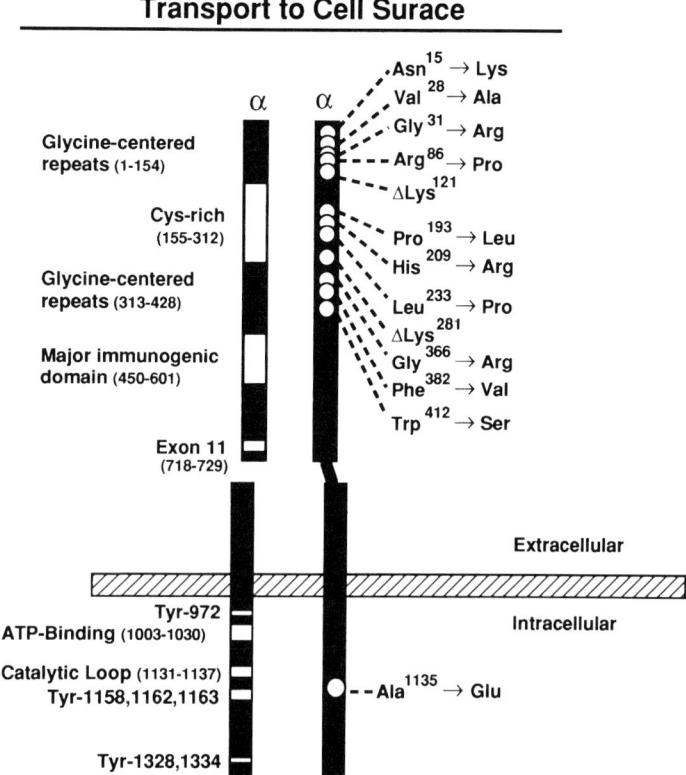

Fig. 4. Mutations that impair transport of receptors to the plasma membrane (class 2). Missense mutations that impair transport of receptors to the plasma membrane are illustrated as described in the legend to Fig. 3.

receptor. In addition, the R735S mutation has been reported to decrease the affinity of insulin binding. Interestingly, Arg^{735} is the last amino acid in the Arg-Lys-Arg-Arg motif at the proteolytic cleavage site between the α- and β-subunits. Thus, substitution of Ser for Arg^{735} also inhibits cleavage of the precursor into two subunits (44,45).

Mutations That Impair Tyrosine Kinase Activity (Class 4). Many mutations have been identified in the tyrosine kinase domain (Fig. 6). The observation that these mutations inhibit receptor tyrosine kinase activity provides strong support for the hypothesis that tyrosine kinase activity is required for the ability of the receptor to mediate the metabolic actions of insulin in humans in vivo. The G1008V mutation was among the first reported mutations in the tyrosine kinase domain (32). Valine is substituted for Gly^{1008}, the third glycine in the highly conserved Gly-X-Gly-X-X-Gly motif in the ATP binding site. This mutation inhibits tyrosine kinase activity—presumably by inhibiting the binding of ATP to the substrate binding site of the enzyme. Unlike most mutations in the extracellular domain of the insulin receptor, mutations in the tyrosine kinase domain appear to cause insulin resistance in a dominant fashion. It has been proposed that the dominant negative effect of mutations in the tyrosine kinase domain is a consequence of the receptor's oligomeric $α_2β_2$ structure. In a heterozygous individual, three distinct oligomeric forms of the receptor are predicted: symmetrical wild-type $[α_2(β_{wt})_2]$ and

Class 1. Premature Termination Mutations

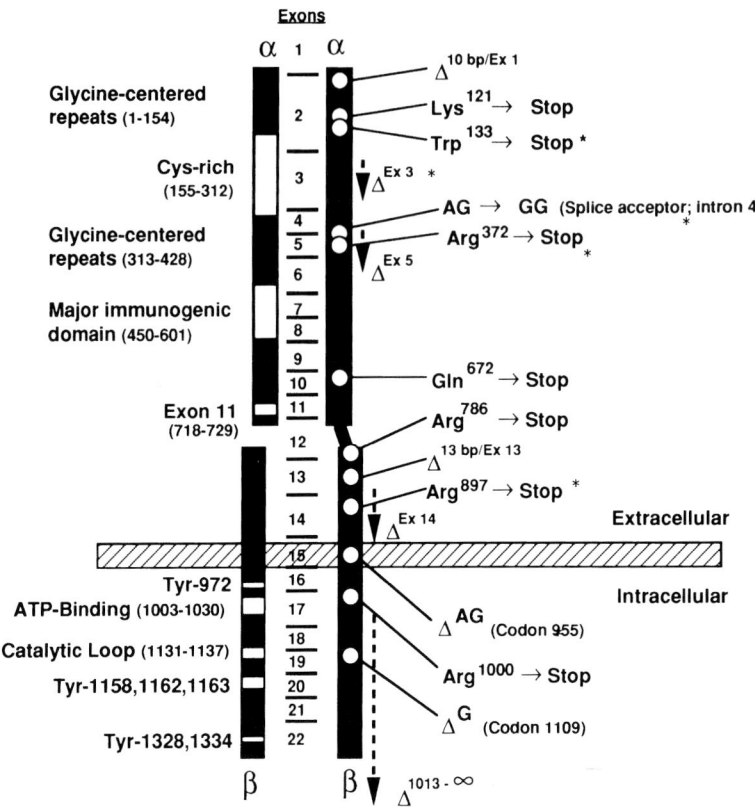

*Decreased mRNA levels

Fig. 3. Mutations that impair receptor biosynthesis (class 1). In this cartoon of the receptor, the structural landmarks are indicated on the left-hand side of the figure (cf. Fig. 1). Examples of class 1 mutations are indicated on the right side of the figure. These include nonsense mutations, as well as frame shifts and splicing mutations. (In the right half of the cartoon, the tetrapeptide connecting the two subunits has not been removed in order to illustrate the sequence of the uncleaved proreceptor.)

binding *(40)*, and the F382V mutation inhibited the ability of insulin binding to activate the receptor tyrosine kinase *(41)*. Most of the published mutations in this class are located in the N-terminal half of the α-subunit. However, the A1135E mutation (located in the cytoplasmic domain of the β-subunit) also inhibited posttranslational processing of the proreceptor and transport of the receptor to the plasma membrane *(31)*.

Mutations That Decrease the Affinity of Insulin Binding (Class 3). When the affinity of insulin binding is decreased, this decreases the quantity of insulin that is bound to the receptor at any given concentration of insulin in the plasma. Therefore, Class 3 mutations decrease the sensitivity with which target cells respond to insulin without causing a decrease in the maximal responsiveness. Two mutations located in the N-terminal half of the α-subunit (N15K and S323L) have been reported to decrease the affinity of insulin receptor binding (Fig. 5) *(35,40,42,43)*. These observations are consistent with the conclusion that the insulin binding domain maps to this region of the mutations in the insulin receptor gene who have insulin resistance in association with acanthosis nigricans, even though they do not have "hyperandrogenism." Furthermore, some obese insulin-resistant patients have been demonstrated to have mutations in one or both alleles of the insulin receptor gene. Thus, the presence of obesity does not totally rule out the possibility that a patient may also have type A insulin resistance. Nevertheless, obesity can predispose to a syndrome of hyperandrogenism, insulin resistance, and acanthosis nigricans (HAIR-AN) even in the absence of mutations in the insulin receptor gene.

Although some patients with type A insulin resistance have been reported to have two mutant alleles of the insulin receptor gene *(27,28)*, some patients (usually those with less severe insulin resistance) are heterozygous for a single mutant allele *(29–32)*. In the patients with two mutant alleles, overt diabetes with fasting hyperglycemia usually develops during childhood or adolescence *(26–28,33)*. In patients with one mutant allele of the insulin receptor gene, there is a variable degree of impairment in glucose tolerance. Some patients have either impaired glucose tolerance or overt diabetes, but glucose tolerance is normal in other individuals *(29,31,32)*.

Rabson-Mendenhall Syndrome. In addition to insulin resistance and acanthosis nigricans, the Rabson-Mendenhall syndrome (Fig. 2) is characterized by short stature, abnormalities of teeth and nails, and (at least in the original description of the syndrome) pineal hyperplasia *(34)*. This syndrome appears to be intermediate in clinical severity between leprechaunism and type A insulin resistance. The clearest distinction beween leprechaunism and the Rabson-Mendenhall syndrome is based on how long the children live. Patients with leprechaunism generally die within the first year or two of life, whereas patients with Rabson-Mendenhall syndrome generally survive at least until adolescence or adulthood. Nevertheless, the molecular defects in the two syndromes are similar. Like patients with leprechaunism, patients with the Rabson-Mendenhall syndrome have been reported to possess two mutant alleles of the insulin receptor gene *(28,35)*. However, unlike leprechaunism *(36,37)*, patients with the Rabson-Mendenhall syndrome have not been reported to have two null alleles of the insulin receptor gene.

Mutations That Impair Receptor Biosynthesis (Class 1). At least two types of mutations have been identified that inhibit receptor biosynthesis (Fig. 3). Premature chain-termination mutations (e.g., either nonsense or "frameshift" mutations) interfere with receptor biosynthesis, thereby leading to the synthesis of truncated receptor fragments. In addition, although the mechanism is not completely understood, most premature chain-termination mutations are associated with a *cis*-acting effect to decrease the level of mRNA transcribed from the mutant allele *(28,38)*. In other patients, the mutations that decrease the level of insulin receptor mRNA have been mapped outside of the protein coding regions of the gene *(38,39)*. It is likely that these mutations are located in regulatory domains of the gene and decrease the rate at which the gene is transcribed.

Mutations That Impair Transport of Receptors to the Cell Surface (Class 2). The insulin receptor precursor undergoes extensive posttranslational processing within the endoplasmic reticulum and Golgi: N- and O-linked glycosylation; formation of intra- and intersubunit disulfide bonds; and proteolytic cleavage into two subunits *(14,23)*. Some mutations inhibit the efficient transport of receptors through the endoplasmic reticulum and Golgi to the plasma membrane, thereby reducing the number of receptors on the cell surface (Fig. 4). In addition, presumably because these mutations prevent the receptor from folding into its normal conformation, these mutations sometimes impair receptor function. For example, the N15K mutation decreased the affinity of insulin

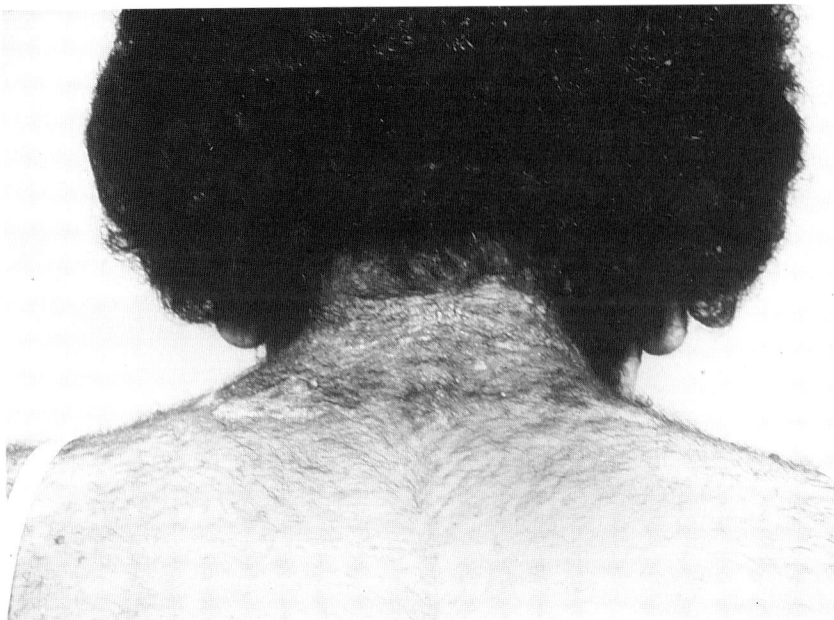

Fig. 2. Photograph of a patient with the syndrome of type A insulin resistance. This is a photograph of a young woman with type A insulin resistance. The syndrome is defined by the triad of insulin resistance, acanthosis nigricans, and hyperandrogenism. Acanthosis nigricans is illustrated by the hyperpigmented, hyperkeratotic skin lesion on the back of the patient's neck. The marked increase in body hair noted in the photograph results from the marked elevation in plasma testosterone (approx 1000 ng/dL in this patient) that resulted from ovarian overproduction of androgens.

Clinical Syndromes Owing to Mutations in the Insulin Receptor Gene. At least three distinct syndromes are caused by mutations in the insulin receptor gene. Two clinical features are common to all the syndromes: acanthosis nigricans and hyperandrogenism (Fig. 2) *(14,23)*. Because acanthosis nigricans and hyperandrogenism correlate with hyperinsulinemia, it has been proposed that they may be caused by "toxic" effects of insulin on the skin and the ovaries, respectively. Because of the impairment in the function of insulin receptors, it is likely that the "toxic" effects of insulin are not mediated by insulin receptors, but by receptors for homologous peptides (e.g., insulin-like growth factor-1). Although all of the syndromes of extreme insulin resistance share some features in common, multiple distinct syndromes can be defined based on the presence or absence of specific clinical features. In general, the severity of the clinical syndrome appears to be determined (at least in part) by the severity of the defect in receptor function caused by the mutation.

Leprechaunism. The most severe clinical syndrome is leprechaunism *(25)*. These patients have multiple abnormalities, including intrauterine growth retardation, fasting hypoglycemia, and glucose intolerance, despite having peak insulin levels that may be increased as much as 100-fold over the normal range. Death usually occurs within the first year of life. Patients with leprechaunism have mutations in both alleles of the insulin receptor gene; in fact, some of these patients are totally deficient in insulin receptors, because they have inherited two null alleles of the insulin receptor gene.

Type A Insulin Resistance. Type A insulin resistance was originally defined by the triad of insulin resistance, acanthosis nigricans, and hyperandrogenism in the absence of obesity or lipoatrophy *(26)*. The diagnosis has been generalized to include men with

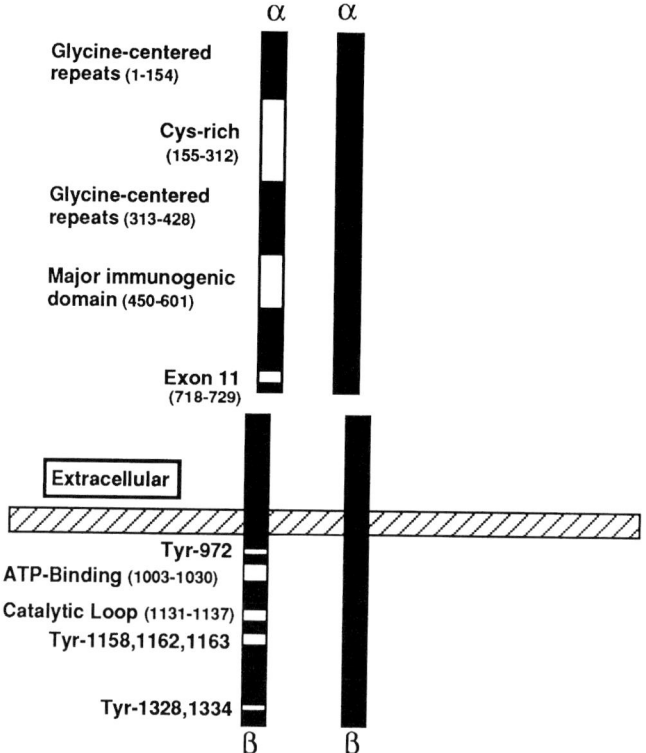

Fig. 1. Oligomeric structure of the insulin receptor. The insulin receptor consists of two α- and two β-subunits. They are derived from a single polypeptide precursor that undergoes extensive post-translational processing (including proteolytic cleavage into two subunits). The N-terminal domain (amino acids 1–154) of the insulin receptor plays an important role in insulin binding. This domain contains five repeats of a loosely conserved motif; the best conserved feature in the motif is a central glycine residue that has been predicted to allow for a "turn" in the polypeptide chain conformation. There are hydrophobic amino acids located at positions 2, 5, and 8 amino acid residues upstream from the glycine residue in most of the repeat motifs. These amino acids are predicted to fold into a short amphipathic α-helical structure with hydrophobic amino acids oriented on the same face of the helix. (Similar glycine-centered repeats are found in the middle of the α-subunit [aa 313–428].) The cysteine-rich domain, which contains 26 cysteine-residues (most of which are probably involved in intrasubunit disulfide bonds), contributes to the determination of of ligand binding specificity.

Insulin receptor mRNA undergoes variable splicing so that there are two isoforms that differ with respect to whether exon 11 has been included or excluded from the mature mRNA. In this chapter, we are using the numbering system corresponding to the isoform of the insulin receptor containing the 12 amino acid residues encoded by exon 11 and located at the C-terminus of the α-subunit *(13, 14)*. The tyrosine kinase domain contains the consensus sequence for ATP binding (Gly-X-Gly-X-X-Gly....Lys, amino acids 1003–1030). The catalytic loop contains several amino acid residues that participate in the catalyzing the phosphotransferase activity of the enzyme. This figure identifies six sites of tyrosine phosphorylation that are described in some detail in the text.

MUTATIONS IN THE INSULIN RECEPTOR GENE

Multiple different mutations have been identified in the insulin receptor genes of insulin-resistant patients. The mutations have been classified into five classes based on the mechanisms whereby the mutations impair insulin receptor function *(14,23)*. This is a modification of the classification scheme proposed by Brown and Goldstein to describe mutations in the low-density lipoprotein receptor gene *(24)*.

dominant trait with incomplete penetrance. This disease is one form of the syndrome "maturity-onset-type diabetes of the young" (i.e., MODY) *(7)*. There are at least four forms of MODY; the form owing to mutations in the glucokinase gene is referred to as MODY2. Natural history studies have demonstrated that MODY2 is a relatively mild form of diabetes, characterized by mild hyperglycemia, onset at an early age, and a relatively low probability of developing the chronic microvascular complications of diabetes *(8)*.

Genes Encoding Transcription Factors

Several other forms of MODY are caused by heterozygosity for mutations in genes encoding transcription factors that are believed to play a role in regulating tissue-specific expression of various genes in pancreatic β-cells: MODY1, resulting from mutations in hepatocyte nuclear factor 4α *(9)*; MODY3, owing to mutations in hepatocyte nuclear factor 1α *(10)*; and MODY4, owing to mutations in insulin promoter factor-1 *(11)*. All of these forms of MODY share in common the characteristics of autosomal-dominant inheritance and onset of diabetes at a relatively early age.

INSULIN RESISTANCE

Insulin Receptors

ROLE OF INSULIN RECEPTORS IN INSULIN ACTION

Over the past 20 years, there has been impressive progress toward understanding the molecular mechanisms of insulin action *(12,13)*. Insulin binds to the insulin receptor, a transmembrane glycoprotein expressed on the surface of the target cell *(14,15)*. The receptor has a heterotetrameric ($\alpha_2\beta_2$) structure (Fig. 1). The α-subunit is entirely extracellular and contains the insulin binding site. The β-subunit spans the plasma membrane, with the extracellular domain anchoring the α-subunit to the membrane and the intracellular domain possessing enzymatic activity as a protein tyrosine kinase. When insulin binds to the α-subunit of the receptor, this induces the phosphorylation of multiple tyrosine residues in the intracellular domain of the β-subunit. The major tyrosine phosphorylation sites are clustered in three sites on the insulin receptor.

1. Tyr^{1158}, Tyr^{1162}, and Tyr^{1163} are located at or near the "activation loop" of the tyrosine kinase domain *(16,17)*. Phosphorylation of the tyrosine residue(s) in the activation loop leads to a conformational change that activates the tyrosine kinase and so facilitates the binding of substrates for the phosphorylation reaction.
2. Tyr^{965} and Tyr^{972} are located in the "juxtamembrane domain" of the receptor. Tyr^{972} is embedded in an Asn-Pro-Xaa-Tyr motif, which, when phosphorylated, provides a binding site for the phosphotyrosine binding domain of several intracellular proteins that are phosphorylated by the insulin receptor, including IRS-1, IRS-2, IRS-3, IRS-4, and Shc (*see* Substrates Containing Phosphotyrosine Binding [PTB] Domains) *(18–20)*.
3. Tyr^{1328} and Tyr^{1334} are located in the C-terminal domain of the receptor. Tyr^{1334} is embedded in a Tyr-Xaa-Xaa-Met motif, which, when phosphorylated, provides a binding site for the SH2 domain of the p85 regulatory subunit of phosphatidylinositol (PI) 3-kinase *(21)*. Although it is well documented that PI 3-kinase can bind to the C-terminal domain of the insulin receptor, the physiological signficance of this binding interaction remains controversial. Indeed, the prevailing view is that the ability of insulin to activate PI 3-kinase is mediated primarily by binding of p85 to IRS-1, 2, 3, or 4, rather than directly to the insulin receptor (*see* The PI 3-Kinase Pathway) *(22)*.

identified in the genes encoding at least five proteins: insulin, glucokinase, and three transcription factors (*see below*). In most cases, the mutations lead to a decrease in the concentration of insulin in the blood. However, in the case of mutations in the insulin gene, immunoreactive insulin levels are frankly elevated, but the majority of the circulating insulin molecules are abnormal and do not retain normal biological activity. Although there is some variation in the clinical presentations, patients with insulin deficiency owing to mutations in single genes seem to share several clinical features in common. First, the syndrome has its onset at a relatively early age (i.e., considerably younger than the majority of patients with type 2 diabetes mellitus). Second, the disease follows an autosomal-dominant pattern of inheritance. Third, the patients are not ketosis-prone. Finally, the patients generally retain normal insulin sensitivity; that is, they do not exhibit clinically signficant insulin resistance.

Insulin Gene

A few families have been identified in which there are mutations in the insulin gene. At least two types of mutations have been identified in the insulin gene *(3)*. Some mutations inhibit the proteolytic processing of the proinsulin molecule. Because excision of the C-peptide from proinsulin is required for full bioactivity, the mutation leads to an impairment in biological activity. Other mutations permit normal posttranslational processing of proinsulin, but directly interfere with normal binding of the mutant insulin molecule to the insulin receptor. In both classes of genetic defects, the mutant alleles direct the synthesis of a molecule with decreased affinity for the receptor. Consequently, the structurally abnormal insulin molecule has reduced biological activity. Furthermore, because receptor-mediated endocytosis (primarily in the liver) provides the principal pathway by which insulin is cleared from the plasma, the defect in insulin binding retards the clearance of the mutant insulin from the circulation. As judged by a radioimmunoassay, insulin levels in the plasma appear to be elevated even though the mutant insulin molecules have markedly reduced biological activity.

Thus far, all of the reported mutations in the insulin gene have been found in the heterozygous state; mutations have not been observed in the homozygous state. The mutations cause elevated levels of insulin-like immunoactivity (i.e., either hyperinsulinemia or hyperproinsulinemia) in a dominant fashion. The phenotype of diabetes is inherited in a dominant pattern, but with incomplete penetrance. Thus, although a mutation in the insulin gene appears to be a risk factor predisposing to diabetes, it is not sufficient to cause diabetes in all patients.

Glucokinase Gene

Glucokinase is an enzyme that catalyzes the phosphorylation of glucose, the first step in glucose metabolism *(4)*. Unlike other hexokinases, it is specific for glucose (in preference to other hexoses). Furthermore, it is expressed primarily in hepatocytes and pancreatic β-cells. Glucokinase is believed to serve a central function in the ability of the β-cell to "measure" levels of glucose in the plasma, and thereby to regulate the rate of insulin secretion. In patients who are heterozygous for mutations in the glucokinase gene, this leads to impaired insulin secretion *(5,6)*. Although patients secrete relatively normal quantities of insulin, they require abnormally high concentrations of plasma glucose (approximately twice normal fasting levels of glucose) to drive the β-cell. Accordingly, diabetes resulting from mutations in the glucokinase gene is inherited as an autosomal-

8 Diabetes Mellitus
Insulin Resistance

Simeon I. Taylor, MD, PhD and Elif Arioglu, MD

CONTENTS

INTRODUCTION
INSULIN DEFICIENCY
INSULIN RESISTANCE
CONCLUSION
REFERENCES

INTRODUCTION

Insulin is the dominant hormone in regulating fuel metabolism. Defects in either insulin secretion or insulin action cause major metabolic abnormalities in the metabolism of glucose, fat, and proteins *(1)*. For example, when insulin secretion is impaired by destruction of the pancreatic β-cell (e.g., autoimmune destruction of the β-cell in type 1 diabetes mellitus), this leads to characteristic metabolic derangements, including hyperglycemia and ketoacidosis. Although insulin deficiency was the first recognized cause of diabetes, the development of radioimmunoassays eventually allowed for the discovery that some diabetic patients retained the ability to secrete insulin. This led to the conclusion that defects in insulin action (i.e., insulin resistance) can also contribute to the pathogenesis of diabetes. Indeed, it is now known that the most common form of diabetes (i.e., type 2 diabetes mellitus) is characterized by at least two defects: insulin deficiency and insulin resistance *(2)*. In some patients (mostly patients with relatively rare variants of noninsulin-dependent diabetes mellitus), molecular defects have been identified as the causes of either deficient insulin secretion or insulin resistance. Nevertheless, the precise molecular causes have not yet been identified in most diabetic individuals.

INSULIN DEFICIENCY

The β-cell of the pancreas is characterized by its ability to synthesize insulin and to secrete the hormone in response to the biological needs of the organism. Complex biochemical mechanisms, involving the products of multiple genes, are required for the β-cell to accomplish these functions. Each of these genes is a potential candidate to harbor mutations that might impair insulin secretion. In fact, mutations have already been

From: *Contemporary Endocrinology: Hormone Resistance Syndromes*
Edited by: J. L. Jameson © Humana Press Inc., Totowa, NJ

80. Forrest D, Hanebuth E, Smeyne RJ, Everds N, Stewart CL, Wehner JM, et al. Recessive resistance to thyroid hormone in mice lacking thyroid hormone receptor β: evidence for tissue-specific modulation of receptor function. EMBO J 1996;15:3006–3015.
81. Forrest D, Erway LC, Ng L, Altschuler R, Curran T. Thyroid hormone receptor β is essential for development of auditory function. Nature Genet 1996;13:354–357.
82. Hayashi Y, Mangoura D, Refetoff S. A mouse model of resistance to thyroid hormone produced by somatic gene transfer of a mutant thyroid hormone receptor. Mol Endocrinol 1996;10:100–106.
83. Wong R, Vasilyev VV, Ting Y-T, Kutler DI, Willingham MC, Weintraub BD, et al. Transgenic mice bearing a human mutant thyroid hormone β1 receptor manifest thyroid function anomalies, weight reduction and hyperactivity. Mol Medicine 1997;3:303–314.
84. Abel ED, Kaulbach HC, Boers M-E, Wondisford FE. Transgenic expression of mutant thyroid hormone receptor isoforms reveal differential effects on pituitary thyroid hormone action in vivo. The Endocrine Society, Bethesda (Abstract P2-38), 1997, p. 294.
85. Wikstrom L, Johansson C, Salto C, Barlow C, Barros AC, Baas F, et al. Abnormal heart rate and body temperature in mice lacking thyroid hormone receptor α1. EMBO J 1998;17:455-461.
86. Fraichard A, Chassande O, Plateroti M, Roux JP, Trouillas J, Dehay C, et al. The T3Rα gene encoding a thyroid hormone receptor is essential for post-natal development and thyroid hormone production. EMBO J 1997;16:4412–4420.
87. Beck-Peccoz P, Piscitelli G, Cattaneo MG, Faglia G. Successful treatment of hyperthyroidism due to nonneoplastic pituitary TSH hypersecretion with 3,5,3'-triiodothyroacetic acid (TRIAC). J Endocrinol Invest 1983;6:217–223.
88. Salmela PI, Wide L, Juustila H, Ruokonen A. Effects of thyroid hormones (T4, T3), bromocriptine and TRIAC on inappropriate TSH hypersecretion. Clin Endocrinol (Oxford) 1988;28:497–507.
89. Crinò A, Borrelli P, Salvatori R, Cortelazzi D, Roncoroni R, Beck-Peccoz P. Anti-iodothyronine autoantibodies in a girl with hyperthyroidism due to pituitary resistance to thyroid hormones. J Endocrinol Invest 1992;15:113–120.
90. Radetti G, Persani L, Molinaro G, Mannavola D, Cortelazzi D, Chatterjee VKK, et al. Clinical and hormonal outcome after two years of triiodothyroacetic acid treatment in a child with thyroid hormone resistance. Thyroid 1997;7:775–778.
91. Bracco D, Morin O, Schutz Y, Liang H, Jequier E, Burger AG. Comparison of the metabolic and endocrine effects of 3,5,3'-triiodothyroacetic acid and thyroxine. J Clin Endocrinol Metab 1993;77: 221–228.
92. Takeda T, Suzuki S, Liu R-T, DeGroot LJ. Triiodothyroacetic acid has unique potential for therapy of resistance to thyroid hormone. J Clin Endocrinol Metab 1995;80: 2033–2040.
93. Ueda S, Takamatsu J, Fukata S, Tanaka K, Shimizu N, Sakata S, et al. Differences in response of thyrotropin to 3,5,3'-triiodothyronine and 3,5,3'-triiodothyroacetic acid in patients with resistance to thyroid hromone. Thyroid 1996;6:563–570.
94. Kunitake JM, Hartman N, Henson LC, Lieberman J, Williams DE, Wong M, et al. 3,5,3'-triiodothyroacetic acid therapy for thyroid hormone resistance. J Clin Endocrinol Metab 1989;69:461–466.
95. Hamon P, Bovier-LaPierre M, Robert M, Peynaud D, Pugeat M, Orgiazzi J. Hyperthyroidism due to selective pituitary resistance to thyroid hormones in 15-month-old boy: efficacy of D-thyroxine therapy. J Clin Endocrinol Metab 1988;67:1089–1093.
96. Dorey F, Strauch G, Gayno JP, Thyrotoxicosis due to pituitary resistance to thyroid hormones. Successful control with D-thyroxine; a study in three patients. Clin Endocrinol (Oxford) 1990;32:221–227.
97. Pohlenz J, Knobl D. Treatment of pituitary resistance to thyroid hormone (PRTH) in an 8 yr old boy. Acta Paediatr 1996;85:387–390.
98. Dulgeroff AJ, Geffner ME, Koyal SN, Wong M, Hershman JM. Bromocriptine and Triac therapy for hyperthyroidism due to pituitary resistance to thyroid hormone. J Clin Endocrinol Metab 1992;75: 1071–1075.
99. Williams G, Kraenzlin M, Sandler L, Burrin J, Law A, Bloom SR, Joplin GF. Hyperthyroidism due to non-tumoural inappropriate TSH secretion: effect of long-acting somatostatin analogue (SMS 201-995. Acta Endocr (Copenh.) 1986;113:42–46.
100. Beck-Peccoz P, Mariotti S, Guillausseau PJ, Medri G, Piscitelli G, Bertoli A, et al. Treatment of hyperthyroidism due to inappropriate secretion of thyrotropin with the somatostatin analog SMS 201-995. J Clin Endocrinol Metab 1989;68:208–214.
101. Weiss RE, Stein MA, Refetoff S. Behavioral effects of liothyronine (L-T3) in children with attention deficit hyperactivity disorder in the presence and absence of resistance to thyroid hormone. Thyroid 1997;7:389–393.

61. Hollenberg AN, Monden T, Wondisford FE Ligand-independent and -dependent function of thyroid hormone receptor isoforms depend upon their distinct amino termini. J Biol Chem 1995;270:14,274–14,280.
62. Tagami T, Madison LD, Nagaya T, Jameson JL. Nuclear receptor corepressors activate rather than suppress basal transcription of genes that are negatively regulated by thyroid hormone. Mol Cell Biol 1997;17:2642–2648.
63. Clifton-Bligh RJ, de Zegher F, Wagner RL, Collingwood TN, Francois I, van Helvoirt M, et al. A novel TRβ mutation (R383H in resistance to thyroid hormone predominantly impairs corepressor release and negative transcriptional regulation. Mol Endocrinol 1998;12:609–621.
64. Hayashi Y, Sunthornthepvarakul T, Refetoff S. Mutations of CpG dinucleotides located in the triiodothyronine (T3)-binding domain of the thyroid hormone receptor (TR) β gene that appears to be devoid of natural mutations may not be detected because they are unlikely to produce the clinical phenotype of resistance to thyroid hormone. J Clin Invest 1994;94:607–615.
65. Hayashi Y, Weiss RE, Sarne DH, Yen PM, Sunthornthepvarakul T, Marcocci C, et al. Do clinical manifestations of resistance to thyroid hormone correlate with the functional alteration of the corresponding mutant thyroid hormone β receptors? J Clin Endocrinol Metab 1995;80:3246–3256.
66. Rodd C, Schwartz HL, Strait KA, Oppenheimer JH. Ontogeny of hepatic nuclear triiodothyronine receptor isoforms in the rat. Endocrinology 1991;131: 2559–2564.
67. Falcone M, Miyamoto T, Fierro-Renoy F, Macchia E, DeGroot LJ. Antipeptide polyclonal antibodies specifically recognize each human thyroid hormone receptor isoform. Endocrinology 1992;131: 2419–2429.
68. Hayashi Y, Janssen OE, Weiss RE, Murata Y, Seo H, Refetoff S. The relative expression of mutant and normal thyroid hormone receptor genes in patients with generalized resistance to thyroid hormone determined by estimation of their specific messenger ribonucleic acid products. J Clin Endocrinol Metab 1993;76:64–69.
69. Mixson AJ, Hauser P, Tennyson G, Renault JC, Bodenner DL, Weintraub BD. Differential expression of mutant and normal beta T3 receptor alleles in kindreds with generalized resistance to thyroid hormone. J Clin Invest 1993;91: 2296–2300.
70. Sasaki S, Nakamura H, Tagami T, Miyoshi Y, Nakao K. Functional properties of a mutant T3 receptor β (R338W) identified in a subject with pituitary resistance to thyroid hormone. Mol Cell Endocrinol 1995;113:109–117.
71. Ando S, Nakamura H, Sasaki S, Nishiyama K, Kitahara A, Nagasawa S, et al. Introducing a point mutation identified in a patient with pituitary resistance to thyroid hormone (Arg 338 to Trp) into other mutant thyroid hormone receptors weakens their dominant negative activities. J Endocrinol 1996; 151:293–300.
72. Flynn TR, Hollenberg AN, Cohen O, Menke JB, Usala SJ, Tollin S, et al. A novel C-terminal domain in the thyroid hormone receptor selectively mediates thyroid hormone inhibition. J Biol Chem 1994; 269:32,713–32,716.
73. Meier CA, Parkison C, Chen A, Ashizawa K, Meier-Heusler SC, Muchmore P, et al. Interaction of human β1 thyroid hormone receptor and its mutants with DNA and retinoid X receptor. T3 response element-dependent dominant negative potency. J Clin Invest 1993;92:1986–1993.
74. Zavacki AM, Harney JW, Brent GA, Larsen PR. Dominant negative inhibition by mutant thyroid hormone receptors is thyroid response element and receptor isoform specific. Mol Endocrinol 1993;7: 1319–1330.
75. Ng L, Forrest D, Haugen BR, Wood WM, Curran T. N-terminal variants of thyroid hormone receptor β: differential function and potential contribution to syndrome of resistance to thyroid hormone. Mol Endocrinol 1995;9:1202–1213.
76. Sjoberg M, Vennstrom B. Ligand-dependent and -independent transactivation by thyroid hormone receptor β2 is determined by the structure of the hormone response element. Mol Cell Biol 1995;15: 4718–4726.
77. Safer JD, Langlois MF, Cohen R, Monden T, John-Hope D, Madura J, et al. Isoform variable action among thyroid hormone receptor mutants provides insight into pituitary resistance to thyroid hormone. Mol Endocrinol 1997;11:16–26.
78. Geffner ME, Su F, Ross NS, Hershman JM, Van Dop C, Menke JB, et al. An arginine to histidine mutation in codon 311 of the *c-erb*Aβ gene results in a mutant thyroid hormone receptor that does not mediate a dominant negative phenotype. J Clin Invest 1993;91:538–546.
79. Weiss RE, Marcocci C, Bruno-Bossio G, Refetoff S. Multiple genetic factors in the heterogencity of thyroid hormone resistance. J Clin Endocrinol Metab 1993;76:257–259.

40. Mixson AJ, Renault JC, Ransom S, Bodenner DL, Weintraub BD. Identification of a novel mutation in the gene encoding the β-triiodothyronine receptor in a patient with apparent selective pituitary resistance to thyroid hormone. Clin Endocrinol (Oxford) 1993;38:227–234.
41. Sasaki S, Nakamura H, Tagami T, Miyoshi Y, Nogimori T, Mitsuma T, Imura H. Pituitary resistance to thyroid hormone associated with a base mutation in the hormone-binding domain of the human 3,5,3'-triiodothyronine receptor β. J Clin Endocrinol Metab 1993;76:1254–1258.
42. Weiss RE, Hayashi Y, Nagaya T, Petty KJ, Murata Y, Tunca H, et al. Dominant inheritance of resistance to thyroid hormone not linked to defects in the thyroid hormone receptor α or β genes may be due to a defective cofactor. J Clin Endocrinol Metab 1996;81:4196–4203.
43. Olson DP, Koenig RJ. Thyroid function in Rubinstein-Taybi syndrome. J Clin Endocrinol Metab 1997;82:3264–3266.
44. Behr M, Loos U. A point mutation (Ala 229 to Thr) in the hinge domain of the c-erbAβ thyroid hormone receptor gene in a family with generalized thyroid hormone resistance. Mol Endocrinol 1992;6: 1119–1126.
45. Onigata K, Yagi H, Sakurai A, Nagashima T, Nomura Y, Nagashima K, et al. A novel point mutation (R243Q) in exon 7 of the c-erbAβ thyroid hormone receptor gene in a family with resistance to thyroid hormone. Thyroid 1995;5:355–358.
46. Pohlenz J, Schonberger W, Wemme H, Winterpacht A, Wirth S, Zabel B. New point mutation (R243W) in the hormone binding domain of the c-erbA β1 gene in a family with generalized resistance to thyroid hormone. Hum Mutat 1996;7:79–81.
47. Matthews C, Collingwood TN, Adams M, Rajanayagam O, Chatterjee VKK. Novel mutations in the hinge domain of the thyroid hormone β receptor and a transactivation mutant in thyroid hormone resistance syndrome. The Endocrine Society, Bethesda (Abstract P3-437), 1995, p. 578.
48. Yagi H, Pohlenz J, Hayashi Y, Sakurai A, Refetoff S. Resistance to thyroid hormone caused by two mutant thyroid hormone receptors β, R243Q and R243W, with marked impairment of function that cannot be explained by altered in vitro 3,5,3'-triiodothyronine binding affinity. J Clin Endocrinol Metab 1997;82:1608–1614.
49. Nagaya T, Madison LD, Jameson JL. Thyroid hormone receptor mutants that cause resistance to thyroid hormone: evidence for receptor competition for DNA sequences in target genes. J Biol Chem 1992;267:13014–13019.
50. Nagaya T, Jameson JL. Thyroid hormone receptor dimerization is required for dominant negative inhibition by mutations that cause thyroid hormone resistance. J Biol Chem 1993;268:15,766–15,771.
51. Rastinejad F, Perlmann T, Evans RM, Sigler P. Structural determinants of nuclear receptor assembly on DNA direct repeats. Nature 1995;375:203–211.
52. Wagner RL, Apriletti JW, McGrath ME, West BL, Baxter JD, Fletterick RJ. A structural role for hormone in the thyroid hormone receptor. Nature 1995;378:690–697.
53. Wurtz J-M, Bourguet W, Renaud J-P, Vivat V, Chambon P, Moras D, et al. A canonical structure for the ligand-binding domain of nuclear receptors. Nature Struct Biol 1996;3:87–94.
54. O'Donnell AL, Koenig RJ. Mutational analysis identifies a new functional domain of the thyroid hormone receptor. Mol Endocrinol 1990;4:715–720.
55. Weiss RE, Weinberg M, Refetoff S. Identical mutations in unrelated families with generalized resistance to thyroid hormone occur in cytosine-guanine-rich areas of the thyroid hormone receptor beta gene. J Clin Invest 1993;91:2408–2415.
56. Collingwood TN, Rajanayagam O, Adams M, Wagner R, Cavailles V, Kalkhoven E, et al. A natural transactivation mutation in the thyroid hormone β receptor: impaired interaction with putative transcriptional mediators. Proc Natl Acad Sci USA 1997;94:248–253.
57. Piedrafita FJ, Bendik I, Ortiz MA, Pfahl M. Thyroid hormone receptor homodimers can function as ligand-sensitive repressors. Mol Endocrinol 1995;9:563–578.
58. Yen PM, Wilcox EC, Hayashi Y, Refetoff S, Chin WW. Studies on the repression of basal transcription (silencing by artificial and natural human thyroid hormone receptor-β mutants. Endocrinology 1995; 136:2845–2851.
59. Kitajima K, Nagaya T, Jameson JL. Dominant negative and DNA-binding properties of mutant thyroid hormone receptors that are defective in homodimerization but not heterodimerization. Thyroid 1995; 5:343–353.
60. Yoh SM, Chatterjee VKK, Privalsky ML. Thyroid hormone resistance syndrome manifests as an aberrant interaction between mutant T3 receptors and transcriptional corepressors. Mol Endocrinol 1997;11:470–480.

60. Quon MJ, Butte AJ, Zarnowski MJ, Sesti G, Cushman SW, Taylor SI. Insulin receptor substrate 1 mediates the stimulatory effect of insulin on GLUT4 translocation in transfected rat adipose cells. J Biol Chem 1994;269:27,920–27,924.
61. Zhou LX, Chen H, Lin CH, Cong L-N, McGibbon MA, Sciacchitano S, et al. Insulin receptor substrate-2 (IRS-2) can mediate the action of insulin to stimulate translocation of GLUT4 to the cell surface in rat adipose cells. J Biol Chem 1997;272:29,829–29,833.
62. Zhou L, Sciacchitano S, Cong LN, Li Y, Xu P, Jacobs AR, et al. Action of insulin receptor substrate-3 (IRS-3) to stimulate translocation of GLUT4 in rat adipose cells. 1998, unpublished observations.
63. Perrotti N, Accili D, Marcus SB, Rees Jones RW, Taylor SI. Insulin stimulates phosphorylation of a 120-kDa glycoprotein substrate (pp120) for the receptor-associated protein kinase in intact H-35 hepatoma cells. Proc Natl Acad Sci USA 1987;84:3137–3140.
64. Margolis RN, Schell MJ, Taylor SI, Hubbard AL. Hepatocyte plasma membrane ECTO-ATPase (pp120/HA4) is a substrate for tyrosine kinase activity of the insulin receptor. Biochem Biophys Res Commun 1990;166, 562–566.
65. Najjar SM, Philippe N, Suzuki Y, Ignacio GA, Formisano P, Accili D, et al. Insulin-simulted phosphorylation of recombinant pp120/HA4, an endogenous substrate of the insulin receptor tyrosine kinase. Biochemistry 1995;34:9341–9349.
66. Fujioka Y, Matozaki T, Noguchi T, Iwamatsu A, Yamao T, Takahashi N, et al. A novel membrane glycoprotein, SHPS-1, that binds the SH2-domain-containing protein tyrosine phosphatase SHP-2 in response to mitogens and cell adhesion. Mol Cell Biol 1996;16:6887–6899.
67. Kharitonenkov A, Chen Z, Sures I, Wang H, Schilling J, Ullrich A. A family of proteins that inhibit signalling through tyrosine kinase receptors. Nature 1997;386:181–186.
68. Imai Y, Fusco A, Suzuki Y, Lesniak MA, D'Alfonso R, Sesti G, et al. Variant sequences of insulin receptor substrate-1 in patients with noninsulin-dependent diabetes mellitus. J Clin Endocrinol Metab 1994;79:1655–1658.
69. Imai Y, Philippe N, Sesti G, Accili D, Taylor SI. Expression of variant forms of insulin receptor substrate-1 identified in patients with noninsulin-dependent diabetes mellitus. J Clin Endocrinol Metab 1997;82:4201–4207.
70. Almind K, Bjorbaek C, Vestergaard H, et al. Aminoacid polymorphism of insulin receptor substrate-1 in non-insulin-dependent diabetes mellitus. Lancet 1993;342:828–832.
71. Hitman GA, Hawrami K, McCarthy MI, et al. Insulin receptor substrate-1 gene mutations in NIDDM; implications for the study of polygenic disease. Diabetologia 1995;38:481–486.
72. Laakso M, Malkki M, et al. Insulin receptor substrate-1 variants in non-insulin-dependent diabetes. J Clin Invest 1994;94:1141–1146.
73. Clausen JO, Hansen T, Bjorbaek C, Echwald SM, Urhammer SA, Rasmussen S, et al. Insulin resistance: interactions between obesity and a common variant of insulin receptor substrate-1. Lancet 1995; 346:397–402.
74. Almind K, Inoue G, Pedersen O, Kahn CR. A common amino acid polymorphism in insulin receptor substrate-1 causes impaired insulin signaling. Evidence from transfection studies. J Clin Invest 1996;97:2569–2575.
75. Brüning JC, Winnay J, Bonner-Weir S, Taylor SI, Accili D, Kahn CR. Development of a novel polygenic model of NIDDM in mice heterozygous for IR and IRS-1 null alleles. Cell 1997;88:561–572.
76. Kohn AD, Summers SA, Birnbaum MJ, Roth RA. Expression of a constitutively active Akt Ser/Thr kinase in 3T3-L1 adipocytes stimulates glucose uptake and glucose transporter 4 translocation. J Biol Chem 1996;271:31,372–31,378.
77. Cong L-N, Chen H, Li YH, Zhou LX, McGibbon MA, Taylor SI, et al. Physiological role of Akt in insulin-stimulated translocation of GLUT4 in transfected rat adipose cells. Mol Endocrinol 1997;11: 1881–1890.
78. Eldar-Finkelman H, Argast GM, Foord O, Fischer EH, Krebs EG. Expression and characterization of glycogen synthase kinase-3 mutants and their effect on glycogen synthase activity in intact cells. Proc Natl Acad Sci USA 1996;93:10,228–10,233.
79. Standaert ML, Galloway L, Karnam P, Bandyopadhyay G, Moscat J, Farese RV. Protein kinase C-zeta as a downstream effector of phosphatidylinositol 3-kinase during insulin stimulation in rat adipocytes. J Biol Chem 1997;272:30,075–30,082.
80. Hansen T, Andersen CB, Echwald SM, Urhammer SA, Clausen JO, Vestergaard H, et al. Identification of a common amino acid polymorphism in the p85alpha regulatory subunit of phosphatidylinositol 3-kinase: effects on glucose disappearance constant, glucose effectiveness, the insulin sensitivity index. Diabetes 1997;46:494–501.

81. Hansen L, Arden KC, Rasmussen SB, Viars CS, Vestergaard H, Hansen T, et al. Chromosomal mapping and mutational analysis of the coding region of the glycogen synthase kinase-3alpha and beta isoforms in patients with NIDDM. Diabetologia 1997;40:940–946.
82. Quon MJ, Chen H, Ing BL, Liu ML, Zarnowski MJ, Yonezawa K, et al. Roles of 1-phosphatidylinositol 3-kinase and ras in regulating translocation of GLUT4 in transfected rat adipose cells. Mol Cell Biol 1995;15:5403–5411.
83. Seip M. Lipodystrophy and gigantism with associated endocrine manifestations. A new diencephalic syndrome? Acta Paediatr Scand 1959;48:555–774.
84. Berardinelli W. An undiagnosed endocrinometabolic syndrome: report of two cases. J Clin Endocrinol 1954;14:193–204.
85. Dunnigan MG, Cochrane MA, Kelly A, Scott JW. Familial lipoatrophic diabetes with dominant transmission. A new syndrome. Q J Med 1974;169:33-48.
86. Kobberling J, Willms B, Katterman R, Creutzfeldt W. Lipodystrophy of the extremities. A dominant inherited syndrome associated with lipoatrophic diabetes. Humangenetik 1975;29:111-120.
87. Lawrence RD. Lipodystrophy and hepatomegaly with diabetes; lipaemia and other metabolic disturbances. Lancet 1946;I:724-741.
88. Spranger S, Spranger M, Tasman AJ, Reith W, Voigtlander T, Voightlander V. Barraquer Simons Syndrome (with sensorineural deafness): a contribution to the differential diagnosis of lipodystrophy syndromes. Am J Med Genet 1997;71:397-400.

9 LH Insensitivity Syndrome

*Axel P. N. Themmen, PhD,
John W. M. Martens, PhD,
and Han G. Brunner, MD, PhD*

CONTENTS

    INTRODUCTION
    PHYSIOLOGICAL FUNCTIONS OF LH AND hCG
    LH RESISTANCE
    MOLECULAR ASPECTS
    CONCLUDING REMARKS
    ACKNOWLEDGMENT
    REFERENCES

INTRODUCTION

In this chapter the syndrome of luteinizing hormone (LH) insensitivity or LH resistance is discussed, and the LH receptor gene mutations underlying this syndrome are described. First, an introduction is given of the physiological functions of LH and human chorion gonadotropin (hCG) during prenatal sex differentiation, puberty, and in the adult state. Subsequently, LH insensitivity syndrome is described in detail in males and females as well as sensitivity to the LH receptor gene mutations that have been found in these patients. In the Molecular Aspects section is discussed the information that the identification of these LH receptor gene mutations provides with respect to the molecular function of the LH receptor and the roles of its subdomains. The chapter is concluded with some remarks on possible future directions of research.

PHYSIOLOGICAL FUNCTIONS OF LH AND hCG

Sex Differentiation

The process of sex differentiation has two major switches. First, the genetic sex of the individual is determined by the presence of a Y chromosome that carries the *SRY* gene, the testis-determining region on the Y chromosome. Once this gene is expressed, the indifferent gonads start to differentiate into testes. In this process, some of the somatic cells that are present in the gonad differentiate into Sertoli cells and organize themselves into testicular cords that envelope the germ cells. Prospective Leydig cells migrate into the gonad from the surrounding mesonephros. The second major switch in sex

From: *Contemporary Endocrinology: Hormone Resistance Syndromes*
Edited by: J. L. Jameson © Humana Press Inc., Totowa, NJ

differentiation is the hormonal control of gonadal and extragonadal development by the Sertoli and Leydig cells. The Sertoli cells of the fetal testis secrete anti-Müllerian hormone (AMH), a member of the transforming growth factor β family of growth and differentiation factors. AMH interacts with an AMH receptor present in the mesenchymal cells that surround the Müllerian ducts, the anlagen of the female urogenital tract. Activation of the AMH receptor by AMH results in regression of the Müllerian ducts in the male. Development of male internal and external genitalia is dependent on the Leydig cell product testosterone, which in most target tissues is converted to the more potent dihydrotestosterone. The Wolffian ducts, the anlagen of the male urogenital tract, develop into the epididymides, vasa deferentia, and seminal vesicles. The other male internal and external genitalia are also formed under the influence of androgens: the prostate, penis, and scrotum.

The androgens that play such an important role during the prenatal period are produced by the testicular Leydig cells. The ontogeny and development of Leydig cells are described in more detail here because their function is changed in male patients with LH insensitivity. The initial differentiation of fetal Leydig cells appears to be under the control of a transcription factor from the orphan nuclear receptor family: steroidogenic factor-1 (SF-1) *(1)*. SF-1 in collaboration with another transcription factor, Wilms' tumor-associated gene (WT1), is essential for the development of the intermediate mesoderm and the formation of the indifferent gonads *(2)*. After the induction of testicular differentiation by *SRY*, SF-1 is expressed in the differentiating Leydig cells and induces steroidogenic enzymes, such as 17-hydroxylase and 3β-hydroxysteroid dehydrogenase, that are necessary for the production of androgens. The initial expression of these enzymatic activities appears not to be dependent on the presence of pituitary LH or hCG from the placenta. Ablation of the βLH gene in mice did not prevent the appearance and function of fetal Leydig cells in the testis, although in the adult mice, low serum androgens levels were observed, indicating that LH is necessary for correct Leydig cell function in the adult *(3)*. In the human fetus, the first signs of functional and morphological differentiation of fetal Leydig cells are discernible at the age of 8 wk *(4)*. Subsequently, fetal Leydig cells go through a series of three developmental stages based on their ultrastructure: differentiation phase: 8–14 wk; maturity phase: 14–18 wk, and involution (apoptotic) phase: 18–40 wk. These phases reflect the concentration of hCG in fetal serum. During the first two phases, hCG is high causing the numbers of Leydig cells to increase until 21–24 wk of age. The involution phase coincides with and may be caused by the decrease in hCG concentration *(4,5)*. After birth, fetal Leydig cells continue to exist, but they are not active in the production of steroid hormones until pituitary LH levels increase at puberty.

Unlike in males, gonadal and urogenital development in the female is largely independent of hormonal activities. Rather, there seems to be a redundancy of these tissues to develop according to the female pattern unless a Y chromosome and accompanying testicular AMH and androgen production are present. In the indifferent gonad destined to become an ovary, the lack of differentiation persists until 11–12 wk of age. Germ cells start to enter meiotic prophase, and the first primordial follicles have formed by 13 wk of age. Prenatally, LH or hCG do not appear to have an effect on the ovaries, and there is no evidence of hormone production by the ovaries before birth.

Puberty and Adult

One of the endocrine changes that occur during puberty (9–14 yr of age for boys; 8–13 yr for girls *[6]*) is the increase in pituitary LH and follicle-stimulating hormone

(FSH) secretion. In boys, LH stimulates the proliferation and, subsequently, the testosterone production of the Leydig cells in the testis. This results in the pubertal changes of secondary sex characteristics. FSH is involved in the initiation and subsequent maintenance of spermatogenesis through its action on the Sertoli cells.

The Leydig cells of the adult testis are not derived through mitotic activity or differentiation from the fetal Leydig cells still present in the testis, but develop independently from undifferentiated precursor cells (7). In experiments in rats using ethane-dimethane sulfonate (EDS), a chemical that specifically removes Leydig cells from the testis, it was found that the mitotic activity of the precursor cells in the testis is not under the control of LH. However, Leydig cell differentiation characteristics, such as 3β-hydroxysteroid dehydrogenase and the recruitment of the precursor cells into the adult population, are strictly LH-dependent (8).

In girls, the pubertal increase in FSH and LH stimulates follicular growth and differentiation of the granulosa and theca cells. This differentiation renders the follicular cells sensitive to the steroidogenic stimulation of LH. The increase in ovarian output of estrogens causes the pubertal changes in female sex characteristics. After puberty, the menstrual cycle is under strict control of both FSH and LH. The cyclic growth and differentiation of follicles causes the high-level estrogens, especially just before the ovulatory peak of LH and the following luteal phase. The corpus luteum of the luteal phase also produces large amounts of progestin. The continuous production of these steroid hormones ensures the support of the female sex characteristics by FSH and LH.

Thus, LH/hCG has different effects pre- and postnatally in males and females. The first differentiation of fetal Leydig cells occurs independently of LH or hCG, while the appearance and development of adult-type Leydig cells from the undifferentiated precursor cells under strict control of LH. In the female, there is no evidence for any role of LH and hCG prenatally. At puberty, LH can regulate follicular function only after the growth and differentiating effects of FSH.

LH RESISTANCE

Leydig Cell Hypoplasia

The first description of Leydig cell hypoplasia can be found in a 1976 publication of a case of male (46, XY) pseudohermaphroditism (9). In this patient, pseudohermaphroditism was not caused by a deficiency in one of the steroidogenic enzymes causing low androgen levels, but resulted from the absence of mature Leydig cells in the testis. Plasma LH was strongly increased, whereas the few androgens that were present were of adrenal origin, since they were suppressed by dexamethasone and increased after treatment with ACTH (9). An iv injection of GnRH agonist did not result in any change in testosterone, notwithstanding the clear increase in serum LH. Absence of receptors for LH was offered as an explanation for this syndrome of Leydig cell dysgenesis. Following this initial report, 10–20 patients were described with similar characteristics (9–21), of which the most important are absence of mature Leydig cells in a testicular biopsy and unresponsiveness to hCG. Furthermore, a distinction was made between severe and mild forms of Leydig cell hypoplasia (22). Complete female external phenotype is characteristic of the severe form, whereas incomplete virilization of the external genitalia, ranging from micropenis to hypospadias, and intact response to hCG were proposed to indicate the milder form of Leydig cell hypoplasia (17,20,22). A sex-limited, autosomal-recessive pattern of inheritance was found (20).

The absence of mature Leydig cells and the unresponsiveness of the testis to LH or hCG indicated either an intrinsic defect in Leydig cell development or an abnormality of the LH signal transduction pathway. In an effort to find the underlying cause of Leydig cell hypoplasia, we used the candidate gene approach and hypothesized that a genetic defect was present in the LH receptor gene. Downstream components of LH signal transduction were less likely to be affected, since the intracellular pathway that involves G_s, adenylyl cyclase, protein kinase A, and the steroidogenic enzymes is shared with many other hormone receptors. Therefore, mutations involving downstream elements would be expected to cause multiple-organ involvement.

LH Receptor Gene Mutations

The LH receptor belongs to the class of GTP binding protein- (G-protein) coupled receptors. These receptors are characterized by a domain that has seven transmembrane helices. The receptors for the glycoprotein hormones, LH, FSH, and TSH, form a separate subfamily that have a large N-terminal hormone binding extracellular domain. Their activity is mediated by coupling via G_s to adenylyl cyclase. In gonadal cells, LH-stimulated cAMP production leads to increased steroid hormone secretion. The LH receptor gene is quite large (>80 kb) and consists of 11 exons. The last exon, exon 11, encodes the transmembrane domain that signals the binding of the hormone to the large extracellular domain (encoded by exons 2–10) to adenylyl cyclase through G_s *(23)*. Exon 1 encodes a signal peptide that targets the LH receptor protein to the plasma membrane.

We investigated exon 11 of the LH receptor gene in two 46,XY sisters with the severe form of Leydig cell hypoplasia *(24)*. They presented with female external genitalia, primary amenorrhea, and lack of breast development. Their parents are first cousins, and there are 14 additional siblings. Both cases had a short blind ending vagina, without uterus or fallopian tubes. Serum levels of testosterone and testosterone precursors were abnormally low, and did not respond to stimulation by hCG. Basal levels of LH were markedly increased, but FSH was within the normal range. The gonads were removed, and on histological examination found to be testes with normal Sertoli cells, but no mature Leydig cells. The patients were found to be homozygous for a missense mutation that resulted in the change of alanine at position 593—located just outside the sixth transmembrane segment—into a proline residue. The mutant receptor was expressed in 293 cells, a human embryonic kidney cell line, and subjected to Scatchard analysis (Fig. 1). A normal K_d, similar to the wild-type LH receptor was found indicating qualitatively normal hormone binding. However, the mutant receptor was expressed at a reduced level compared to the wild-type receptor. More importantly, the proline residue changed the LH receptor in such a way that in spite of the high-affinity hCG binding, no coupling to adenylyl cyclase could be demonstrated in the absence or presence of hCG (Fig. 1) *(24)*. These findings established this inactivating mutation in the LH receptor gene as the cause of Leydig cell hypoplasia.

The complete absence of G-protein coupling of the LH receptor A593P mutant explains the severe Leydig cell hypoplasia phenotype. This raises the question of whether patients with the milder form of Leydig cell hypoplasia have LH receptor gene mutations that compromise receptor function to a lesser degree. Two mutations, S616Y and I625K in the seventh transmembrane segment, that were found in patients with the mild Leydig cell hypoplasia phenotype were compared *(25,26)*. Expressed in 293 cells, these mutant LH receptors showed a diminished, but not absent response to hCG (Fig. 2). Not only was

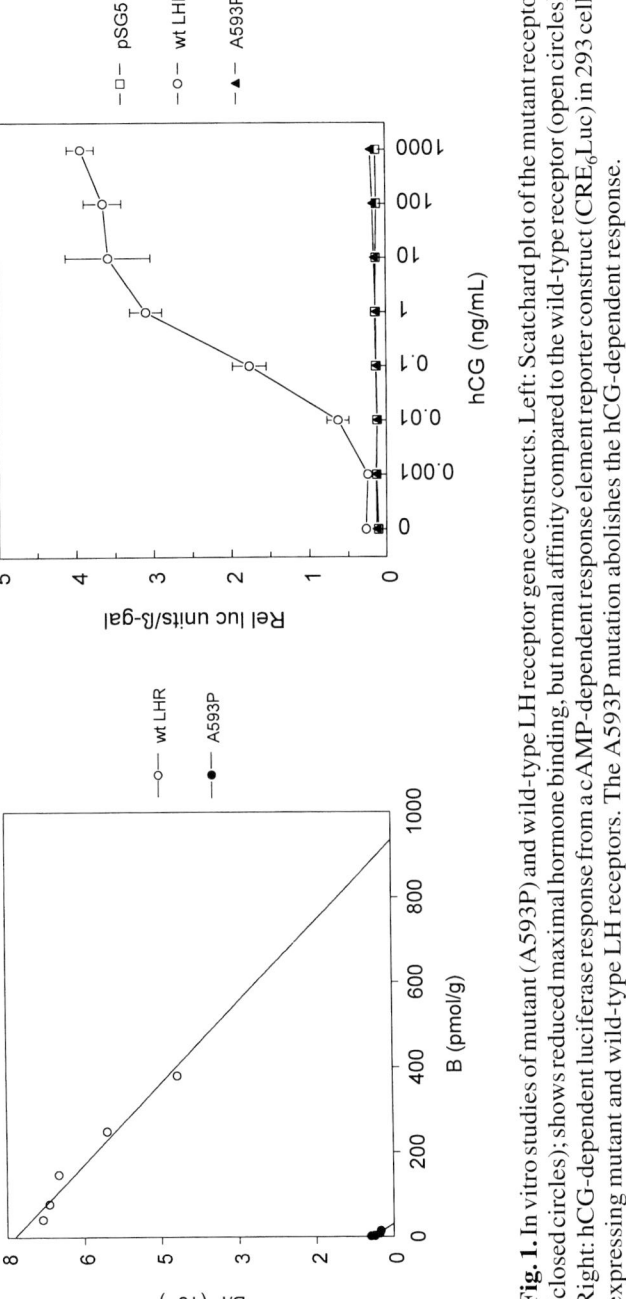

Fig. 1. In vitro studies of mutant (A593P) and wild-type LH receptor gene constructs. Left: Scatchard plot of the mutant receptor (closed circles); shows reduced maximal hormone binding, but normal affinity compared to the wild-type receptor (open circles). Right: hCG-dependent luciferase response from a cAMP-dependent response element reporter construct (CRE_6Luc) in 293 cells expressing mutant and wild-type LH receptors. The A593P mutation abolishes the hCG-dependent response.

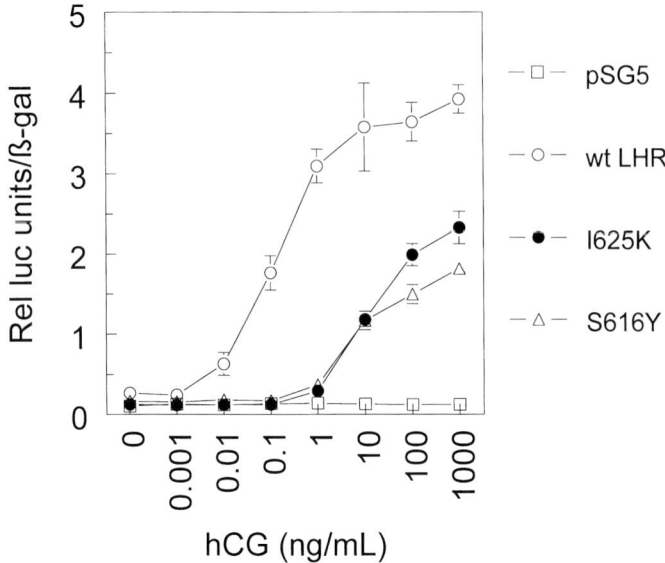

Fig. 2. Effect of LH receptor mutations found in less severely affected Leydig cell hypoplasia patients. The mutant receptors display a diminished and right-shifted response to hCG.

the maximal response at high hCG concentrations lower, but also the ED_{50} was shifted to higher hCG concentrations. These results show that the effect of the mutations on LH receptor function is well correlated with the severity of the phenotype in the patients.

Primary Amenorrhea

As discussed in the Puberty and Adult section, LH plays a pivotal role in female reproduction. After induction of LH receptors in the growing follicles by FSH, LH stimulates estrogen and progestin production by these follicles, and ultimately triggers ovulation of the Graafian follicle through the midcycle ovulatory LH peak. In contrast, LH does not appear to be important for female sex differentiation. Therefore, it was of great interest that one of the female siblings of the patients with 46,XY pseudohermaphroditism owing to Leydig cell hypoplasia described by Kremer et al. *(24)* presented with infertility resulting from primary amenorrhea. Indeed, checking the LH receptor gene of this sister revealed that she also was a homozygous carrier of the A593P mutation *(27)*. The ovarian LH insensitivity in this case was associated with normal development of female internal and external genitalia, amenorrhea, and infertility. Follicular development occurred normally, at least up to the antral stage, but ovulation did not occur. A small size of the uterus, a thin-walled hyposecretory vagina, and reduced bone mass all suggest that this patient had long-standing hypoestrogenization. Indeed, estradiol never reached levels normally found during the ovulatory or luteal phase of the cycle. The effects of LH insensitivity on female external sex characteristics is less severe compared to patients with ovarian dysgenesis caused by a mutation in the FSH receptor *(28)*. The absence of follicular stimulation in these FSH receptor-deficient patients results in much lower estradiol production. Other cases of primary amenorrhea caused by different LH receptor gene mutations confirm the effects of LH insensitivity on female secondary sex characteristics *(25,29)*.

Familial Male-Limited Precocious Puberty (FMPP)

LH receptor mutations have been described that have an opposite affect on function, since they render the receptor constitutively active. Such mutations are present in patients who have FMPP, also named testotoxicosis, first described in 1981 *(30)*. This autosomal-dominant hereditary form of precocious puberty is characterized by excessive secretion of testosterone by Leydig cells in the absence of elevated levels of pituitary gonadotropins and is found only in boys *(31,32)*. It is different from true (central) precocious puberty, which involves premature activation of the hypothalamic GnRH pulse generator *(33)*. In FMPP, excessive androgen production occurs probably also in the fetus, because enlargement of the external genitalia has been noted at birth in some cases *(34)*. Since the increased testosterone production might be the result of an inherent change in the Leydig cells themselves, the LH receptor gene of these patients was investigated *(35,36)*. Mutations in the LH receptor were predicted to change the behavior of the receptor in such a manner that it would signal continuously through G_s to the adenylyl cyclase system, activating the Leydig cells without the ligand being present. In their investigations of the α_1-adrenergic receptor, Lefkowitz and coworkers *(37)* had shown that changing an alanine residue in the third intracellular loop of this receptor to any other amino acid residue resulted in different levels of constitutive activation of the phosphatidylinositide pathway without ligand being present. Initially, two LH receptor gene mutations were reported both in the sixth transmembrane segment, methionine at position 571 to isoleucine (M571I) *(36)* and aspartate at position 578 to glycine (D578G) *(35,36)*. When transfected into HEK293 cell lines, the cells that expressed the mutant receptor species indeed showed an increased basal cAMP level when compared to cells expressing the wild-type receptor (Fig. 3). The effect of the mutations found in FMPP is the opposite of those identified in Leydig cell hypoplasia. In these cases, the mutations render the LH receptor molecule constitutively active, that explains the dominant hereditary pattern of the syndrome. No phenotype is found in girls that can be explained by the absence of prepubertal ovarian LH receptor expression in girls. Rosenthal et al. *(38)* described a clinical study of an adult woman who carries an activating LH receptor mutation (D578G). The hormonal responses to a GnRH-agonist challenge were found to be normal, and based on these findings, the authors suggest that the mutation did not activate LH receptor function beyond the pubertal level and that the negative feedback systems that regulate ovarian function are intact.

MOLECULAR ASPECTS

The mutations that are identified in the LH receptor gene of the patients also provide essential information about the importance of the individual amino acids that have been changed and about the subdomains in which they reside. In Fig. 4 are indicated the inactivating and activating mutations in the LH receptor gene that have been found thus far. Transmembrane segment 6 and the C-terminus of the third intracellular loop appears to be a hot spot for amino acid changes that cause LH receptor constitutive activity. The third intracellular loop and its flanking transmembrane segments have been indicated as major interactive regions with G_s, and this might explain the hyperactivity in the absence of ligand. Thus, changes in this region may directly activate G_s, in the absence of ligand-induced changes in the transmembrane domain. Alternatively, in addition to the single

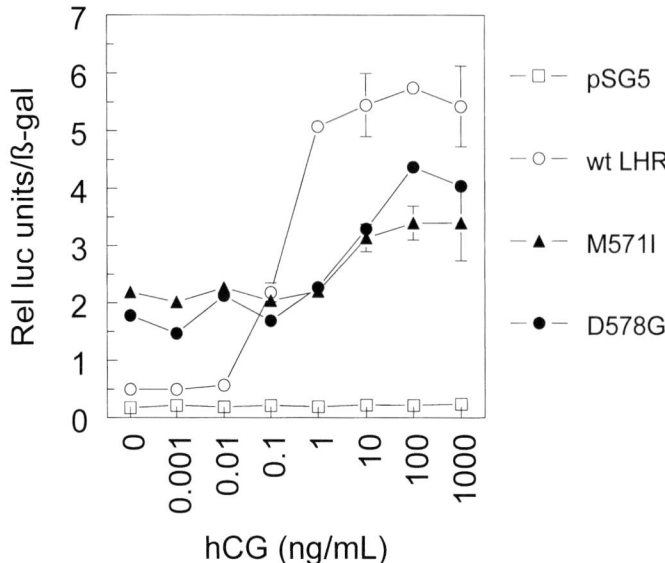

Fig. 3. Effect of activating LH receptor mutations identified in FMPP patients. The FMPP mutant receptors show enhanced activity in the absence of added hCG, but show a diminished maximal response.

amino acid change identified in transmembrane segment 2, these mutations may destroy interactions between transmembrane segments that keep the LH receptor in an inactive state in the absence of ligand.

Some of the LH receptor gene alterations that were identified in Leydig cell hypoplasia or primary amenorrhea patients were stop codons or gene deletions (not shown) that completely inactivate the LH receptor molecule by truncating it or by preventing its expression completely. Two amino acid changes were found in the extracellular domain (Fig. 4). However, one of these, N291S (not shown), does not change LH receptor activity *(25)* and may therefore be a rare neutral DNA polymorphism. Inhibition of LH binding may be caused by C133R, but this has not been formally tested in vitro *(39)*. In this respect, it is of interest that the patient with this mutation has a mild phenotype characterized by hypospadias and a micropenis. Not only are LH receptor molecules carrying the A593P mutation completely unable to transduce the hCG signal, but they also express very poorly when transfected in cells in vitro *(24)*. Patients homozygous for this mutation are complete pseudohermaphrodites. The presence of a proline just outside the sixth transmembrane domain possibly inhibits the proper insertion of the receptor into the plasma membrane. The I625K mutation was found in two brothers with a mild phenotype, micropenis, and reduced fertility *(17,26)*. Two patients were reported with the S616Y mutation. In Latronico et al. *(29)*, this mutation was found to be homozygous in a patient with a mild phenotype. The other patient presenting with a micropenis and severe hypospadias was a compound heterozygous carrier of the S616Y allele with an allele that carries a deletion of exon 8 of the LH receptor, which completely disables receptor function *(25)*. The degree of masculinization may depend on the combined residual LH receptor activity of the two alleles, which is lower in the latter patient.

Based on homology studies of a number of G-protein-coupled receptors, Baldwin *(40,41)* has suggested a model of the most probable orientation of the seven-transmem-

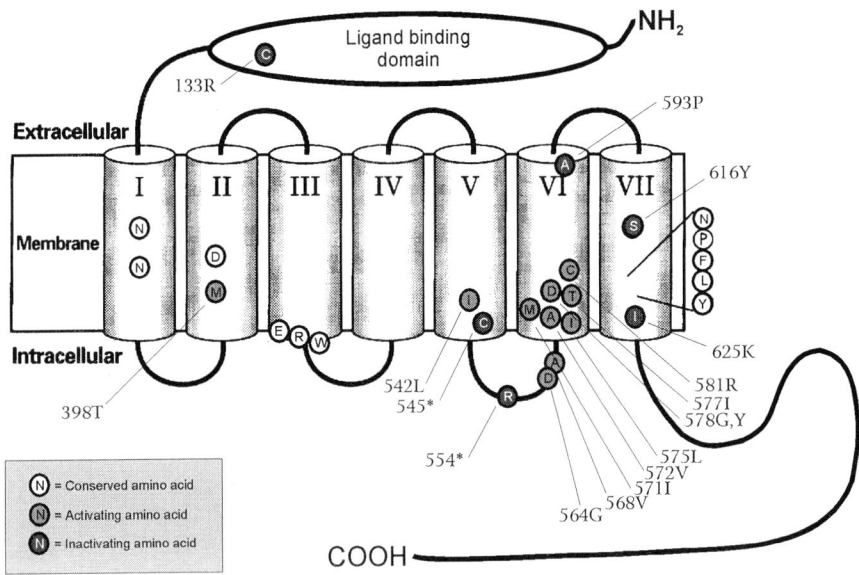

Fig. 4. Activating and inactivating mutations identified in the LH receptor gene. Activating mutations are indicated by light gray circles, and inactivating amino acid substitutions by dark gray circles. This figure is an adaptation of a model of the GnRH receptor *(42)*.

brane α-helices in the membrane. In this model, isoleucine 625 and serine 616 are located close to the conserved NPXXY motif *(see also* Fig. 4). This motif is found in all G-protein-coupled receptors, and is involved in receptor activation and sequestration *(43)*. A shift or a turn of the seventh transmembrane segment caused by an amino acid change at position 616 or 625 may alter the position of the NPXXY motif in relation to the other transmembrane segments of the LH receptor and diminish its response to LH or hCG.

In conclusion, the effects of the amino acid changes found in LH receptor genes of Leydig cell hypoplasic (LCH) and FMPP patients confirm the importance of the third intracellular loop and flanking regions for G-protein coupling. Furthermore, the inactivating effect of LH receptor mutations may reflect absence or reduction of LH receptor expression and/or a direct effect on signal transduction. Finally, the degree of inactivation, or more precisely, the level of residual activity determines the severity of the LCH phenotype.

CONCLUDING REMARKS

The finding of LH receptor gene mutations in patients with FMPP, Leydig cell hypoplasia or primary amenorrhea has not only increased our understanding of these rare conditions, but has also provided new insights in normal gonadal development and function. However, in some patients an LH receptor gene alteration has not been found *(44)* (our unpublished observations). Changes in the LH receptor gene promoter or in the 3'-untranslated region may be present. Because many of the introns of the human LH receptor gene are quite large, variants caused by mutated splice donor or acceptor sites have not been investigated yet. Lymphocytes that are used as a source of genomic DNA do not express the LH receptor mRNA, so screening for mutations using cDNA is not feasible. The possibility that other genes are affected cannot be excluded. Candidate alternative genes might be involved in the transport of the LH receptor protein to the plasma membrane, specific receptor kinases that affect receptor desensitization, or other proteins involved in LH receptor function. It is clear that continuing studies using patient material will unravel the signaling pathway that LH employs in the developing and adult gonadal cells.

ACKNOWLEDGMENT

J. W. M. Martens is supported by a grant from the Netherlands Organization for Scientific Research through MW.

REFERENCES

1. Parker KL, Schimmer BP. Steroidogenic factor 1: key determinant of endocrine development and function. Endocr Rev 1997;18:361–377.
2. MacLean HE, Warne GL, Zajac JD. Intersex disorders: shedding light on male sexual differentiation beyond SRY. Clin Endocrinol (Oxford) 1997;46:101–108.
3. Kendall SK, Samuelson LC, Saunders TL, Wood RI, Camper SA. Targeted disruption of the pituitary glycoprotein hormone alpha-subunit produces hypogonadal and hypothyroid mice. Genes Dev 1995;9:2007–2019.
4. Pelliniemi LJ, Kuopio T, Fröjdman K. The cell biology and function of the fetal Leydig cell. In: Payne AH, Hardy MP, Russel LD, eds. The Leydig Cell. Cache River Press. Vienna, IL, 1996, pp. 143–158.
5. Zondek LH, Zondek T. Fetal hilar cells and Leydig cells in early pregnancy. Biol Neonate 1976;30:193–199.
6. Styne DM. Puberty. In: Greenspan FS, Strewler GJ, eds. Basic and Clinical Endocrinology. Prentice-Hall International, London, 1997, pp. 521–547.
7. Ge R-S, Shan L-X, Hardy MP. Pubertal development of Leydig cells. In: Payne AH, Hardy MP, Russel LD, eds. The Leydig Cell. Cache River Press. Vienna, IL, 1996, pp. 143–158.
8. Molenaar R, de Rooij DG, Rommerts FF, van der Molen HJ. Repopulation of Leydig cells in mature rats after selective destruction of the existent Leydig cells with ethylene dimethane sulfonate is dependent on luteinizing hormone and not follicle-stimulating hormone. Endocrinology 1986;118:2546–554.
9. Berthezène F, Forest MG, Grimaud JA, Claustrat B, Mornex R. Leydig-cell agenesis: a cause of male pseudohermaphroditism. N Engl J Med 1976;295:969–972.
10. Brown DM, Markland C, Dehner LP. Leydig cell hypoplasia: a cause of male pseudohermaphroditism. J Clin Endocrinol Metab 1978;46:1–7.
11. Perez-Palacios G, Scaglia HE, Kofman-Alfaro S, Saavedra D, Ochoa S, Larraza O, et al. Inherited male pseudohermaphroditism due to gonadotrophin unresponsiveness. Acta Endocrinol 1981;98:148–155.
12. Schwartz M, Imperato-McGinley J, Peterson RE, Cooper G, Morris PL, MacGillivray M, et al. Male pseudohermaphroditism secondary to an abnormality in Leydig cell differentiation. J Clin Endocrinol Metab 1981;53:123–127.
13. Lee PA, Rock JA, Brown TR, Fichman KM, Migeon CJ, Jones H Jr. Leydig cell hypofunction resulting in male pseudohermaphroditism. Fertil Steril 1982;37:675–679.
14. Wu RH, Rosenfeld R, Fukushima D. Hypogonadism and Leydig cell hypoplasia unresponsive to human luteinizing hormone (hLH). Am J Med Sci 1984;287:23–25.
15. Eil C, Austin RM, Sesterhenn I, Dunn J, Cutler GB Jr, Johnsonbaugh RE. Leydig cell hypoplasia causing male pseudohermaphroditism: diagnosis 13 years after prepubertal castration. J Clin Endocrinol Metab 1984;58:441–448.
16. Arnhold IJ, Mendonca BB, Bloise W, Toledo SP. Male pseudohermaphroditism resulting from Leydig cell hypoplasia. J Pediatr 1985;106:1057.
17. Toledo SP, Arnhold IJ, Luthold W, Russo EM, Saldanha PH. Leydig cell hypoplasia determining familial hypergonadotropic hypogonadism. Prog Clin Biol Res 1985;200:311–314.
18. Arnhold IJ, de Mendonca BB, Toledo SP, Madureira G, Nicolau W, Bisi H, et al. Leydig cell hypoplasia causing male pseudohermaphroditism: case report and review of the literature. Rev Hosp Clin Fac Med Sao Paulo 1987;42:227–232.
19. El-Awady MK, Temtamy SA, Salam MA, Gad YZ. Familial Leydig cell hypoplasia as a cause of male pseudohermaphroditism. Hum Heredity 1987;37:36–40.

20. Saldanha PH, Arnhold IJ, Mendonca BB, Bloise W, Toledo SP. A clinico-genetic investigation of Leydig cell hypoplasia. Am J Med Genet 1987;26:337– 344.
21. Martinez-Mora J, Saez JM, Toran N, Isnard R, Perez-Iribarne MM, Egozcue J, et al. Male pseudohermaphroditism due to Leydig cell agenesia and absence of testicular LH receptors [see comments]. Clin Endocrinol 1991;34:485–491.
22. Toledo SP. Leydig cell hypoplasia leading to two different phenotypes: male pseudohermaphroditism and primary hypogonadism not associated with this [Letter; comment]. Clin Endocrinol (Oxford) 1992;36:521–522.
23. Segaloff DL, Ascoli M. The gonadotrophin receptors: insights from the cloning of their cDNAs [Review]. Oxford Rev Reprod Biol 1992;14:141–168.
24. Kremer H, Kraaij R, Toledo SPA, Post M, Fridman JB, Hayashida CY, et al. Male pseudohermaphroditism due to a homozygous missense mutation of the luteinizing hormone receptor gene. Nature Genet 1995;9:160–164.
25. Laue LL, Wu SM, Kudo M, Bourdony CJ, Cutler GB Jr, Hsueh AJ, et al. Compound heterozygous mutations of the luteinizing hormone receptor gene in Leydig cell hypoplasia. Mol Endocrinol 1996; 10:987–997.
26. Martens JWM, Verhoef-Post M, Abelin N, Ezabella M, Toledo SPA, Brunner HG, et al. A homozygous mutation in the Luteinizing Hormone receptor causes partial Leydig cell hypoplasia: correlation between receptor activity and phenotype. Mol Endocrinol 1998;12:775–784.
27. Toledo SP, Brunner HG, Kraaij R, Post M, Dahia PL, Hayashida CY, et al. An inactivating mutation of the luteinizing hormone receptor causes amenorrhea in a 46,XX female. J Clin Endocrinol Metab 1996;81:3850–3854.
28. Aittomaki K, Dieguez Lucena JL, Pakarinen P, Sistonen P, Tapanainen J, Gromoll J, et al. Mutation in the follicle-stimulating hormone receptor gene causes hereditary hypergonadotropic ovarian failure. Cell 1995;82:959–968.
29. Latronico AC, Anasti J, Arnhold I, Rapaport R, Mendonca BB, Bloise W, et al. Testicular and ovarian resistance to luteinizing hormone caused by inactivating mutations of the luteinizing hormone-receptor gene. N Engl J Med 1996;334:507–512.
30. Schedewie HK, Reiter EO, Beitins IZ, Seyed S, Wooten VD, Jimenez JF, et al. Testicular leydig cell hyperplasia as a cause of familial sexual precocity. J Clin Endocrinol Metab 1981;52:271–278.
31. Egli CA, Rosenthal SM, Grumbach MM, Montalvo JM, Gondos B. Pituitary gonadotropin-independent male-limited autosomal dominant sexual precocity in nine generations: familial testotoxicosis. J Pediatr 1985;106:33–40.
32. Gondos B, Egli CA, Rosenthal SM, Grumbach MM. Testicular changes in gonadotropin-independent familial male sexual precocity. Familial testotoxicosis. Arch Pathol Lab Med 1985;109:990–995.
33. Grumbach MM, Styne DM. Puberty: ontogeny, neuroendocrinology, physiology, and disorders. In: Wilson JD, Foster DW, eds. Textbook of Endocrinology, 8th ed. W.B. Saunders, Philadelphia, 1992, pp. 733–798.
34. Rosenthal SM, Grumbach MM, Kaplan SL. Gonadotropin- independent familial sexual precocity with premature Leydig and germinal cell maturation (familial testotoxicosis): Effects of a potent luteinizing hormone-releasing factor agonist and methoxyprogesterone acetate therapy in four cases. J Clin Endocrinol Metab 1983;57:571.
35. Shenker A, Laue L, Kosugi S, Merendino J Jr, Minegishi T, Cutler GB Jr. A constitutively activating mutation of the luteinizing hormone receptor in familial male precocious puberty. Nature 1993;365: 652–654.
36. Kremer H, Mariman E, Otten BJ, Moll G Jr, Stoelinga GB, Wit JM, et al. Cosegregation of missense mutations of the luteinizing hormone receptor gene with familial male-limited precocious puberty. Hum Mol Genet 1993;2:1779–1783.
37. Kjelsberg MA, Cotecchia S, Ostrowski J, Caron MG, Lefkowitz RJ. Constitutive activation of the alpha 1B-adrenergic receptor by all amino acid substitutions at a single site. Evidence for a region which constrains receptor activation. J Biol Chem 1992;267:1430–1433.
38. Rosenthal IM, Refetoff S, Rich B, Barnes RB, Sunthornthepvarakul T, Parma J, et al. Response to challenge with gonadotropin-releasing hormone agonist in a mother and her two sons with a constitutively activating mutation of the luteinizing hormone receptor - a clinical research center study. J Clin Endocrinol Metab 1996;81:3802–3806.
39. Misrahi M, Pissard S, Meduri G, Bouvattier C, Chaussain JL, Milgrom E, et al. Hypospadias and micropenis associated with a homozygous mutation of the luteinizing hormone receptor (LH/CGR)

gene. 35th Annual Meeting of the European Society for Paediatric Endocrinology (ESPE). Abstract 12, 1996.
40. Baldwin JM. The probable arrangement of the helices in G protein-coupled receptors. Embo J 1993;12:1693–1703.
41. Baldwin JM. Structure and function of receptors coupled to G proteins. Curr Opinion Cell Biol 1994;6: 180–190.
42. Arora KK, Cheng Z, Catt KJ. Dependence of agonist activation on an aromatic moiety in the DPLIY motif of the gonadotropin-releasing hormone receptor. Mol Endocrinol 1996;10:979–986.
43. Barak LS, Menard L, Ferguson SS, Colapietro AM, Caron MG. The conserved seven-transmembrane sequence NP(X)2,3Y of the G-protein-coupled receptor superfamily regulates multiple properties of the beta 2-adrenergic receptor. Biochemistry 1995;34:15,407–15,414.
44. Arnhold IJ, Latronico AC, Batista MC, Carvalho FM, Chrousos GP, Mendonca BB. Ovarian resistance to luteinizing hormone: a novel cause of amenorrhea and infertility. Fertil Steril 1997;67:394–397.

10   FSH Resistance

*Kristina Aittomäki, MD, PhD,
and Ilpo T. Huhtaniemi, MD, PhD*

CONTENTS

> INTRODUCTION
> THE ROLE OF FSH IN REGULATION OF NORMAL OVARIAN
>   AND TESTICULAR FUNCTION
> FSH BINDING AND SIGNALING PATHWAYS OF THE FSH RECEPTOR
>   (FSHR)
> MUTATIONS OF THE FSH β GENE
> FSHR GENE
> CLINICAL FEATURES OF FSH RESISTANCE
> CONCLUSIONS
> REFERENCES

INTRODUCTION

The essential role of the two pituitary gonadotropins, follicle-stimulating hormone (FSH) and luteinizing hormone (LH), as important regulators of ovarian and testicular function is well known. As a consequence, the ability of gonadotropins to stimulate ovarian follicular growth has not only been scientifically studied, but also vigorously exploited in the treatment of infertile women, although the exact molecular mechanisms of gonadotropin action are not yet known in detail in either sex. Despite the importance of gonadotropins in normal reproductive function, their role in infertility and reproductive disorders has long remained an open question. The recent development in molecular genetics has brought about not only the cloning of the genes encoding the glycoprotein hormones and their receptors, but has also provided new tools for the study of their actions at the molecular level. This development has enabled the identification of genetic mutations altering the function of both FSH and its receptor. Through these spontaneous mutations, effects of abnormal gonadotropin action can now be observed in humans as well as in experimental models using genetically modified animals. The phenotype caused by each mutation not only reveals the functional role of each specific gene, but also highlights the role of FSH hormone–receptor interaction in normal reproductive development and functions.

From: *Contemporary Endocrinology: Hormone Resistance Syndromes*
Edited by: J. L. Jameson © Humana Press Inc., Totowa, NJ

THE ROLE OF FSH IN REGULATION OF NORMAL OVARIAN AND TESTICULAR FUNCTION

The target cells of FSH in the ovary are the estrogen-producing granulosa cells. The ovaries of a female fetus contain up to 7 million oocytes, the majority of which undergo atresia already during the embryonic period and during childhood. This atresia is preceded by follicular development up to the preantral stage. The notion that this process is independent of FSH stimulation *(1)* is supported by the persistence of initial follicular growth in gonadotropin-deficient mice *(2)* and its occurrence in the perinatal rat ovary before the appearance of gonadotropin-receptors *(3,4)*. At puberty, the interrupted process of follicular growth is overcome, and cyclic ovarian function ensues. FSH is the crucial factor that rescues the follicles from atresia, and its receptors are exclusively expressed in the granulosa cells of developing follicles *(5)*. FSH activates the production of estrogen in granulosa cells through aromatization of androgens produced by theca cells, and by stimulating follicular growth, it initiates normal ovulatory cycles *(6,7)*.

In the testis, the target cells of FSH are the Sertoli cells. The significance of FSH in the regulation of spermatogenesis is much more controversial than its role in regulation of ovarian function *(8)*. It has been suggested that FSH is needed for the pubertal proliferation of Sertoli cells and for the initiation of spermatogenesis, as well as for maintaining qualitatively normal spermatogenesis in adult testes. It has also been postulated that FSH action might not be an absolute requirement for spermatogenesis *(9,10)*. Thus, the role of FSH in the regulation of spermatogenesis has remained a contentious issue.

FSH BINDING AND SIGNALING PATHWAYS OF THE FSH RECEPTOR (FSHR)

Both in the ovary and testis, FSH mediates its actions through the FSHR expressed in granulosa *(11,12)* and Sertoli cells *(13,14)*. FSH is a heterodimer consisting of two glycoprotein subunits; an α-subunit shared by FSH, LH, thyroid-stimulating hormone (TSH) and chorionic gonadotropin (CG) and a unique β-subunit. The two subunits are coupled together by noncovalent interactions *(15,16)* and probably both subunits participate in receptor binding *(17,18)*.

The FSHR belongs to the G-protein-coupled receptors, and together with luteinizing hormone receptor (LHR) and thyroid-stimulating hormone receptor (TSHR), it constitutes the subgroup of glycoprotein hormone receptors. As in the other receptors of this family, it consists of three domains: extracellular, transmembrane, and intracellular (Fig. 1). The seven-times plasma membrane spanning α-helices form the transmembrane domain and are typical of all G-protein-coupled receptors *(19,20)*. It is also the most conserved domain of the FSHR. The glycoprotein hormone receptors differ from the other receptors of the family by their exceptionally large extracellular domain, the site of hormone binding. It includes several repeated leucine-rich motifs *(21)*, which have also been identified in other proteins participating in protein–protein interaction *(22,23)*.

On ligand binding, the FSHR is activated. This activation takes place through conformational changes that allow the receptor to bind to the stimulatory G-protein (G_s), resulting in stimulation of adenylyl cyclase and production of cAMP *(24)*, the main intracellular secondary messenger of FSH action. Subsequently, protein kinase A (PKA) is activated by the accumulation of cAMP, and the function of target proteins, such as enzymes or transcription factors, is then regulated by phosphorylation of serine/threo-

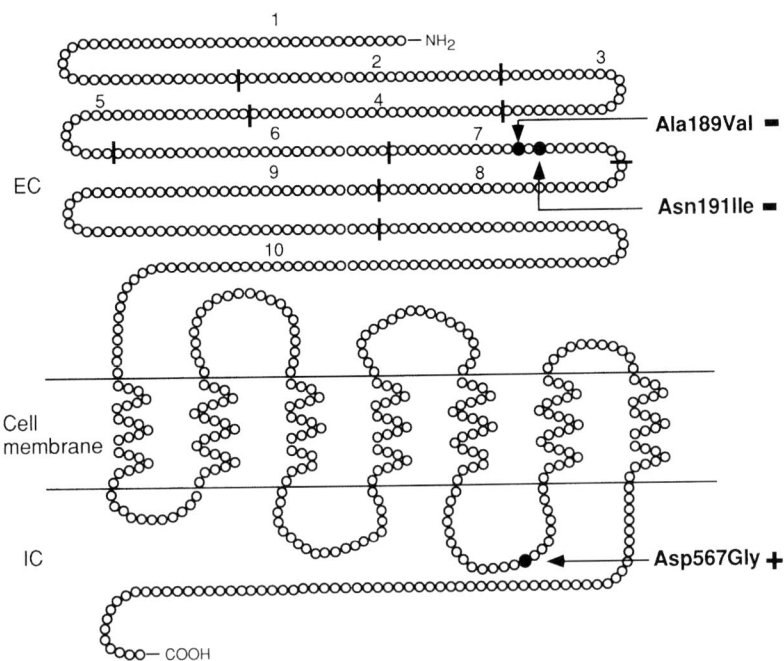

Fig. 1. A schematic presentation of the FSHR and gene structure. The location of each of the three known genomic mutations is designated with a black circle. An activating mutation is marked with (+) and inactivating with (−). The 10 exons of the FSHR gene are numbered and separated by a short line (EC, extracellular domain, IC, intracellular domain).

nine residues *(25)*. FSH also increases the intracellular free calcium concentrations in granulosa and Sertoli cells *(26,27)*. The mechanisms, signaling pathways, and the significance of the calcium changes still remain unknown. The possible role of the protein kinase C (PKC) signaling pathway, although supported by several studies *(28)*, also remains unclear together with some other suggested signaling pathways *(29)*.

MUTATIONS OF THE FSHβ GENE

The two subunits of FSH, α and β, are both encoded by separate genes. The gene encoding the α-subunit is shared by FSH, LH, TSH, and CG, and resides in chromosome 6 *(30)*. A loss-of-function mutation in this gene would cause resistance to all these hormones and a complex phenotype with symptoms. Although such mutations have not yet been found in humans, the knockout mouse model is viable *(31)*, suggesting that such mutations need not necessarily be lethal. A different gene encodes the specific β-subunit of each glycoprotein hormone. Inactivating homozygous mutations have been reported in TSHβ and LHβ, resulting in hypothyroidism and hypogonadism *(32–34)*. The gene encoding the FSH specific β-subunit has three exons and is located in chromosome 11p *(35,36)*. A loss-of-function mutation of FSHβ should cause FSH deficiency and resistance to endogenous, but not to exogenous FSH. The first of four such mutations known to date was identified in a female with primary amenorrhea, infertility, and isolated FSH deficiency *(37)*. She was shown to be homozygous for a deletion of 2 bp (TG) in exon 3 of FSHβ (Val61X), resulting in a frameshift and predicting a premature termination with truncated protein lacking the amino acids from 87 to 111. This truncated

FSH is inactive, because the carboxy-terminal end of FSHβ is necessary for both receptor binding and coupling with the α-subunit *(38)*. With exogenous FSH treatment, the patient ovulated and later underwent a normal pregnancy.

A similar phenotype with primary amenorrhea, incomplete pubertal development, and isolated FSH deficiency prompted the search for mutations in FSHβ in another female *(39)*. With sequencing the patient was shown to be a compound heterozygote, that is, to have different mutations in each of FSHβ alleles. One of these two mutations was identical to the one previously identified by Matthews et al. *(37)*, whereas the other was a missense point mutation (C to T) predicting a Cys51Gly change in the protein structure. Relatives of the patient, shown to be heterozygous for the Cys51Gly mutation, did not show any evidence of hormone deficiency suggesting that this mutation is inherited as an autosomal recessive trait: heterozygotes do not show any abnormalities, and the phenotype is limited to individuals with a loss-of-function mutation in both alleles of the gene.

The Val61X deletion mutation has also been identified in a third patient *(40)* previously reported for isolated FSH deficiency. Many other reports of male and female patients with isolated FSH deficiency have been published at the time when molecular genetic studies of FSHβ were not yet available *(41–43)*. In some reported cases, the males had spermatogenic failure, whereas others have reported FSH deficiency with normal fertility. Of interest is therefore the very recent report of Lindstedt et al. describing a male with azoospermia and isolated FSH deficiency *(44)*. He was shown to have a missense point mutation predicting a Cys82Arg change in FSHβ. However, exogenous FSH treatment did not correct the spermatogenic failure in this patient. Although the female phenotype caused by isolated FSH deficiency owing to FSHβ mutations seems to be similar in all reported cases, the significance of FSHβ mutations for male phenotype awaits further studies.

FSH RECEPTOR GENE

The Structure of the FSHR Gene

The FSHR gene is known to reside in chromosome 2p *(45,46)* in the same chromosomal area as the gene encoding LHR. The two genes show 70% homology with each other and have most probably arisen from the same ancestral gene during evolution *(21)*. Human FSHR was cloned in the early 1990s *(47,48)*, and the entire gene seems to span some 54 kb of genomic DNA. The FSHR gene has 10 exons and 9 introns (Fig. 1). The length of exons 1–9 varies between 69, and 251 bp whereas exon 10 spans 1251 bp constituting about half of the entire cDNA. The extracellular domain is encoded by exons 1-9, whereas the transmembrane as well as the intracellular domains are encoded by the large exon 10. The entire gene encodes 695 amino acids, a signal peptide of 17 amino acids, and a protein of 678 amino acids with mol wt of 75 kDa *(49)*. The FSHR gene expression is highly tissue-specific, hormone-dependent, and confined to granulosa and Sertoli cells. The promoter has several transcriptional start sites, but does not have a consensus TATA or CAAT start site *(50)*. These features are typical of the constitutively active housekeeping genes.

The Mutations of the FSHR Gene

To date, several mutations have been identified in the genes encoding G-protein-coupled receptors. These include both constitutively activating and inactivating mutations with known clinical phenotypes. Although a number of such mutations are known in the glycoprotein hormone receptor genes, only three genomic mutations are known in the FSHR gene. One of them is a constitutively activating mutation *(51)*, while two are inactivating, causing FSH resistance in a homozygous individual. In addition, the FSHR gene harbors several polymorphisms.

The first genomic mutation of the FSHR gene was identified in a Finnish study concerning the genetic etiology of hypergonadotropic ovarian failure *(52)*. For this purpose, 75 females with hypergonadotropic primary or early secondary amenorrhea were first identified in a population-based study in Finland *(53)*. This patient series included 18 familial and 57 sporadic cases. Genealogic studies were initiated to identify the possible mode of inheritance for this condition. The results, namely segregation analysis (0.23), parental consanguinity in 12%, uneven geographical distribution of ancestral birthplaces, and kinship between families, predicted autosomal-recessive inheritance at least in a proportion of patients. The multiplex families, that is families with two or more affected female sibs, were then chosen for a linkage study to map the gene locus for hypergonadotropic ovarian failure *(52)*. As a result, it was mapped to chromosome 2p, to the same chromosomal area where both the FSHR *(45,46)* and LHR gene *(54)* had previously been mapped. Of these two already cloned *(47,48,55)* candidate genes, the FSHR was considered more promising, since there was no known phenotype in male sibs of the affected females. When the FSHR was screened for mutations by DGGE and sequencing, a 566C→T point mutation was identified in exon 7. The mutation was shown to segregate perfectly with the phenotype as an autosomal-recessive trait in the study families. The 566C→T mutation predicts an Ala189Val change in the extracellular domain of the receptor protein (Fig. 1). To study its functional significance, FSH signaling was tested.

Mouse Sertoli cells (MSC-1) that do not express endogenous FSHR were transfected with wild-type FSHR plasmids and mutated FSHR plasmids *(52)*. Receptor function was then studied by measuring the cAMP production on stimulation with recombinant human FSH and by FSH binding. A clear-cut reduction of both cAMP production and FSH binding was observed, although the affinity of binding was similar in both cell lines. This, together with recent observations, suggests that the mutation affects folding of the receptor protein and its trafficking to the cell surface (Rannikko et al, unpublished).

Interestingly, the second inactivating point mutation of the FSHR gene identified so far is located adjacent to the Ala189Val *(56)* and predicts an Asn191Ile change in the protein structure. Both mutations involve the same conserved stretch of five amino acids shared by FSHR, LHR, and TSHR, implying that this conserved region, which also contains an N-linked glycosylation site, is structurally important. The Asn191Ile mutation was detected in a heterozygous, healthy, and fertile female, the situation therefore being comparable to the heterozygous mothers and female sibs of the Finnish families with the Ala189Val mutation. These females are healthy and fertile, as already demonstrated by the large kindreds of the multiplex families. Although the Asn191Ile mutation has not been found in homozygous individuals, its possible consequences have been functionally tested in vitro. Clearly reduced production of cAMP was observed in transfected COS-7 cells when compared to the cells transfected with the wild-type FSHR. Since these studies showed a loss of function and correspond to those of the Ala189Val mutation, the same mutation in a homozygous individual would most probably cause a phenotype similar to the one caused by the Ala189Val mutation.

It has been suggested that mutations of the FSHR might play a role in certain ovarian malignancies, such as granulosa cell tumors. Since many of the tumors have functional FSHR *(57)*, it could be important in tumor development. Although constitutively activating mutations could exert such an effect, the first mutation identified in ovarian tumors was functionally inactivating *(58)*. A heterozygous point mutation predicting a Phe591Ser change was found in several sex cord tumors and ovarian small-cell carcinomas, but not in control specimens. Since there was no other tissue available from the same patients, it could not be studied whether the mutation was somatic or genetic. The significance of this finding is unclear, as other studies have shown contradictory results *(59)*.

CLINICAL FEATURES OF FSH RESISTANCE

Female Phenotype

The resistance to gonadotropins was first suggested as a clinical syndrome during the late 1960s by Jones and de Moraes-Ruehsen *(60)*, although it was not possible to verify the cause of ovarian failure in the suggested patients at that time. The data on both female and male phenotypes for FSH resistance are mostly derived from Finnish patients homozygous for the Ala189Val mutation. In a clinical study, the phenotype of female patients was compared to the patients with hypergonadotropic ovarian failure of unknown origin, designated as ODG for ovarian dysgenesis *(61)*. All the patients with FSHR mutation had primary amenorrhea, whereas 8 out of 30 ODG patients had early secondary amenorrhea (Table 1). In hormonal analyses, the two groups did not differ and showed results typical of ovarian failure. However, when ovarian histology was compared between these two groups, a clear-cut difference was noted. All the patients with FSH-resistant ovaries (FSHRO) had ovarian follicles, whereas in 3/4 patients with hypergonadotropic ovarian failure, the ovaries were devoid of oocytes (Table 1). Follicles, up to the stage of preantral–antral follicle, were seen in ovaries of patients with the FSHR mutation. However, the most typical finding in histological sections was an ovary with many primordial or primary follicles (Fig. 2). This finding supports the previous notion that initial follicular development is independent of FSH regulation *(62)*. In one instance, a mature-looking follicle was seen in a microscopic slide. The possible explanation for this is the persistence of some residual function of the FSHR that might, with extremely low likelihood, allow some follicles to mature.

Male Phenotype of FSH Resistance

The effects of FSH resistance have also been studied in males homozygous for the same Ala189Val mutation as the females with FSH resistance *(63)*. When five such males were studied for fertility, two of them had two children each, whereas three had no children. All five men were normally masculinized, but had small to normal testicular size, elevated serum FSH concentrations, and low inhibin B levels, whereas their androgen production was normal (Table 2). Semen analysis of all men was abnormal, although with variable abnormal characteristics. The fact that the two males that had fathered two children as young adults had very low sperm counts in their 50s might be explained by a time factor. Although these males may have been fertile in their early 20s when the children were conceived, the spermatogenic capacity may have been lost by the time they were studied. Therefore, FSH deficiency might be accentuated by time and, in fact, the best semen quality of the five homozygous males was seen in the youngest patient.

Table 1
Clinical and Hormonal Data of Patients
with FSHRO Compared with Patients with Ovarian Failure
of Unknown Origin (ODG) Studied at Diagnosis[a]

	FSHRO, n = 22	ODG, n = 30
Amenorrhea		
Primary	22	22
Secondary	0	8
Adult height (cm)		
Median	160.5	165.8
Range	150–172	154–183
FSH (IU/L)		
Median	61	84
Range	42–176	42–146
LH (IU/L)		
Median	40	36
Range	17–1196	25–82
Estradiol (nmol/L)		
Median	0.06	0.05
Range	0.03–0.3	0.01–0.2
Follicles[b]		
Absent (stroma only)	0	3
Present	9	1

[a] Modified from ref. 61.
[b] Nine patients with FSHRO and four patients with ODG.

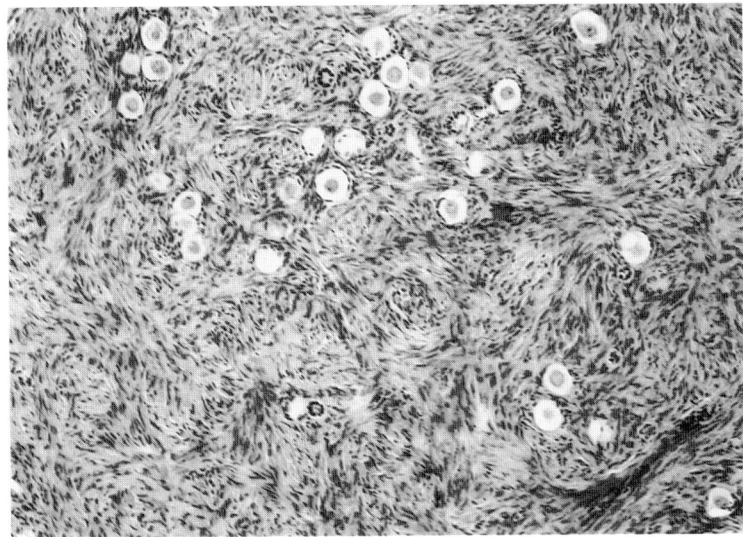

Fig. 2. An ovarian biopsy from a patient with FSH resistance owing to the inactivating Ala189Val mutation showing several primordial follicles. (Courtesy of Riitta Herva.)

Animal Models

Relevant to the findings in humans are the studies on knockout animals. An FSHβ knockout mouse was recently produced by Kumar et al. *(64)*. The findings in both sexes coincided with the clinical phenotype of the inactivating FSHR mutation: the females were

Table 2
Clinical and Hormonal Data of Five Males Homozygous for the Ala189Val Mutation[a]

Age	Testis size mL, right/left	Sperm count, /mL	FSH IU/L	LH IU/L	Testosterone, mU/L	Inhibin B, ng/L
45	4.0/4.0	$<0.1 \times 10^{6}$ [b]	23.5	16.3	14.5	<15.0
47	15.0/13.8	5.6×10^{6}	12.5	5.6	8.8	33.0
55	13.5/15.8	$<0.1 \times 10^{6}$	15.1	4.2	15.8	62.0
42	8.0/8.0	$<1.0 \times 10^{6}$	20.6	16.2	26.2	54.0
29	8.6/6.0	42×10^{6}	39.6	11.1	14.7	53.0
Reference range	≥15 ml	$\geq 20 \times 10^{6}$	1–10.5	1-8.4	8.2-34.6	76-447 [c]

[a] Modified from ref. 63.
[b] Studied at the age of 30 yr.
[c] Range of the mean ± 2 SD of 10 normal men.

infertile, but the males had reduced testis size and some impairment of spermatogenesis, but were fertile. The results of the clinical phenotype of FSH resistance in humans and animal models therefore support each other both for the female and male phenotypes.

CONCLUSIONS

A recent development in molecular genetic methods has enabled the identification of mutations underlying FSH deficiency and resistance. The phenotype caused by mutations of the genes encoding the two subunits of FSH and its receptor will not only explain the disturbances themselves, but also enlighten their role in human reproduction. Presently, few mutations and their phenotypes have been described. The frequency of such mutations in different reproductive disorders and different populations has not yet been studied. The Ala189Val mutation of FSHR has been identified in 24 females with hypergonadotropic ovarian failure, and it shows an uneven geographical distribution in Finland. It has not been reported elsewhere, and neither have we been able to detect it in mutation screening of patients with hypergonadotropic hypogonadism from other populations (64a). An enrichment of mutations for certain recessively inherited disorders is a well-known feature of the Finnish population, and regional subisolates often show even higher frequencies than the population as a whole (65,66). Since these are also features of the Ala189Val mutation of FSHR, it is included in the Finnish disease heritage as the first gynecological disorder with some 30 other inherited disorders that are common in Finland, but rare elsewhere (66). The identification of another inactivating FSHR mutation in a German female, however, suggests that similar mutations do exist in other populations (56). Moreover, in addition to the known mutations, the FSHR gene also harbors polymorphisms (52,67), which by definition are not clinically important. Transfection studies and functional testing are thus invaluable when determining the significance of newly found genetic changes, especially since all mutations cannot be studied in large families.

A clear phenotype is caused in females homozygous for an inactivating FSHR mutation. This includes hypergonadotropic primary amenorrhea and infertility with ovaries showing mostly primordial and primary follicles. Such a histological finding in an ovarian biopsy therefore suggests FSH resistance owing to an inactivating mutation of FSHR gene. Although it is evident that there are yet new mutations to be identified, they probably will not explain but a fraction of hypergonadotropic ovarian failure. Others might turn out to be caused by mutations of other genes, some perhaps by genes encoding other proteins important in the FSH signal transduction. However, it is already evident from the existing data that FSH is absolutely necessary for the normal cyclic ovarian function and female fertility.

Regarding the males, the role of FSH in spermatogenesis has been uncertain. Judged by the male phenotype in inactivating FSHR mutations, it does play a role in determination of testis size and both quantitatively and qualitatively contributes to spermatogenesis. But contrary to the females, FSH is not an absolute requirement for male fertility. What then is the proportion of male infertility caused by FSHR mutations? Probably not very high. This was studied by screening 151 infertile males originating from the geographical area where the Ala189Val mutation is known to be enriched with estimated carrier frequency of 1:80–1:100 *(63)*. Two males out of 151 where shown to be heterozygous carriers, but there were no homozygotes among these men, implying that for male infertility, FSHR mutations are a rare cause.

REFERENCES

1. Baker TG, Scrimgeour JB. Development of the gonads in anencephalic human fetuses. In: Coutts JRT, ed. Functional Morphology of the Human Ovary. MTP Press, Lancaster, 1981, pp. 13–25.
2. Halpin DM, Charlton HM, Faddy MJ. Effects of gonadotrophin deficiency on follicular development of hypogonadal (hpg) mice. J Reprod Fertil 1986;78:119–125.
3. Sokka T, Huhtaniemi I. Ontogeny of the gonadotrophin receptors and gonadotrophin stimulated cAMP production in neonatal rat ovary. J Endocrinol 1990;127:297–303.
4. Rannikko AS, Zhang FP, Huhtaniemi I. Ontogeny of follicle-stimulating hormone receptor gene expression in the rat testis and ovary. Mol Cell Endocrinol 1995;107:196–208.
5. Richards JS. Hormonal control of gene expression in the ovary. Endocr Rev 1994;15:725–751.
6. Dorrington JH, Moon YS, Armstrong DT. Estradiol 17α biosynthesis in cultured granulosa cells from hypophysectomized immature rats: stimulation by follicle-stimulating hormone. Endocrinology 1975; 97:1328–1331.
7. Gore-Langton RE, Armstrong DT. Follicular steroidogenesis and its control. In: Knobil E, Neill JD, Ewing LL, Greenwald GS, Markert SL, Pfaff DW, eds. The Physiology of Reproduction. Raven, New York, 1988, pp. 331–386.
8. Zirkin BR, Awoniyi C, Griswold MD, Russell LD, Sharpe RM. Is FSH required for adult spermatogenesis? J Androl 1994;15:273–276.
9. Matsumoto AM. Hormonal control of human spermatogenesis. In: Burger H, de Kretser D, eds. The Testis. Raven, New York, 1989, pp. 181–196.
10. Weinbauer GF, Nieschlag E. Hormonal control of spermatogenesis. In: de Kretser D, ed. Molecular Biology of the Male Reproductive System. Academic, New York, 1993, pp. 99–142.
11. Richards JS, Midgley AR. Protein hormone action: a key to understanding ovarian follicular and luteal cell development. Biol Reprod 1976; 14:82–94.
12. Richards JS. Maturation of ovarian follicles: actions and interaction of pituitary and ovarian hormones on follicular cell differentiation. Physiol Rev 1980;60:51–89.
13. Fritz IB. Sites of action of androgens and follicle-stimulating hormone on cells of the seminiferous tubule. In: Litwack G, ed. Biochemical Actions of Hormones. Academic, New York, 1978, pp. 240–281.
14. Kangasniemi M, Kaipia A, Toppari J, Perheentupa A, Huhtaniemi I, Parviainen M. Cellular regulation of follicle-stimulating hormone (FSH) binding in rat seminiferous tubules. J Androl 1990;11:336–343.
15. Bousfield GR, Perry WM, Ward DN. Gonadotropins; chemistry and biosynthesis. In: Knobil E, Neill JD, eds. The Physiology of Reproduction. 2nd ed. Raven, New York, 1994, pp. 1749–1792.
16. Moyle WR, Campbell RK. Gonadotropins. In: Adashi EY, Rock JA, Rosenwaks Z, eds. Reproductive Endocrinology, Surgery, and Technology. Lippincott-Raven, Philadelphia, 1996, pp. 683–724.
17. Chopineau M, Maurel MC, Combarnous Y, Durand Y. Topography of equine chorionic gonadotropin epitopes relative to the luteinizing hormone and follicle-stimulating hormone receptor interaction sites. Mol Cell Endocrinol 1993;92:229–239.
18. Liu C, Dias JA. Long loop residues 33-53 in the human glycoprotein hormone common α-subunit contain structural components for subunit heterodimerization and human follitropin receptor binding. Arch Biochem Biophys 1996;329:127–135.

19. Baldwin JM Structure and function of receptors coupled to G proteins. Curr Opinion Biol 1994;6:180–190.
20. Oliveira L, Paiva ACM, Vriend G. A common motif in G-protein-coupled seven transmembrane helix receptors. J Computer-Aided Molec Design 1993;7:649–658.
21. Vassart G, Parmentier M, Libert F, Dumont J. Molecular genetics of the thyrotropin receptor. Trends Endocrinol Metab 1991;2:151–156.
22. McFarland KC, Sprengel R, Phillips HS, Köhler M, Rosemblit N, Nikolics K, et al. Lutropin-choriogonadotropin receptor: an unusual member of the G protein-coupled receptor family. Science 1989;245:494–499.
23. Kobe B, Deisenhofer J. Crystal structure of porcine ribonuclease inhibitor, a protein with leucine-rich repeats. Nature 1993;366:751–756.
24. Birnbaumer L. Receptor-to-effector signaling through G proteins: roles for beta gamma dimers as well as alpha subunits. Cell 1992;71:1069–1072.
25. Lalli E, Sassone-Corsi P. Signal transduction and gene regulation: the nuclear response to cAMP. J Biol Chem 1994;269:17,359–17,362.
26. Flores JA, Vledhuis JD, Leong DA. Follicle-stimulating hormone evokes an increase in intracellular free calcium ion concentrations in single ovarian cells. Endocrinology 1990;127:3172–3179.
27. Gorczynska E, Handelsman DJ. The role of calcium in follicle-stimulating hormone signal transduction in Sertoli cells. J Biol Chem 1991;266:23,739–23,744.
28. Berridge MJ. Inositol trisphosphate and calcium signalling. Nature 1993;361(6410):315–325.
29. Richards JS. Hormonal control of gene expression in the ovary. Endocr Rev 1994;15(6):725–751.
30. Naylor SL, Chin WW, Goodman HM, Lalley PA, Grzeschnik K-H, Sakaguchi AY. Chromosome assignment of the genes encoding the alpha and beta subunits of the glycoprotein hormones in man and mouse. Somat Cell Genet 1983;9:757–770.
31. Kendall SK, Samuelson LC, Saunders TL Wood RI, Camper SA. Targeted disruption of the pituitary glycoprotein hormone (α-subunit produces hypogonadal and hypothyroid mice. Genes Dev 1995;9: 2007–2019.
32. Hayashizaki Y, Hiraoka Y, Endo Y, Matsubara K. Thyroid-stimulating hormone (TSH) deficiency caused by a single base substitution in the CAGYC region of the beta-subunit. EMBO J 1989;8:2291–2296.
33. Medeiros-Neto G, Herodotou DT, Rajan S, Kommareddi S, de Lacerda L, Sandrini R, et al. A circulating, biologically inactive thyrotropin caused by a mutation in the beta subunit gene. J Clin Invest 1996;97:1250–1256.
34. Weiss J, Axelrod L, Whitcomb RW, Harris PE, Crowley WF, Jameson JL. Hypogonadism caused by a single amino acid substitution in the beta subunit of luteinizing hormone. N Engl J Med 1992;326:179–183.
35. Watkins P, Eddy R, Beck A, Vellucci V, Gusella J, Shows T. Assignment of the human gene for the beta subunit of follicle-stimulating hormone (FSHB) to chromosome 11. Cytogenet Cell Genet 1985;40:773.
36. Watkins PC, Eddy R, Beck AK, Vellucci V, Leverone B, Tanzi RE, et al. DNA sequence and regional assignment of the human follicle-stimulating hormone beta-subunit gene to the short arm of human chromosome 11. DNA 1987;6:205–212.
37. Matthews CH, Borgato S, Beck-Peccoz P, Adams M, Tone Y, Gambino G, et al. Primary amenorrhea and infertility due to a mutation in the beta-subunit of follicle-stimulating hormone. Nature Genet 1993; 5:83–86.
38. Dias JA. Progress and approaches in mapping the surfaces of human follicle-stimulating hormone. Trends Endocrinol Metab 1992;3:24–29.
39. Layman LC, Lee EJ, Peak DB, Namnoum AB, Vu KV, van Lingen BL, et al. Delayed puberty and hypogonadism caused by mutations in the follicle-stimulating hormone beta-subunit gene. N Engl J Med 1997;337:607–611.
40. Matthews C, Chatterjee VK. Isolated deficiency of follicle-stimulating hormone re-visited. N Engl J Med 1997;337:642.
41. Maroulis GB, Parlow AF, Marshall JR. Isolated follicle-stimulating hormone deficiency in a man. Fertil Steril 1977;28:818–822.
42. Al-Ansari AA-K, Khalil TH, Kelani Y, Mortimer CH. Isolated follicle-stimulating hormone deficiency in a men: successful long-term gonadotropin therapy. Fertil Steril 1984;42:618–626.
43. Diez JJ, Iglesias P, Sastre J, Salvador J, Gomez-Pan A, Otero I, et al. Isolated deficiency of follicle-stimulating hormone in man: a case report and literature review. Int J Fertil Menopausal Stud 1994;39: 26–31.
44. Lindstedt G, Ernest I, Nyström E, Janson PO. Fall av manlig infertilitet. Kliniskt Kemi i Norden 1997;3: 81–87.

45. Rousseau-Merck MF, Atger M, Loosfelt H, Milgrom E, Berger R. The chromosomal localization of the human follicle-stimulating hormone receptor gene (FSHR) on 2p21-p16 is similar to that of the luteinizing hormone receptor gene. Genomics 1993;15:222–224.
46. Gromoll J, Ried T, Holtgreve-Grez H, Nieschlag E, Gudermann T. Localization of the human FSH receptor to chromosome 2p21 using a genomic probe comprising exon 10. J Mol Endocr 1994;12:265–271.
47. Minegishi T, Nakamura K, Takakura Y, Ibuki Y, Igarashi M. Cloning and sequencing of human FSH receptor cDNA. Biochem Biophys Res Commun 1991;175:1125–1130.
48. Kelton CA, Cheng SVY, Nugent NP, Schweickhardt RL, Rosenthal JL, Overton SA, et al. The cloning of the human follicle-stimulating hormone receptor and its expression in COS-7, CHO, and Y-1 cells. Mol Cell Endocr. 1992;89:141–151.
49. Gromoll J, Pekel E, Nieschlag E. The structure and organization of the human follicle-stimulating hormone receptor (FSHR) gene. Genomics 1996;35:308–311.
50. Griswold MD, Heckert L, Linder C. The molecular biology of the FSH receptor. J Steroid Biochem Mol Biol 1995;53:215–218.
51. Gromoll J, Simoni M, Nieschlag E. An activating mutation of the follicle-stimulating hormone receptor autonomously sustains spermatogenesis in a hypophysectomized man. J Clin Endocrinol Metab 1996;81:1367–1370.
52. Aittomäki K, Dieguez Lucena JL, Pakarinen P, Sistonen P, Tapanainen J, Gromoll J, et al. Mutation in the follicle-stimulating hormone receptor gene causes hereditary hypergonadotropic ovarian failure. Cell 1995;82:959–968.
53. Aittomäki K. The genetics of XX gonadal dysgenesis. Am J Hum Genet 1994;54:844–851.
54. Rousseau-Merck MF, Mirashi M, Atger M, Loosfelt H, Milgrom E, Berger R. Localization of the human LH (luteinizing hormone) receptor gene to chromosome 2p21. Cytogenet Cell Genet 1990;54:77–79.
55. Minegishi T, Nakamura K, Takakura Y, Miyamoto K, Hasegawa Y, Ibuki Y, et al. Cloning and sequencing of human LH/hCG receptor cDNA. Biochem Biophys Res Commun 1990;172:1049–1054.
56. Gromoll J, Simoni M, Nordhoff V, Behre HM, De Geyter C, Nieschlag E. Functional and clinical consequences of mutations in the FSH receptor. J Clin Endocrinol Metab 1996;125(1–2):177–182.
57. Stouffer RL, Gordin MS, Davis JS, Surwit EA. Investigation of binding sites for follicle-stimulating hormone and chorionic gonadotropin in human ovarian cancers. J Clin Endocrinol Metab 1984;59:441–446.
58. Kotlar TJ, Young RH, Albanese C, Crowley WF Jr, Scully RE, Jameson JL. A mutation in the follicle-stimulating hormone receptor occurs frequently in human ovarian sex cord tumors. J Clin Endocrinol Metab 1997;82(4):1020–1026.
59. Ichikawa Y, Yoshida S, Suzuki H, Nishida M, Tsunoda H, Kubo T, et al. Mutation analysis of gonadotropin receptor and G protein genes in various types of human ovarian tumors. Jap J Clin Oncol 1996;26:298–302.
60. Jones GS, de Moraes-Ruehsen M. A new syndrome of amenorrhea in association with hypergonadotropism and apparently normal ovarian follicular apparatus. Am J Obstet Gynecol 1969;104:597–600.
61. Aittomäki K, Herva R, Stenman U-H, Juntunen K, Ylöstalo P, Hovatta O, et al. Clinical features of ovarian failure caused by a point mutation in the follicle-stimulating hormone receptor gene. J Clin Endocrinol Metab 1996;81:3722–3726.
62. Hillier SG. Current concepts of the roles of follicle stimulating hormone and luteinizing hormone in folliculogenesis. Hum Reprod 1994;9:188–191.
63. Tapanainen J, Aittomäki K, Min J, Vaskivuo T, Huhtaniemi I. Men homozygous for an inactivating mutation of the follicle-stimulating hormone receptor gene present variable suppression of spermatogenesis and infertility. Nature Genet 1997;15:205–206.
64. Kumar TR, Wang Y, Lu N, Matzuk MM. Follicle stimulating hormone is required for ovarian follicle maturation but not male fertility. Nature Genet 1997;15(2):201–204.
64a. Jiang M, Aittomäki K, Nilsson C, Pakarinen P, Iitiä A, Torresani T, Simonsen H, Goh V, Pettersson K, de la Chapelle A, Huhtaniemi I. The frequency of an inactivating point mutation (566C->T) of the human FSH receptor gene in four populations using allele-specific hybridization and time-resolved fluorometry. J Clin Endocrinol Metab, in press.
65. Norio R. Diseases of Finland and Scandinavia. In: Rotschild H, ed. Biocultural Aspects of Disease. Academic, New York, 1981, pp. 359–415.
66. de la Chapelle A. Disease gene mapping in isolated human populations: the example of Finland. J Med Genet 1993;30:857–865.
67. Whitney EA, Layman LC, Chan PJ, Lee A, Peak DB, McDonough GP. The follicle-stimulating hormone receptor gene is polymorphic in premature ovarian failure and normal controls. Fertil Steril 1995; 64:581–524.

11 Genetic Alterations of Androgen Receptor Function

Ken Brantley, MD, PhD, Tianshu Gao, MD, and Michael J. McPhaul, MD

CONTENTS

    INTRODUCTION
    THE SPECTRUM OF PHENOTYPES CAUSED BY DEFECTS
      OF THE HUMAN ANDROGEN RECEPTOR
    METHODS TO ANALYZE THE DEFECTS OF THE ANDROGEN RECEPTOR
    THE HUMAN ANDROGEN RECEPTOR AND ITS FUNCTIONAL DOMAINS
    TYPES OF AR MUTATIONS
    ABSENCE OF SPECIFIC LIGAND BINDING
    ANDROGEN RECEPTOR MUTATIONS WITH NORMAL LEVELS
      OF LIGAND BINDING
    MUTATIONS OF THE ANDROGEN RECEPTOR CAUSING MINIMAL
      DEFECTS OF VIRILIZATION
    THE RELATIONSHIP BETWEEN MUTATIONS OF THE AR
      AND THE OBSERVED PHENOTYPE
    VARIATIONS IN PHENOTYPE
    HOMOPOLYMERIC REPEATS AND HETEROGENEITY WITHIN
      THE ANDROGEN RECEPTOR GENE
    SPINAL AND BULBAR MUSCULAR ATROPHY
    MUTATIONS OF THE ANDROGEN RECEPTOR IN PROSTATE CANCER
    DISTRIBUTION OF MUTATIONS WITHIN THE ANDROGEN
      REECEPTOR GENE
    SUMMARY
    ACKNOWLEDGMENT
    REFERENCES

INTRODUCTION

In vertebrates, the morphogenesis of tissues is a complex developmental process that requires the interaction of many different cell types via endocrine and paracrine pathways. Although these processes are only just beginning to be unraveled for many tissues, a number of the genes involved in the development and differentiation of the vertebrate

From: *Contemporary Endocrinology: Hormone Resistance Syndromes*
Edited by: J. L. Jameson © Humana Press Inc., Totowa, NJ

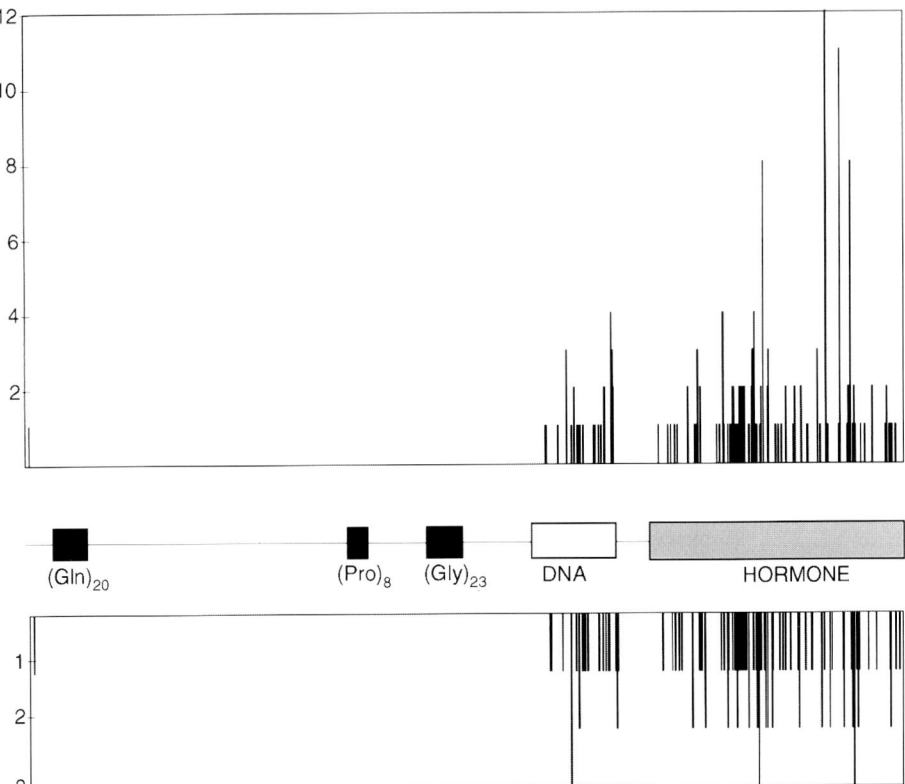

Fig. 1. Overview of effectors participating in the process of male sexual development. This diagram demonstrates several key features of male sexual development. The initial processes accounting for the development of the male phenotype stem from the establishment of embryonic genotype (46, XY genotype). Presence of the Y-chromosome containing a functional copy of the SRY gene dictates testicular differentiation. As a result of this, specific cell types within the developing testis produce androgens (testosterone and 5α-dihydrotestosterone) and MIS. These hormones act to inhibit the development of female structures (uterus, fallopian tubes) and to promote the development of structures characteristic of the male phenotype. An important feature is that defects in specific proteins involved in testicular differentiation, testosterone synthesis, or androgen metabolism (5α-reductase deficiency) can all give rise to defects of virilization. It should also be noted that defects of androgen receptor function can act to alter the development of structures dependent on both testosterone and on 5α-dihydrotestosterone for their formation.

urogenital tract have been identified, and information pertaining to their biology and expression is becoming available.

In mammals, the complement of sex chromosomes in the fertilized ovum establishes the genotypic sex, which is the principal determinant of the sequence of events that take place during sexual development. In particular, the presence or absence of the Y chromosome determines whether the primordial gonad—morphologically identical in males and females in its indifferent state—will develop as a testis or as an ovary *(1)*. It is this event—the functional differentiation of the gonad—that determines the subsequent events in mammalian sexual development (Fig. 1) *(2)*.

In the male embryo, as a result of the differentiation of the testis, the gonad begins to secrete steroid and polypeptide hormones. At approx 9 wk of development, the testes begin to secrete androgen. Testosterone, in combination with its 5α-reduced metabolite,

5α-dihydrotestosterone, acts to induce the virilization of both the internal and external genitalia. The external genitalia virilize with the enlargement of the phallus and fusion of the genital ridges to form the scrotum. In parallel, the Wolffian ducts grow to form the pelvic portion of the urogenital sinus, and give rise to the seminal vesicles and the epididymis. Within this same time frame, the Sertoli cells of the testes secrete a polypeptide hormone, Müllerian inhibiting substance (MIS), which acts to induce a regression of the Müllerian duct-derived structures, such as the uterus and fallopian tubes. In the absence of these testicular hormones, the uterus and fallopian tubes will develop, and the upper vaginal segments and urogenital swellings fail to fuse.

As can be inferred from the preceding comments, normal male development requires the combined actions of testosterone, 5α-dihydrotestosterone, and MIS. Although defects in a number of genes have been identified that act to alter normal sexual development, it is clear that defects in male phenotypic development can be caused by any abnormality that acts to impair the development of the testes, normal androgen synthesis, or the capacity of tissues to respond to androgens *(2)*.

THE SPECTRUM OF PHENOTYPES CAUSED BY DEFECTS OF THE HUMAN ANDROGEN RECEPTOR

Androgen receptor (AR) defects have been identified more frequently than defects in the other members of the steroid/thyroid hormone receptor family. This is most likely owing to several characteristics of the androgen receptor. First, the androgen receptor gene is located on the X chromosome. For this reason, males—in whom androgens exert their most obvious effects—inherit only one copy of the gene and will display phenotypic effects if the AR gene is defective. Second, though the androgen receptor mediates many important functions, including virilization of external genitalia, maintenance of bone density, and normal spermatogenesis, these functions are not absolutely required for survival. As a result, it appears that even marked abnormalities of AR function have minimal effects on fetal viability, and individuals in which AR function is completely absent will be viable and available for ascertainment. This may not be true of abnormalities of other receptors, where pronounced defects may be underrepresented owing to *in utero* deaths. Finally, the phenotypes resulting from androgen insensitivity or resistance are often obvious at birth (ambiguities of sexual development) or detected at puberty (complete testicular feminization).

One of the most remarkable features of androgen resistance caused by mutation in the androgen receptor gene is the remarkable variation in phenotype that is available for study *(3)*.

In the complete absence of receptor function, both the internal and external male structures fail to develop. This clinical picture has been given a variety of names, including complete testicular feminization and complete androgen insensitivity. These patients externally are normally developed females with developed breasts and normal external female genitalia. In some instances, decreased axillary and pubic hair provides a diagnostic clue to the underlying disorder. Careful examination of such individuals will detect testes either within the labia majora or within the abdominal cavity. Owing to the action of MIS produced by the functional testes, the uterus and Fallopian tubes are absent, and the vagina is blind ending.

Less severely affected individuals can display a range of intermediate phenotypes *(3)*. These have been referred as partial androgen insensitivity, incomplete testicular

Table 1
Phenotypes and Ligand Binding Assays in a Collection
of Patients With Androgen Resistance Caused by AR Defects[a]

	Complete testicular feminization	Incomplete testicular feminization	Reifenstein syndrome	Infertile male	Undervirilized, infertile male
Ligand binding negative	24	6	3		
Qualitative abnormality of ligand binding	13	9	23	8	3
Reduced ligand binding	2	4	7	5	
Normal ligand binding	3	6	12	2	

[a] The phenotypes and monolayer binding assay results from 130 patients are depicted. The criteria employed—both clinically and experimentally—are described in ref. (3), from which this table is adapted.

feminization, and Reifenstein syndrome. Such individuals exhibit varying degrees of virilization, and may display either a predominantly female phenotype (incomplete androgen insensitivity, incomplete testicular feminization) or may display a predominantly male phenotype with severe urogenital abnormalities (Reifenstein phenotype). Some authors have suggested a more detailed system by which to categorize patients with incomplete forms of androgen resistance (4).

Finally, a small group of patients display phenotypes in which male sexual development is normal or near normal, but in whom some processes mediated by androgen are not normal. Such patients may display subtle signs of undervirilization—such as gynecomastia—or may simply demonstrate infertility as a manifestation of their defective receptor function. A number of such families have been described, and are discussed in more detail below.

METHODS TO ANALYZE THE DEFECTS OF THE ANDROGEN RECEPTOR

A wide spectrum of receptor abnormalities have been defined based on ligand binding studies performed on genital skin fibroblasts established from 46, XY patients with androgen insensitivity or resistance syndromes. Table 1 displays the results of studies of genital skin fibroblasts from 130 families with androgen resistance syndromes (3). There is no clear relationship between the clinical phenotype and the character of the receptor abnormalities determined by assays of qualitative and quantitative ligand binding. For example, receptor abnormalities that are indistinguishable on the basis of ligand binding assays are associated with complete androgen resistance in some families and only mild undervirilization in others. In addition, in a considerable number of families, no abnormality of the androgen receptor is detected using ligand binding assays, yet a defect of the androgen receptor is suggested by family history or by endocrine testing.

In recent years, the tools for analyzing the defects in the AR have undergone dramatic changes. Following the isolation of cDNAs* encoding the human androgen receptor

*The polymorphisms of the glutamine repeat within the androgen receptor amino-terminus led to differences in numbering between the coordinates of each of the cDNAs encoding the human androgen receptor that were isolated (5–8). Since the Androgen Receptor Mutation Database (available at http://www.mcgill.ca/androgendb) has adopted the numbering protocol of Lubahn et al. (6), this numbering system has been employed in reference to nucleotide and amino acid coordinates in this chaper as well.

(5–8), the characterization of the AR gene made it possible to develop techniques that permitted the identification of mutations within the androgen receptor gene from patients with the various classes of androgen resistance. In a small proportion of cases, techniques such as Southern blotting, were sufficient to detect large-scale changes in the AR gene. In most instances, however, the genetic changes were found to be much subtler, requiring the use of methods to amplify and analyze the sequence of individual segments of the AR gene.

In parallel, methods have also been developed to study the effects of these mutations on the abundance and function of the AR. Methods to measure the AR mRNA using ribonuclease protection assays have been utilized to identify mutations that may not change the function of the AR protein, but may alter the steady-state levels of the AR mRNA. Likewise, the development of immunoblot techniques using specific antibodies that recognize the AR have made it possible to measure the levels of immunoreactive receptor protein in fibroblasts and transfected cells *(9–11)*.

It was anticipated from the ligand binding studies that some mutations of the AR would have effects on AR function not reflected in measurements of ligand binding or receptor abundance. For this reason, the development of assays capable of measuring receptor function independent of effects on binding and abundance was critical. Androgen-response elements from a number of different androgen-responsive genes (e.g., the C3, probasin, PSA genes) *(12–15)* have been identified and used to devise tests that could be applied to normal and mutant receptors to measure their capacities to regulate gene transcription. Although arguably one of the least physiologic response elements, the DNA fragment containing the steroid responsive elements of the mouse mammary tumor virus long terminal repeat (MMTV-LTR, ref. *16,17*) has been widely used to characterize mutations that impair AR function regardless of the effect that the mutation has on ligand binding. This element, when linked to an easily assayable enzymatic activity, such as chloramphenicol acetyltransferase (CAT) or luciferase, has proven to be a valuable tool to measure receptor function.

THE HUMAN ANDROGEN RECEPTOR AND ITS FUNCTIONAL DOMAINS

The predicted amino acid sequence for the AR cDNA reveals the AR to be a member of the steroid/thyroid hormone receptor gene superfamily (ref. *18*, Fig. 2). In keeping with its similarity to other members of this family, the human androgen receptor is comprised of discrete functional domains responsible for the high-affinity binding of androgen ligands (residues 676–917) *(19)* and the recognition of specific DNA sequences (residues 559–624). In addition, a number of studies have demonstrated the importance of segments within the amino-terminus for the transcriptional activation of responsive genes *(20–22)*. Distinct sequences within and adjacent to the DNA binding domain serve roles in the nuclear localization of the AR *(23,24)*.

In addition to the preceding structural features (which are similar to other members of this receptor family), the androgen receptor also possesses several unusual motifs. In particular is the presence of several repeated elements within the predicted open reading frame of the amino-terminus. In the human androgen receptor, these repeats are comprised of three repeating elements: glutamine, proline, and glycine (Fig. 2). These elements are particularly significant with respect to their function and the pathogenesis of specific disease states (*see below*).

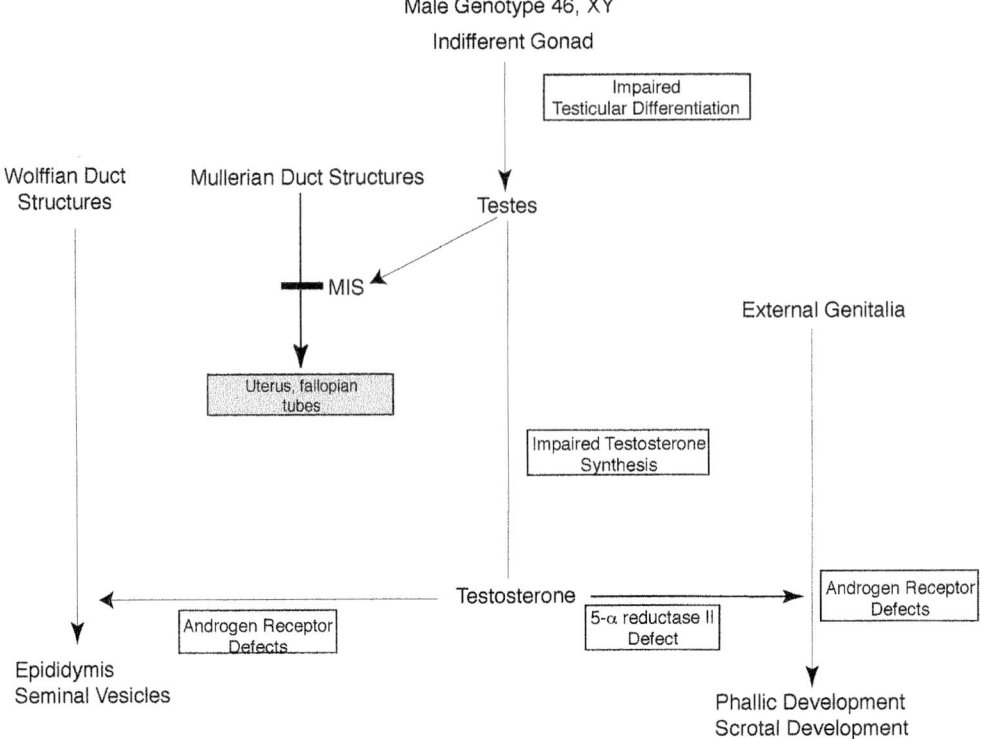

Fig. 2. Schematic structure of the androgen receptor gene. This figure represents the organization of the human androgen receptor gene and open reading frame. The gene itself is divided into eight coding exons. Exon 1 encodes the amino-terminus of receptor, exons 2 and 3 encode the DNA binding domains of the receptor, and exons 5–8 encode the hormone binding domain of the receptor protein. The homopolymeric segments (composed of repeated glutamine, proline, or glycine residues) within the amino-terminus of the receptor protein are indicated. The transcription of androgen receptor messenger RNA begins approx 1 kb 5' to the site of translation initiation and gives rise to two mRNAs that differ in the length of their 3'-untranslated segments. The upper panel demonstrates the positions of amino acid substitutions within the AR open reading frame. The height of the vertical bar indicates the frequency that substitutions have been reported at the individual amino acid residues. The lower panel represents this same data, but indicates the number of different amino acid substitutions that have been reported at each residue.

TYPES OF AR MUTATIONS

Mutations in the androgen receptor gene causing androgen resistance phenotypes can be categorized in many different ways. Although superseded by more advanced techniques of analysis, the results of binding assays remain a useful perspective from which to review the genetic lesions causing defective receptor function *(3)*.

ABSENCE OF SPECIFIC LIGAND BINDING

Large-Scale Deletions

Many patients who lack detectable levels of AR in ligand binding assays performed using genital skin biopsy cultures (ligand binding negative) have the phenotype of complete testicular feminization. The initial patient in whom a defect of the AR was identified was the only individual in this category in whom studies using Southern blotting demonstrated a partial deletion of the androgen receptor gene. Although further characterization was not reported, restriction mapping suggested that large segments of the gene encoding most or all of the hormone-binding domain (HBD) had been deleted. Given the large segments of the gene that were removed, it is unlikely that an androgen receptor was produced *(25)*.

As larger numbers of patients have been examined, additional families have been identified that harbor deletions of segments of the androgen receptor gene. Two patients have been reported in whom a complete deletion of the coding sequence of the androgen receptor gene has occurred. In the first instance, the phenotype was complete testicular feminization *(26)*. The second patient had additional abnormalities, including mental retardation *(27)*. These patients, although representing gene alterations that are relatively infrequent, are important because they establish the "null" phenotype. This information has particular importance with respect to understanding the pathogenesis of spinal and bulbar muscular atrophy (*see below*). In this regard, it seems likely that the additional abnormalities in the second patient *(27)* are caused by deletion of adjacent genetic loci.

A number of patients have been identified with less than complete deletions of the androgen receptor gene, including those with both single and multiple exon deletions. The structures of several of these partially deleted androgen receptor gene alleles have not been completely defined. An exception to this is the pedigree described by MacLean et al. *(28)*. In this family, deletions were detected that appeared to have distinct breakpoints within the androgen receptor gene. In some individuals, a deletion of exon 5 was identified, but in others, a larger deletion was detected that included both exons 5 and 6. The authors speculated that the results might imply the existence of a mutation "hot spot" in the region of exon 5 in this allele. Another possible explanation could be the movement of some sort of transposable element into or from the region *(28)*. The description of two different families with similar large-scale deletions involving multiple exons suggests that this mechanism may be operative in selected populations or under specific circumstances *(29)*.

Small Deletions and Insertions

In some patients, deletions have been detected that are substantially smaller consisting of the deletion of one or several nucleotides. When the reading frame is maintained, such changes result in the deletion of a single amino acid. If the deletion alters the reading frame, the result is the premature termination of the receptor protein *(30–39)*. Based on the studies reported, insertions appear to occur less frequently than deletions *(40–43)*. The one exception to this generalization is the lengthening of the glutamine repeat that occurs in patients with X-linked spinal and bulbar muscular atrophy (Kennedy's disease; *see below*).

Alterations in AR Structure Caused by Changes in AR mRNA Splicing

Single-nucleotide substitutions can give rise to large-scale structural changes of the androgen receptor as a result of the effects on the synthesis or processing of the AR mRNA. This is most frequently traced to base substitutions that alter the splice donor or acceptor sites used in the processing of AR mRNA.

In one patient with complete androgen resistance *(44)*, a single nucleotide change altered the splice donor site at the 5'-end of intron 4 and altered the splicing of the androgen receptor mRNA. As a result of this change, an alternate splicing event occurred (using a cryptic splice donor in intron 4), which resulted in the production of an AR protein lacking 41 amino acids of the normal AR coding sequence, including a portion at the 5'-boundary of the AR HBD *(19)*. This protein was not functional in assays of ligand binding or transactivation. A number of other examples have been described *(45–48)*.

Probably one of the most interesting mutations of this type was described as the etiology of androgen resistance in a patient with partial androgen insensitivity *(49)*. Analysis of this patient's AR gene revealed a large (>6 kb deletion) of intron 2 that began 18 bp upstream of the 5'-boundary of exon 3. As a result of this change, aberrant splicing of the AR mRNA occurred and led to the accumulation of an mRNA that encoded an AR protein that lacked exon 3 but preserved the open reading frame. Although this protein had no function in reporter gene assays, approx 10% of the AR mRNA was correctly spliced using a cryptic site. This lower abundance message resulted in the low-level expression of a normal AR protein. It is postulated that this low-level expression of normal AR accounts for the partial AIS phenotype.

These findings suggest that alterations of AR mRNA splicing that result in the synthesis of a truncated AR protein will be associated with complete forms of androgen resistance, as in the case of mutations that lead to the premature termination of the receptor protein. In those unusual circumstances where a partial androgen-resistant phenotype is observed, it will likely be traced to the function of a reduced amount of intact AR that is still synthesized.

Premature Termination Codons

Single-nucleotide substitutions that result in the introduction of premature termination codons into the androgen receptor gene were among the first defects in the AR gene to be defined. For the most part, these mutant receptors have no activity in assays of androgen binding and transcriptional activation. Such mutations have been detected in most of the coding exons of the AR gene, and there appears to be little in the way of bias as to the regions of the AR gene in which these types of mutation have been identified. In the overwhelming majority of cases that have been described, the mutant androgen receptor molecules produced by these patients are truncated, lack the carboxy-terminal hormone binding domain, and have no activity in assays of androgen binding or transcriptional activation.

To date, only a single exception to this generalization has been reported. In this instance, a single nucleotide substitution resulted in the premature termination of the AR protein early in the open reading frame (at amino acid 60). Owing to this amino-terminal location of the termination codon, downstream initiation occurred (at methionine 189) and resulted in reduced, but measurable levels of the shortened AR protein *(50,51)*.

Amino Acid Substitutions

When categorized on the basis of ligand binding studies performed using patient fibroblasts, mutations resulting in absent ligand binding are frequently traced to single-nucleotide substitutions within the open reading frame that result in amino acid substitutions.** In all instances, these substitution mutations are localized to the hormone binding domain of the receptor protein.

** The number of amino acid substitutions that have been identified in the androgen receptor causing various forms of androgen resistance has grown dramatically over the last several years. Although the reference is provided for most of the published literature, owing to the breadth of mutations reported, this is not true for amino acid substitutions in the hormone binding domain. Readers are referred to the Androgen Receptor Mutation Database (http://www.mcgill.ca/androgendb) for a more comprehensive listing of such mutations.

Amino acid substitutions in the hormone binding domain of the androgen receptor that result in absent ligand binding (as measured in patient fibroblasts) appear to fall into two major categories. One category, typified by the mutation (W739R) described by McPhaul and coworkers *(52,53)*, involves amino acid substitution mutations within the hormone binding domain that induce such major changes in the structure of the hormone binding domain that it is no longer capable of binding ligand. This type of mutation has been encountered in only a very small number of instances, but is likely to reflect major alterations in the conformation of the ligand binding pocket.

More frequently, although binding studies performed in genital skin fibroblasts indicate that specific ligand binding is absent, when cDNAs encoding the mutant receptors predicted from the analysis of such AR genes are expressed in heterologous cells, it is found that the mutant receptor is still capable of binding ligand, although with reduced ability and or affinity. For example, it was demonstrated in one subject that a single nucleotide substitution resulted in an amino acid replacement, R774C *(54)*. When assayed in fibroblasts, this single amino acid replacement resulted in levels of ligand binding that were below the limits of detection. When expressed in heterologous cells, however, although the expressed receptor was shown to be capable of binding ligand specifically, it was also found to exhibit marked lability to thermal denaturation *(54)*. The end result in this instance, as in most mutants within this class, is normal—or near normal—levels of immunoreactive receptor, but decreased ligand binding *(55)*. In many cases, this type of effect is the result of replacement of a charged amino acid by an uncharged amino acid. These residues are likely involved in the formation of ionic bridges that are important to the maintenance of the three-dimensional structure of the hormone binding domain.

This discordance between the results of the fibroblast binding assays and those performed in cells transfected with mutant cDNAs does not appear to indicate an inherent difference in how the mutant receptor behaves in the different cellular environments. Instead, it appears simply to reflect differences in the sensitivity of the assays employed and differences in the levels of receptor expressed.

Androgen Receptor Mutations Causing Qualitative Abnormalities of Ligand Binding

The comparison of results describing qualitative abnormalities from different research groups is often difficult, since the range and methods employed differ from laboratory to laboratory. On the whole, however, the qualitative tests that have been applied have examined the affinity of the receptor for its ligand, the stability of the AR protein expressed (e.g., to thermal denaturation), and the stability of the hormone–receptor complexes that are formed. Owing to the number of mutations that have now been described, the following discussion is not encyclopedic, and is instead focused on illustrative studies in which the mutation responsible for causing the androgen-resistant phenotype has been identified.

The first mutation of this type was reported by Lubahn et al. *(56)*. They identified a single amino acid substitution (V866M) in the hormone binding region of the androgen receptor in a family with complete testicular feminization. This produced an increased K_d in ligand binding studies. Subsequent analysis of the function of this mutated androgen-receptor demonstrated a reduced capacity to stimulate a model androgen-responsive reporter gene *(57)*. This decreased capacity of the mutant androgen receptor to induce activity of the responsive gene was minimized at high ligand concentrations and maximized at lower concentration ranges.

That high doses of androgen could cause discernible different effects on the activity of mutant androgen receptors has been observed in other studies. In a patient with Reifenstein phenotype, studies performed in fibroblast cultures indicated a normal level of ligand binding, but showed accelerated dissociation *(58)*. The androgen receptor gene in this patient contained two mutations *(59)*: (1) a tyrosine-to-cysteine substitution in exon 7 and (2) a shortened glutamine homopolymeric region in the amino-terminus. One or both of these mutations were inserted in an androgen receptor cDNA using site-directed mutagenesis. The resulting plasmids were then inserted into eukaryotic cells, and the expressed receptor protein was assessed. The alteration to the glutamine homopolymeric region caused no changes in the receptor stability or ligand dissociation. However, the tyrosine-to-cysteine substitution caused the receptor to be qualitatively abnormal with accelerated dissociation rate and thermal lability *(59)*. The abnormalities of the receptor containing both mutations were more severe than the receptor containing only the mutation in exon 7 or the receptor containing the glutamine truncation. The cooperative effects of the two mutations were evident in studies of the dissociation rate, transcriptional activation, and thermal stability, suggesting that the receptor and the phenotype were the result of both mutations. Subsequent investigations *(60,61)* have supported a functional interaction between the amino-terminal and carboxy-terminal segments of the AR protein.

Subsequent analyses have identified a number of different amino acid substitutions that result in mutant ARs that exhibit qualitative abnormalities of androgen receptor function. Virtually all of these mutations are localized to the hormone binding domain, and their distribution within the HBD is similar compared to the distribution of the mutations that result in the lack of detectable ligand binding *(53)*. This suggests that the degree of disruption of the structure of the ligand binding is related to the abnormality that is identified: more severe leading to absent ligand binding, and less severe alterations leading to qualitative defects of the AR. This inference has been reinforced most dramatically in the studies of Prior et al. *(62)*, who identified families in which distinct mutations of the androgen receptor in the same amino acid residue (774) led to distinctive abnormalities of the androgen receptor protein. In one instance, replacement of an arginine residue by cysteine residue (R774C, equivalent to the R774C mutation described above) led to absent ligand binding. Replacement of this same arginine residue by a histidine, by contrast, led to normal levels of androgen binding that displayed marked thermal instability.

Other investigations have identified mutant receptors in which a different substitution mutation at the same residue led to identifiably different effects on ligand binding and on receptor function *(63–65)*. In those instances in which replacement of a residue with different amino acids has resulted in discernibly different phenotypes, the level of AR function measured for the mutant receptors has varied in concert with the phenotype that is apparent.

An additional interesting aspect of the studies of mutant ARs harboring amino acid substitutions in the HBD are those described by Marcelli et al. (66). This study examined the functional responsiveness of a number of the androgen receptor mutants that exhibited different types of qualitative abnormalities. When assayed in cells capable of metabolizing the androgens testosterone and 5α-dihydrotestosterone, it was observed that the presentation and type of androgen used in functional assay experiments had a dramatic effect on the levels of androgen receptor function that were observed. These findings were relatively consistent for all the mutant receptors that were capable of binding androgen. In each case, testosterone was the least potent, but dihydrotestosterone and mibolerone exhibited higher potencies. These experiments permitted three important conclusions. First, they suggested that mutant receptors that were able to bind hormone— however weakly—could be manipulated pharmacologically to exhibit near-normal levels of androgen receptor function. This possibility has been tested in only a limited number of circumstances *(58,67,68)*. Second, these experiments demonstrated the importance of the stability of the hormone–androgen receptor complex. Conditions that favored the formation and stability of these complexes could be seen to have major effects on the function of the mutant receptors in functional assays. Finally, these experiments demonstrate that extreme caution must be used in attempting to correlate the results of functional assays performed using transfected cells with the phenotype observed in vivo, since minor alterations in protocol can lead to major differences in the receptor function that is measured.

ANDROGEN RECEPTOR MUTATIONS WITH NORMAL LEVELS OF LIGAND-BINDING

In about 10% of patients with family histories and endocrine studies suggesting androgen resistance, no qualitative or quantitative abnormality of ligand binding can be detected. In these families, the observed phenotypes can range from complete testicular feminization to the most mildly affected phenotypes. This constellation of features had suggested the presence of subtle abnormalities of androgen receptor or mutations in other genes affecting the action of androgens.

Analyses of such families have now demonstrated that in pedigrees in which the family history is positive, mutations within the conserved DNA binding domain of the receptor are the most frequently implicated. In four subjects with androgen resistance, the androgen receptor was characterized, and in each case, amino acid substitutions were identified in the DNA binding region of the androgen receptor. When these mutations were introduced into the androgen receptor cDNA, the receptors produced bound ligand with normal or near-normal kinetics. However, the mutant receptors were severely impaired in transcriptional activation of a model androgen-responsive reporter gene. Detailed studies indicated that these receptors displayed an impaired capacity to bind to target DNA sequences (androgen response elements) *(69)*. These conclusions have been borne out by the results of studies conducted on patients with complete and partial forms of androgen resistance in a number of different laboratories *(70–74)*.

Alterations other than amino acid substitutions within the DNA binding domain (DBD) can affect the function of the androgen receptor in a similar fashion. A deletion of exon 3 was detected in a patient with complete testicular feminization in which levels of ligand binding were measured and considered to be normal. Exon 3 of the androgen receptor is unique in that its size is a multiple of 3 nucleotide residues. Therefore, deletion of this exon leads to an in frame deletion of the sequence encoding the segment of the protein encoded by exon 3. As expected from these facts, this mutant receptor binds hormone normally, but is unable to activate the transcription of a target androgen-responsive reporter gene. Based on the preceding studies, it appears that whatever the mutation is, alterations restricted to the DNA binding domain of the receptor result in a form of androgen insensitivity by interfering with a capacity of the receptor to recognize its specific target DNA sequences. The available studies would suggest that in general the degree to which DNA binding by the mutant receptors is impaired correlates with the degree of phenotypic abnormality observed (75).

Although a number of mutations within the AR DBD have now been reported (76–87), a subset of patients with partial forms of androgen resistance caused by mutations within the DBD are particularly intriguing. In one pedigree, reported by Wooster et al. (88), two brothers with Reifenstein syndrome were described that had developed breast cancer. Analysis of the androgen receptor revealed a mutation in the second zinc finger (exon 3 position 607), but functional studies were not reported at that time. The segment that was mutated is a part of the segment implicated in determining the specificity of DNA binding by the receptor (89). Although the development of breast cancer in such patients may simply reflect the gynecomastia observed in such individuals, a second report by Lobaccaro and coworkers (90) would seem to make this possibility less likely. In this second report, an unrelated individual with the Reifenstein phenotype who developed breast cancer was found to harbor an amino acid substitution at amino acid residue (residue 608) adjacent to that described by Wooster et al. The fact that in these two pedigrees adjacent amino acid residues were mutated suggests that the pattern of gene activation that occurs in response to androgens in these patients may somehow directly contribute to the development of breast cancer. Subsequent studies by this second group of investigators (91) demonstrated that the mutant AR displayed altered DNA binding specificity, such that promoters with isolated single androgen-responsive elements were no longer activated. Given the setting in which these mutations were identified, it is possible that these substitutions may alter gene regulation in a fashion that is distinct from that which is seen for other mutant ARs causing androgen insensitivity and not associated with male breast cancer.

Androgen Receptor Mutations Resulting in Decreased Levels of Ligand Binding

This category appears to be the most heterogeneous of the categories in terms of the number of different mechanisms implicated. The first mutation of this type reported was reported by Zoppi et al. (50). Fibroblast samples from affected individuals within this pedigree were found to express reduced amounts of androgen receptor as assayed using monolayer binding assays. Surprisingly, immunoblot experiments performed using antibodies directed at the amino-terminus of the androgen receptor did not detect the receptor protein. This paradox was explained by the identification of a termination codon resulting from a single nucleotide substitution at codon 60 within the receptor protein. Subsequent investigations established that the low level of binding detected in the initial screening assays was the result of the downstream initiation at methionine 189. More recent studies have demonstrated that this receptor protein lacking amino acids 1–189 is synthesized in normal cells and is analogous to the A-form of the progesterone receptor (51). Studies in heterologous cells established that the phenotype (complete testicular feminization) that is observed in affected individuals within this family is owing to a combination of reduced receptor expression and a reduced function of the receptor protein that is synthesized (92).

More recently, the report by Choong et al. *(93)* has described a patient with partial Androgen Insensitivity Syndrome (AIS) in which reduced levels of apparently normal AR were synthesized. In this pedigree, the genetic lesion was traced to a single nucleotide substitution that resulted in the replacement of the second amino acid residue of the androgen receptor open reading frame (aspartate to lysine). Although it was not possible to examine the effects of this mutation in cultured fibroblasts, the authors concluded that the alterations causing the AIS phenotype were the result of the reduced levels of AR that were expressed, as well as a subtle increase in the rate of ligand dissociation.

It is clear that this category will be heterogeneous at least in part because of the types of assays that are applied. Patients classified in this category may express reduced amounts of normal receptor or may express reduced levels of AR that exhibits subtle alterations that may be missed in initial measurements of AR expression examined in fibroblast cultures.

MUTATIONS OF THE ANDROGEN RECEPTOR CAUSING MINIMAL DEFECTS OF VIRILIZATION

The preceding discussion has focused on the more severely affected phenotypes. In addition to the wealth of information relating to the pathogenesis of androgen receptor defects identified in such patient populations, information is now becoming available on the types of genetic lesions that cause more subtle phenotypic defects, particularly those associated with the undervirilized male syndrome and those causing infertility. The first publication to identify a mutation in such a pedigree was that described by Tsukada et al. *(94)*, who identified the genetic lesion causing androgen resistance in an undervirilized man. The receptor defect identified, which in this individual resulted in undervirilization with preserved fertility, was traced to amino acid substitution in exon 5 of the receptor that resulted in the substitution of a phenylalanine in place of the normal leucine residue at position 790. This mutant receptor displayed thermal instability, as well as slightly altered kinetics of androgen binding. Functional assays demonstrated only slightly reduced levels of receptor function when assayed using the MMTV reporter gene. As would be expected for a mutation causing a qualitative abnormality of the androgen receptor, the defect of receptor function was most pronounced at lower hormone concentrations and less evident at higher concentrations. In addition to the defect described in this patient with undervirilization, the structure of a small number of other mutant androgen receptors has been associated with a defect of AR function that appears to be largely or exclusively restricted to the process of spermatogenesis *(68,95)*. In each instance, when assayed in transactivation assays, the mutations associated with these phenotypes have resulted in relatively small alterations of androgen receptor function. To this point, it remains unclear whether these minimally affected phenotypes will be found to reflect defects of androgen receptor function, or whether these abnormalities are reflected disproportionately in selected tissues.

THE RELATIONSHIP BETWEEN MUTATIONS OF THE AR AND THE OBSERVED PHENOTYPE

From the time that androgen receptor binding assays were used to classify patients with defects in the androgen receptor gene, it was clear that the results of these assays did not show a precise alignment with the types of phenotypes exhibited by affected individuals. It was reasonable to expect that the information derived from the analysis of mutant androgen receptor genes might allow a more direct comparison of the genetic defect and the resulting phenotype.

The range of AR mutations that has now been identified permits a number of generalizations to be made. First, truncations of the AR protein result—with few exceptions—in a phenotype of complete androgen resistance. This is owing to the fact that most truncations of the receptor protein either remove essential functional domains (i.e., portions of the hormone binding domain), result in levels of AR protein that display diminished function, or are expressed at reduced levels. This is in contrast to alterations that result in amino acid substitutions, a class of mutations that has been shown to result in the complete range of phenotypes, from the most severely affected to the most mildly affected individuals. Second, although there is broad agreement between the degree of defect that is defined in assays of receptor function and the phenotype that is observed in individual patients, it is clear that this relationship can at times be difficult to discern. In instances where no receptor is expressed or the genetic defect completely inactivates the receptor, the degree of deficiency is not difficult to associate with the clinical phenotypes of complete testicular feminization. In situations where the receptor is not completely defective, however, quantitation of the degree of deficiency is more difficult, and it is more subject to variations introduced by the conditions under which such assays are performed. This was demonstrated by Marcelli et al., who demonstrated that variations in hormone dosing or the exact ligands employed could dramatically alter the levels of receptor function that were measured *(66)*. The use of an adenovirus to deliver a model androgen-responsive reporter gene into patient fibroblasts to measure the level of androgen receptor function *in situ* is one way to minimize the artifacts inherent in such assays performed in heterologous cells *(96,97)*.

VARIATIONS IN PHENOTYPE

In some instances of hormone resistance, such as the generalized resistance to thyroid hormone *(98,99)*, similar genotypes have been associated with considerable differences in phenotype. In the case of androgen resistance, however, similar phenotypes are usually observed in pedigrees (presumably independent) in which the same mutation is observed. There are exceptions to the statement, however. In most such instances, the phenotypic variation has been relatively minor and has been attributed to the potential contribution of factors, such as levels of 5α-reductase expression, but this is not known with any certainty. The recent publication by Holterhus et al. *(100)*, however, describes a patient in whom a termination codon had been detected in a patient affected with a partial form of androgen resistance. This apparent paradox was resolved when detailed studies established that the patient was a mosaic and the partial phenotype was in fact owing to the expression of a proportion of normal androgen receptor mRNA in this patient's samples. How frequently this occurs in other instances in which the phenotype varies from that which is expected has not yet been determined.

HOMOPOLYMERIC REPEATS AND HETEROGENEITY WITHIN THE ANDROGEN RECEPTOR GENE

As noted above, the structure of the human androgen receptor is unusual in that it contains three motifs located within the amino-terminus of the receptor, each composed of direct repeats of single amino acid residues: glutamine, glycine, and proline. Although unusual, these homopolymeric repeats are not unique to the androgen receptor and are found in a number of other transcription factors, including other members of the nuclear receptor family.

From the data available presently, it appears that the proline homopolymeric repeat is relatively constant in size. No instances have been identified in which variations in the length of this segment exist in the normal population. Likewise, the glycine homopolymeric repeat is also relatively invariant in size, although some length polymorphisms have been identified in some analyses *(101)*.

The glutamine homopolymeric domain, however, shows substantial variation. This has been established as a consequence of the analyses of many alleles in the course of analyzing the structure of individual androgen receptor genes, as well as in studies focused on determining the extent with which such polymorphisms are present in the general population *(59,102)*. Population studies have established, however, that a wide range of homopolymeric lengths can be detected in the normal population, although it appears that in most instances, this repeat encodes approx 20–23 glutamine residues *(103)*. These polymorphisms have been employed in family studies to track the inheritance of specific androgen receptor alleles.

In addition to their importance as polymorphic markers, it appears that variations in the length of the glutamine homopolymeric segment of the human androgen receptor have functional implications as well. As noted, expansion of this repeated element is involved in the pathogenesis of spinal and bulbar muscular atrophy, a disease characterized by the progressive development of androgen resistance phenotype and the progressive deterioration of anterior motor neurons. Functional assays have established that the expansion of the glutamine homopolymeric domain results in a decrease of androgen receptor function *(22,95,104–107; see below)*. The mechanism by which the motor neuron degeneration occurs has not yet been determined, although a toxic gain of function is suspected.

Shortening of the glutamine homopolymeric domain also results in a discernible phenotype. Although a shortened glutamine repeat was identified in a single prostate cancer sample *(108)*, recent studies by several groups have suggested that individuals possessing such shortened glutamine repeats have an inherently increased incidence of developing aggressive forms of prostate cancer *(109–111)*. Although some authors have suggested that the mechanism underlying this increased risk can be traced to it increased potency of androgen receptors containing such shortened glutamine repeats *(105)*, this type of observation has not been obtained by all groups *(22)*, and thus, other explanations may lie at its root.

When considering variations between individual men, it is clear that the substantial amount of variability exists from individual to individual. It was possible that such variations might be traced to alterations in the structure of the androgen receptor itself. For this reason, many investigators were quite cautions in interpreting in the sequence variations encountered during the course of androgen receptor gene analyses. It has become evident at that this point, however, that such sequence variations are quite infrequent in the general population. In our own series of patients, we have not encountered sequence variations that do not cause androgen resistance in regions outside mutations that do cause androgen resistance *(112)*. This is been the experience of other groups engaged in similar types of analysis.

SPINAL AND BULBAR MUSCULAR ATROPHY

Probably one of the most unexpected findings relating to mutations of the androgen receptor gene relates to the identification of the genetic alterations that cause X-linked spinal and bulbar muscular atrophy. This disorder, also known as Kennedy's syndrome, is inherited as an X-linked trait. Affected males display normal genotypic development and normal sexual function early in life. Later in life, however, a phenotype characteristic of mild androgen resistance emerges (gynecomastia). In association with these clinical findings is the appearance of neurological symptoms caused by a progressive loss of motor neurons from spinal and bulbar nuclei. Seminal work by LaSpada and colleagues *(113)* identified the expansion of the glutamine homopolymeric domain within the amino-terminus of the androgen receptor gene. Subsequent studies have demonstrated that this expansion of the glutamine homopolymeric domain is invariably associated with the Kennedy's syndrome phenotype. Although the mechanisms for the androgen resistance and the neurological symptoms have not been defined, it is clear that this glutamine expansion is somehow responsible for both a partial loss of function of (with respect to the function of the receptor) and for a toxic gain of function responsible for the appearance of the neurological symptoms *(114,115)*.

MUTATIONS OF THE ANDROGEN RECEPTOR IN PROSTATE CANCER

All of the mutations that have been described from families with various forms of androgen resistance have been found to result from a partial or complete loss of androgen receptor function. Studies of the anomalous responses observed for the LNCaP prostatic epithelial cancer cell line (response to an androgen receptor antagonists in a fashion characteristic of androgen receptor agonists) led to the discovery that mutations within the hormone binding domain of the human androgen receptor could cause a relaxation in the ligand responsiveness of the androgen receptor *(116)*. In the case of the mutations identified in the LNCaP cell line, although responsiveness to androgen agonists (activation of androgen-responsive reporter genes) was preserved, the androgen receptor was capable of activating model androgen-responsive reporter genes after stimulation with a variety of agents that would be incapable of activating the normal human androgen receptor.

The potential relevance of such somatic mutations as explanations for the emergence of "hormone refractory" (androgen-independent or hormone-independent) growth of advanced prostate cancer is obvious. Although no consensus has yet emerged, it appears that such mutations are less abundant in early forms of prostate cancer *(117,118)* and are more frequent in more advanced *(119,120)* stages of the disease (reviewed in ref. *121*). Mutant receptors of this type that have been studied in detail display altered specificities and can be activated by ligands that will not activate the normal androgen receptor *(119,120,122–124)*. A number of additional reports have been described in which additional alterations of the AR gene have been identified, including amplification and alterations in the untranslated segments of the AR *(125–129)*. This continues to be an area of active investigation.

DISTRIBUTION OF MUTATIONS WITHIN THE ANDROGEN RECEPTOR GENE

The considerable amount of information that is now available permits some tentative conclusions to be drawn regarding the types of genetic alterations in the androgen receptor that results in androgen resistance. First, it is clear that as has been found for many other genes causing human disease, a variety of types of mutation may cause defects of androgen receptor function. These include complete or partial gene deletions, insertions, premature termination, and abnormalities of androgen receptor mRNA splicing. In each of these different circumstances, despite the diversity of lesions (and their locations), the mechanism is the production of a defective receptor protein or the accumulation of the reduced quantity of functional AR. Despite this heterogeneity, the phenotype that results is that of complete testicular feminization.

Mutations that result in the substitution of the single amino acid residue within the AR protein are the most frequent—and the most interesting—defects that cause androgen resistance. With a single exception, these mutations are all localized to the hormone or DNA binding domains of the receptor. For the most part, the locations of these mutations within the AR protein do not reflect sites subject to increased rates of mutagenesis, but instead identify residues within the AR that are critical and discrete enough to be affected by a single amino acid replacement.

In addition to drawing conclusions from the locations of mutations that have been identified in different forms of androgen resistance, it is also interesting to consider the paucity of mutations that have been identified within the amino-terminus of the receptor protein. Those few mutations that have been localized to the amino-terminus result—directly or indirectly—in the premature termination or inefficient synthesis of the receptor protein. This fact, coupled with the results of in vitro mutagenesis studies conducted by a number of investigators, clearly demonstrates the importance of the amino-terminus for full receptor function, and suggests that those functions exerted by the amino-terminus are so diffusely encoded that single amino acid substitutions are unable to affect AR function significantly.

SUMMARY

A large amount of information is now available from the analysis of patients with different disorders of AR function. This information has considerable importance concerning the identification of elements that are crucial for the normal function of the AR and their relevance to androgen physiology.

The behavior of most, if not all, of the mutations encountered in patients presenting with disorders of sexual development are consistent with a simple "loss-of-function" type of mechanism. Although precise quantitation is not possible at present, it appears that there is broad agreement between the phenotypes observed in patients and measures of AR abundance and function in cell culture and in in vitro assays. Understanding those instances in which discordance is observed between pedigrees or between affected members of a single pedigree should provide important information regarding the physiological modifiers of androgen action.

As our knowledge of the disorders of the androgen receptor has increased, it has become apparent that mutations of the AR other than those associated with classic forms of androgen resistance also have clinical significance. In two diseases—spinal and bulbar muscular atrophy and prostate cancer—alterations in the length of the glutamine homopolymeric domain within the amino-terminus have been implicated in the pathogenesis of these disorders. Although the mechanisms by which these changes in AR structure contribute to the biology of these diseases have not been completely elucidated, it appears that in both instances, the genetic alteration has conferred on the mutant receptor some type of novel activity ("gain-of-function" mutation). In the same vein, the identification of somatic mutations of the androgen receptor in advanced prostatic malignancies that display altered ligand responsiveness suggests that the appearance of such genetic alterations in the tumor cells may play a role in the progression of this disorder as well.

ACKNOWLEDGMENT

Support was provided by NIH grant DK03892.

REFERENCES

1. Jost A, Vigier B, Prepin J, Perchellet JP. Studies on sex differentiation in mammals. Recent Prog Hormone Res 1973;29:1–41.
2. Quigley CA. Disorders of sex determination and differentiation. In: Jameson JL, ed. Principles of Molecular Medicine. Totowa, NJ, Humana Press, 1998.
3. Griffin JE, McPhaul MJ, Russell DW, Wilson JD. The androgen resistance syndromes: Steroid 5α-reductase 2 deficiency, testicular feminization, and related disorders. In: Scriver CR, Beaudet AL, Sly WS, Valle D, eds. The Metabolic and Molecular Bases of Inherited Disease, 7th ed., Vol. II. McGraw-Hill, New York, 1995, pp. 2967–2998.
4. Quigley CA, De Bellis A, Marschke KB, el-Awady MK, Wilson EM, French FS. Androgen receptor defects: historical, clinical, and molecular perspectives Endocr Rev 1995;16:271–321.
5. Chang CS, Kokontis J, Liao ST. Structural analysis of complementary DNA and amino acid sequences of human and rat androgen receptor. Proc Natl Acad Sci USA 1988;85:7211–7215.
6. Lubahn DB, Joseph DR, Sar M, Tan J, Higgs HN, Larson RE, et al. The human androgen receptor: complementary deoxyribonucleic acid cloning, sequence analysis and gene expression in prostate. Mol Endocrinol 1988;2:1265–1275.
7. Faber PW, Kuiper GG, van Rooij HC, van der Korput JA, Brinkmann AO, Trapman J. The N-terminal domain of the human androgen receptor is encoded by one, large exon. Mol Cell Endocrinol 1989; 61:257–262.
8. Tilley WD, Marcelli M, Wilson JD, McPhaul MJ. Characterization and expression of a cDNA encoding the human androgen receptor. Proc Natl Acad Sci USA 1989;86:327–331.
9. Chang CS, Whelan CT, Popovich TC, Kokontis J, Liao ST. Fusion proteins containing androgen receptor sequences and their use in the production of poly- and monoclonal anti-androgen receptor antibodies. Endocrinology 1989;125:1097–1099.
10. Zegers ND, Claassen E, Neelen C, Mulder E, van Laar JH, Voorhorst MM, et al. Epitope prediction and confirmation for the human androgen receptor: generation of monoclonal antibodies for multi-assay performance following the synthetic peptide strategy. Biochim Biophys Acta 1991;1073:23–32.
11. Husmann DA, Wilson CM, McPhaul MJ, Tilley WD, Wilson JD. Antipeptide antibodies to two distinct regions of the androgen receptor localize the receptor protein to the nuclei of target cells in the rat and human prostate. Endocrinology 1990;126:2359–2368.
12. Celis L, Claessens F, Peeters B, Heyns W, Verhoeven G, Rombauts W. Proteins interacting with an androgen-responsive unit in the C3(1) gene intron. Mol Cell Endocrinol 1993;94:165–172.
13. Allison J, Zhang YL, Parker MG. Tissue-specific and hormonal regulation of the gene for rat prostatic steroid-binding protein in transgenic mice. Mol Cell Biol 1989;9:2254–2257.
14. Claessens F, Alen P, Devos A, Peeters B, Verhoeven G, Rombauts W. The androgen-specific probasin response element 2 interacts differentially with androgen and glucocorticoid receptors J Biol Chem 1996;271:19,013–19,016.
15. Cleutjens KB, van der Korput HA, Ehren-van Eekelen CC, Sikes RA, Fasciana C, Chung LW, Trapman J. A 6-kb promoter fragment mimics in transgenic mice the prostate-specific and androgen-regulated expression of the endogenous prostate-specific antigen gene in humans. Mol Endocrinol 1997;11:1256–1265.
16. Chavez S, Beato M. Nucleosome-mediated synergism between transcription factors on the mouse mammary tumor virus promoter. Proc Natl Acad Sci USA 1997;94:2885–2890.
17. Rundlett SE, Miesfeld RL. Quantitative differences in androgen and glucocorticoid receptor DNA binding properties contribute to receptor-selective transcriptional regulation. Mol Cell Endocrinol 1995;109:1–10.

18. Mangelsdorf DJ, Thummel C, Beato M, Herrlich P, Schutz G, Umesono K, et. al. The nuclear receptor superfamily: the second decade. Cell 1995;83:835–839.
19. Cooper B, Gruber JA, McPhaul MJ. Hormone-binding and solubility properties of fusion proteins containing the ligand-binding domain of the human androgen receptor J Steroid Biochem Mol Biol 1996;57:251–257.
20. Simental JA, Sar M, Lane MV, French FS, Wilson EM. Transcriptional activation and nuclear targeting signals of the human androgen receptor J Biol Chem 1991;266:510–518.
21. Jenster G, van der Korput HA, van Vroonhoven C, van der Kwast TH, Trapman J, Brinkmann AO. Domains of the human androgen receptor involved in steroid binding, transcriptional activation, and subcellular localization. Mol Endocrinol 1991;5:1396–1404.
22. Gao TS, Marcelli M, McPhaul MJ. Transcriptional activation and transient expression of the human androgen receptor J Steroid Biochem Mol Biol 1996;59:9–20.
23. Zhou ZX, Sar M, Simental JA, Lane MV, Wilson EM. A ligand-dependent bipartite nuclear targeting signal in the human androgen receptor. Requirement for the DNA-binding domain and modulation by NH2-terminal and carboxyl-terminal sequences. J Biol Chem 1994;269:13,115–13,123.
24. Jenster G, Trapman J, Brinkmann AO. Nuclear import of the human androgen receptor. Biochem J 1993;293:761–768.
25. Brown TR, Lubahn DB, Wilson EM, Joseph DR, French FS, Migeon CJ. Deletion of the steroid-binding domain of the human androgen receptor gene in one family with complete androgen insensitivity syndrome: evidence for further genetic heterogeneity in the syndrome. Proc Natl Acad Sci USA 1988;85:8151–8155.
26. Quigley CA, Friedman KJ, Johnson A, Lafreniere RG, Silverman LM, Lubahn DB, et al. Complete deletion of the androgen receptor gene: definition of the null phenotype of the androgen insensitivity syndrome and determination of carrier status. J Clin Endocrinol Metab 1992;74:927–933.
27. Trifiro M, Gottlieb B, Pinsky L, Kaufman M, Prior L, Belsham DD, et al. The 56/58 kDa androgen-binding protein in male genital skin fibroblasts with a deleted androgen receptor gene. Mol Cell Endocrinol 1991;75:37–47.
28. MacLean HE, Chu S., Warne GL, Zajac JD. Related individuals with different androgen receptor gene deletions. J Clin Invest 1993;91:1123–1128.
29. Jakubiczka S, Nedel S, Werder EA, Schleiermacher E, Theile U, Wolff G, et al. Mutations of the androgen receptor gene in patients with complete androgen insensitivity Hum Mutat 1997;9:57–61.
30. Ris-Stalpers C, Verleun-Mooijman MC, de Blaeij TJ, Degenhart HJ, Trapman J, Brinkmann AO. Differential splicing of human androgen receptor pre-mRNA in X-linked Reifenstein syndrome, because of a deletion involving a putative branch site Am J Hum Genet 1994;54(4):609–617.
31. Baldazzi L, Baroncini C, Pirazzoli P, Balsamo A, Capelli M, Marchetti G, et al. Two mutations causing complete androgen insensitivity: a frame-shift in the steroid binding domain and a Cys→Phe substitution in the second zinc finger of the androgen receptor. Hum Mol Genet 1994;3:1169–1170.
32. Batch JA, Williams DM, Davies HR, Brown BD, Evans BA, Hughes IA, et al. Androgen receptor gene mutations identified by SSCP in fourteen subjects with androgen insensitivity syndrome. Hum Mol Genet 1992;1:497–503.
33. Brown TR, Scherer PA, Chang YT, Migeon CJ, Ghirri P, Murono K, Zhou Z. Molecular genetics of human androgen insensitivity. Eur. J. Pediatr. 1993;152 Suppl 2:S62–69.
34. Lobaccaro JM, Lumbroso S, Poujol N, Georget V, Brinkmann AO, Malpuech G, et al. Complete androgen insensitivity syndrome due to a new frameshift deletion in exon 4 of the androgen receptor gene: functional analysis of the mutant receptor Mol Cell Endocrinol 1995;111:21–28.
35. Beitel LK, Prior L, Vasiliou DM, Gottlieb B, Kaufman M, Lumbroso R, et al. Complete androgen insensitivity due to mutations in the probable alpha-helical segments of the DNA-binding domain in the human androgen receptor. Hum Mol Genet 1994;3:21–27.
36. Hiort O, Wodtke A, Struve D, Zollner A, Sinnecker GH. Detection of point mutations in the androgen receptor gene using non-isotopic single strand conformation polymorphism analysis. German Collaborative Intersex Study Group. Hum Mol Genet 1994;3:1163–1166.
37. Imai A, Ohno T, Iida K, Ohsuye K, Okano Y, Tamaya T. A frame-shift mutation of the androgen receptor gene in a patient with receptor-negative complete testicular feminization: comparison with a single base substitution in a receptor-reduced incomplete form. Ann Clin Biochem 1995;32:482–486.
38. Bruggenwirth HT, Boehmer AL, Verleun-Mooijman MC, Hoogenboezem T, Kleijer WJ, Otten BJ, et al. Molecular basis of androgen insensitivity. J Steroid Biochem Mol Biol 1996;58:569–575.
39. Hiort O, Sinnecker GH, Holterhus PM, Nitsche EM, Kruse K. The clinical and molecular spectrum of androgen insensitivity syndromes. Am J Med Genet 1996;63:218–222.

40. Brinkmann AO, Jenster G, Ris-Stalpers C, van der Korput JA Bruggenwirth HT, Boehmer AL, et al. Androgen receptor mutations. J Steroid Biochem Mol Biol 1995;53:443–448.
41. Batch JA, Williams DM, Davies HR, Brown BD, Evans BA, Hughes IA, et al. Androgen receptor gene mutations identified by SSCP in fourteen subjects with androgen insensitivity syndrome. Hum Mol Genet 1992;1:497–503.
42. Hiort O, Wodtke A, Struve D, Zollner A, Sinnecker, GH. Detection of point mutations in the androgen receptor gene using non-isotopic single strand conformation polymorphism analysis. German Collaborative Intersex Study Group. Hum Mol Genet 1994;3:1163–1166.
43. Bruggenwirth HT, Boehmer AL, Verleun-Mooijman MC, Hoogenboezem T, Kleijer WJ, Otten BJ, et al. Molecular basis of androgen insensitivity. J Steroid Biochem Mol Biol 1996;58:569–575.
44. Ris-Stalpers C, Kuiper GG, Faber PW, Schweikert HU, van RHC, Zegers ND, et al. Aberrant splicing of androgen receptor mRNA results in synthesis of a nonfunctional receptor protein in a patient with androgen insensitivity. Proc Natl Acad Sci USA 1990;87:7866–7870.
45. Ris-Stalpers C, Turberg A, Verleun-Mooyman MC, Romalo G, Schweikert HU, Trapman J, et al. Expression of an aberrantly spliced androgen receptor mRNA in a family with complete androgen insensitivity. Ann NY Acad Sci 1993;684:239–242.
46. Trifiro MA, Lumbroso R, Beitel LK, Vasiliou DM, Bouchard J, Deal C, et al. Altered mRNA expression due to insertion or substitution of thymine at position +3 of two splice-donor sites in the androgen receptor gene. Eur J Hum Genet 1997;5:50–58.
47. Pinsky L, Trifiro M, Kaufman M, Beitel LK, Mhatre A, Kazemi-Esfarjani P, et al. Androgen resistance due to mutation of the androgen receptor. Clin Invest Med 1992;15:456–472.
48. Yong EL, Chua KL, Yang M, Roy A, Ratnam S. Complete androgen insensitivity due to a splice-site mutation in the androgen receptor gene and genetic screening with single-stranded conformation polymorphism. Fertil Steril 1994;61:856–862.
49. Ris-Stalpers C, Verleun-Mooijman MC, de BTJ, Degenhart HJ, Trapman J, Brinkmann AO. Differential splicing of human androgen receptor pre-mRNA in X-linked Reifenstein syndrome, because of a deletion involving a putative branch site. Am J Hum Genet 1994;54:609–617.
50. Zoppi S, Wilson CM, Harbison MD, Griffin JE, Wilson JD, McPhaul MJ, et al. Complete testicular feminization caused by an amino-terminal truncation of the androgen receptor with downstream initiation. J Clin Invest 1993;91:1105–1112.
51. Wilson CM, McPhaul MJ. A and B forms of the androgen receptor are present in human genital skin fibroblasts. Proc Natl Acad Sci USA 1994;91:1234-1238.
52. McPhaul MJ, Marcelli M, Zoppi S, Griffin JE, Wilson JD. The spectrum of mutations in the androgen receptor gene that causes androgen resistance. J. Clin Endocrinol Metab 1993;76:17–23.
53. McPhaul MJ, Marcelli M, Zoppi S, Wilson CM, Griffin JE, Wilson JD. Mutations in the ligand-binding domain of the androgen receptor gene cluster in two regions of the gene. J Clin Invest 1992; 90:2097-2101.
54. Marcelli M, Tilley WD, Zoppi S, Griffin JE, Wilson JD, McPhaul MJ. Androgen resistance associated with a mutation of the androgen receptor at amino acid 772 (Arg→Cys) results from a combination of decreased messenger ribonucleic acid levels and impairment of receptor function. J Clin Endocrinol Metab 1991;73:318-325.
55. Wilson CM, Griffin JE, Wilson JD. A survey of immunoreactive androgen receptor (AR) expression in genital skin fibroblasts established from patients with androgen resistance and absent ligand binding. 75th Endocrine Society Annual Meeting, Las Vegas, Nevada, 1993; Abstract #12, p. 53.
56. Lubahn DB, Brown TR, Simental JA, Higgs HN, Migeon CJ, Wilson EM, et al. Sequence of the intron/exon junctions of the coding region of the human androgen receptor gene and identification of a point mutation in a family with complete androgen insensitivity. Proc Natl Acad Sci USA 1989;86:9534-9538.
57. Brown TR, Lubahn DB, Wilson EM, French FS, Migeon CJ, Corden JL. Functional characterization of naturally occurring mutant androgen receptors from subjects with complete androgen insensitivity. Mol Endocrinol 1990;4:1759-1772.
58. Grino PB, Isidro-Gutierrez RF, Griffin JE, Wilson JD. Androgen resistance associated with a qualitative abnormality of the androgen receptor and responsive to high dose androgen therapy. J Clin Endocrinol Metab 1989;68:578-584.
59. McPhaul MJ, Marcelli M, Tilley WD, Griffin JE, Isidro-Gutierrez RF, Wilson JD. Molecular basis of androgen resistance in a family with a qualitative abnormality of the androgen receptor and responsive to high-dose androgen therapy. J Clin Invest 1991;87:1413-1421.
60. Langley E, Zhou ZX, Wilson EM. Evidence for an anti-parallel orientation of the ligand-activated human androgen receptor dimer. J Biol Chem 1995;270:29,983-29,990.

61. Doesburg P, Kuil CW, Berrevoets CA, Steketee K, Faber PW, Mulder E, et al. Functional in vivo interaction between the amino-terminal, transactivation domain and the ligand binding domain of the androgen receptor. Biochemistry 1997;36:1052-1064.
62. Prior L, Bordet S, Trifiro MA, Mhatre A, Kaufman M, Pinsky L, et al. Replacement of arginine 773 by cysteine or histidine in the human androgen receptor causes complete androgen insensitivity with different receptor phenotypes. Am J Hum Genet 1992;51:143–155.
63. Beitel LK, Kazemi-Esfarjani P, Kaufman M, Lumbroso R, DiGeorge AM, Killinger DW, et al. Substitution of arginine-839 by cysteine or histidine in the androgen receptor causes different receptor phenotypes in cultured cells and coordinate degrees of clinical androgen resistance. J Clin Invest 1994; 94:546–554.
64. Kazemi-Esfarjani P, Beitel LK, Trifiro M, Kaufman M, Rennie P, Sheppard P, et al. Substitution of valine-865 by methionine or leucine in the human androgen receptor causes complete or partial androgen insensitivity, respectively with distinct androgen receptor phenotypes. Mol Endocrinol 1993;7:37–46.
65. Ris-Stalpers C, Trifiro MA, Kuiper GG, Jenster G, Romalo G, Sai T, et al. Substitution of aspartic acid-686 by histidine or asparagine in the human androgen receptor leads to a functionally inactive protein with altered hormone-binding characteristics. Mol Endocrinol 1991;5:1562–1569.
66. Marcelli M, Zoppi S, Wilson CM, Griffin JE, McPhaul MJ. Amino acid substitutions in the hormone-binding domain of the human androgen receptor alter the stability of the hormone receptor complex. J Clin Invest 1994;94:1642–1650.
67. Tincello DG, Saunders PT, Hodgins MB, Simpson NB, Edwards CR, Hargreaves TB, et al. Correlation of clinical, endocrine and molecular abnormalities with in vivo responses to high-dose testosterone in patients with partial androgen insensitivity syndrome. Clin Endocrinol 1997;46:497–506.
68. Yong EL, Ng SC, Roy AC, Yun G, Ratnam SS. Pregnancy after hormonal correction of severe spermatogenic defect due to mutation in androgen receptor gene Lancet 1994;344: 826–827.
69. Zoppi S, Marcelli M, Deslypere JP, Griffin JE, Wilson JD, McPhaul MJ. Amino acid substitutions in the DNA-binding domain of the human androgen receptor are a frequent cause of receptor-binding positive androgen resistance. Mol Endocrinol 1992;6:409–415.
70. Beitel LK, Prior L, Vasiliou DM, Gottlieb B, Kaufman M, Lumbroso R, et al. Complete androgen insensitivity due to mutations in the probable alpha-helical segments of the DNA-binding domain in the human androgen receptor. Hum Mol Genet 1994;3:21–27.
71. De Bellis, A, Quigley CA, Marschke KB, el-Awady MK, Lane MV, Smith EP, et al. Characterization of mutant androgen receptors causing partial androgen insensitivity syndrome. J Clin Endocrinol Metab 1994;78:513–522.
72. Sultan C, Lumbroso S, Poujol N, Belon C, Boudon C, Lobaccaro JM. Mutations of androgen receptor gene in androgen insensitivity syndromes. J Steroid Biochem Mol Biol 1993;46:519–530.
73. Lumbroso S, Lobaccaro JM, Belon C, Martin D, Chaussain JL, Sultan C. A new mutation within the deoxyribonucleic acid-binding domain of the androgen receptor gene in a family with complete androgen insensitivity syndrome. Fertil Steril 1993;60:814–819.
74. Mowszowicz I, Lee HJ, Chen HT, Mestayer C, Portois MC, Cabrol S, et al. A point mutation in the second zinc finger of the DNA-binding domain of the androgen receptor gene causes complete androgen insensitivity in two siblings with receptor-positive androgen resistance. Mol Endocrinol 1993; 7:861–869.
75. Quigley CA, Evans BAJ, Sinemetal JA. Complete androgen insensitivity due to deletion of exon C of the androgen receptor gene hilights the functional importance of the second zinc-finger of the androgen receptor in vivo. Mol Endocrinol 1992;6:1103–1112.
76. Beitel LK, Prior L, Vasiliou DM, Gottlieb B, Kaufman M, Lumbroso R, et al. Complete androgen insensitivity due to mutations in the probable alpha-helical segments of the DNA-binding domain in the human androgen receptor. Hum Mol Genet 1994;3:21–27.
77. Ris-Stalpers C, Hoogenboezem T, Sleddens HF, Verleun-Mooijman MC, Degenhart HJ, Drop SL, et al. A practical approach to the detection of androgen receptor gene mutations and pedigree analysis in families with x-linked androgen insensitivity. Pediatr Res 1994;36:227–234.
78. Saunders PT, Padayachi T, Tincello DG, Shalet SM, Wu FC. Point mutations detected in the androgen receptor gene of three men with partial androgen insensitivity syndrome. Clin Endocrinol 1992;37: 214–220.
79. Hiort O, Wodtke A, Struve D, Zollner A, Sinnecker GH. Detection of point mutations in the androgen receptor gene using non-isotopic single strand conformation polymorphism analysis. Hum Mol Genet 1994;3:1163–1166.

80. Allera A, Herbst MA, Griffin JE, Wilson JD, Schweikert HU, McPhaul MJ. Mutations of the androgen receptor coding sequence are infrequent in patients with isolated hypospadias. J Clin Endocrinol Metab 1995;80:2697–2699.
81. Gast A, Neuschmid-Kaspar F, Klocker H, Cato AC. A single amino acid exchange abolishes dimerization of the androgen receptor and causes Reifenstein syndrome. Mol Cell Endocrinol 1995;111:93–98.
82. Sutherland RW, Wiener JS, Hicks JP, Marcelli M, Gonzales ET Jr, Roth DR, et al. Androgen receptor gene mutations are rarely associated with isolated penile hypospadias. J Urol 1996;156:828–831.
83. Imasaki K, Okabe T, Murakami H, Tanaka Y, Haji M, Takayanagi R, et al. Androgen insensitivity syndrome due to new mutations in the DNA-binding domain of the androgen receptor. Mol Cell Endocrinol 1996;120:15–24.
84. Bruggenwirth HT, Boehmer AL, Verleun-Mooijman MC, Hoogenboezem T, Kleijer WJ, Otten BJ, et al. Molecular basis of androgen insensitivity. J Steroid Biochem Mol Biol 1996;58:569–575.
85. Lobaccaro JM, Poujol N, Chiche L, Lumbroso S, Brown TR, Sultan C. Molecular modeling and in vitro investigations of the human androgen receptor DNA-binding domain: application for the study of two mutations. Mol Cell Endocrinol 1996;116:137–147.
86. Weidemann W, Linck B, Haupt H, Mentrup B, Romalo G, Stockklauser K, et al. Clinical and biochemical investigations and molecular analysis of subjects with mutations in the androgen receptor gene. Clin Endocrinol 1996;45:733–739.
87. Tincello DG, Saunders PT, Hodgins MB, Simpson NB, Edwards CR, Hargreaves TB, et al. Correlation of clinical, endocrine and molecular abnormalities with in vivo responses to high-dose testosterone in patients with partial androgen insensitivity syndrome. Clin Endocrinol 1997;46:497–506.
88. Wooster R, Mangion J, Eeles R, Smith S, Dowsett M, Averill D, et al. A germline mutation in the androgen receptor gene in two brothers with breast cancer and Reifenstein syndrome. Nat Genet 1992;2:132–134.
89. Evans RM. The steroid and thyroid hormone receptor superfamily. Science 1988;240:889–895.
90. Lobaccaro JM, Lumbroso S, Belon C, Galtier-Dereure F, Bringer J, Lesimple T, et al. Androgen receptor gene mutation in male breast cancer. Hum Mol Genet 1993;2:1799–1802.
91. Poujol N, Lobaccaro JM, Chiche L, Lumbroso S, Sultan C. Functional and structural analysis of R607Q and R608K androgen receptor substitutions associated with male breast cancer. Mol Cell Endocrinol 1997;130:43–51.
92. Gao TS, McPhaul MJ. Functional activities of the A- and B- forms of the human androgen receptor in response to androgen receptor agonists and antagonists, Mol Endocrinol 1998;12:654–663.
93. Choong CS, Quigley CA, French FS, Wilson EM. A novel missense mutation in the amino-terminal domain of the human androgen receptor gene in a family with partial androgen insensitivity syndrome causes reduced efficiency of protein translation. J Clin Invest 1996;98:1423–1431.
94. Tsukada T, Inoue M, Tachibana S, Nakai Y, Takebe H. An androgen receptor mutation causing androgen resistance in undervirilized male syndrome. J Clin Endocrinol Metab 1994;79:1202–1207.
95. Tut TG, Ghadessy FJ, Trifiro MA, Pinsky L, Yong EL. Long polyglutamine tracts in the androgen receptor are associated with reduced trans-activation, impaired sperm production, and male infertility. J Clin Endocrinol Metab 1997;82:3777–3782.
96. McPhaul MJ, Deslypere JP, Allman DR, Gerard RD. The adenovirus mediated delivery of a reporter gene permits the assessment of androgen receptor function in genital skin fibroblast cultures. J Biol Chem 1993;268:26,063–26,066.
97. McPhaul MJ, Schweikert H-U, Allman DR. Assessment of androgen receptor function in genital skin fibroblasts using a recombinant adenovirus to deliver an androgen-responsive reporter gene. J Clin Endocrinol Metab 1997;82:1944–1948.
98. Refetoff S. Resistance to thyroid hormone. Curr Therapy Endocrinol Metabol 1997;6:132–134.
99. Kopp P, Kitajima K, Jameson JL. Syndrome of resistance to thyroid hormone: insights into thyroid hormone action. Proc Soc Exp Biol Med 1996;211:49–61.
100. Holterhus PM, Bruggenwirth HT, Hiort O, Kleinkaufhoucken A, Kruse K, Sinnecker GHG, et al. Mosaicism due to a somatic mutation of the androgen receptor gene determines phenotype in androgen insensitivity syndrome J Clin Endocrinol Metab 1997;82:3584–3589.
101. Sleddens HF, Oostra BA, Brinkmann AO, Trapman J. Trinucleotide (GGN) repeat polymorphism in the human androgen receptor (AR) gene. Hum Mol Genet 1993;2:493.
102. Sleddens HF, Oostra BA, Brinkmann AO, Trapman J. Trinucleotide repeat polymorphism in the androgen receptor gene (AR). Nucleic Acids Res 1992;20:1427.
103. Edwards A, Hammond HA, Jin L, Caskey CT, Chakraborty R. Genetic variation at five trimeric and tetrameric tandem repeat loci in four human population groups Genomics 1992;12:241–253.

104. Mhatre AN, Trifiro MA, Kaufman M, Kazemi-Esfarjani P, Figlewicz D, Rouleau G, et al. Reduced transcriptional regulatory competence of the androgen receptor in X-linked spinal and bulbar muscular atrophy Nature Genet 1993;5:184–188.
105. Chamberlain NL, Driver ED, Miesfeld RL. The length and location of CAG trinucleotide repeats in the androgen receptor N-terminal domain affect transactivation function. Nucleic Acids Res 1994;22: 3181–3186.
106. MacLean HE, Warne GL, Zajac JD. Spinal and bulbar muscular atrophy: androgen receptor dysfunction caused by a trinucleotide repeat expansion. J Neurol Sci 1996;135:149–157.
107. Choong CS, Kemppainen JA, Zhou ZX. Wilson EM. Reduced androgen receptor gene expression with first exon CAG repeat expansion. Mol Endocrinol 1996;10:1527–1535.
108. Schoenberg MP, Hakimi JM, Wang S, Bova GS, Epstein JI, Fischbeck KH, et al. Microsatellite mutation (CAG24→18) in the androgen receptor gene in human prostate cancer. Biochem Biophys Res Commun 1994;198:74–80.
109. Giovannucci E, Stampfer MJ, Krithivas K, Brown M, Dahl D, Brufsky A, et al. The CAG repeat within the androgen receptor gene and its relationship to prostate cancer. Proc Natl Acad Sci USA 1997;94: 3320–3323.
110. Stanford JL, Just JJ, Gibbs M, Wicklund KG, Neal CL, Blumenstein BA, et al. Polymorphic repeats in the androgen receptor gene: molecular markers of prostate cancer risk. Cancer Res 1997;57: 1194–1198.
111. Ingles SA, Ross RK, Yu MC, Irvine RA, La Pera G, Haile RW, et al. Association of prostate cancer risk with genetic polymorphisms in vitamin D receptor and androgen receptor. J Natl Cancer Inst 1997;89:166–170.
112. McPhaul MJ, Marcelli M, Zoppi S, Griffin JE, Wilson JD. Genetic basis of endocrine disease. 4. The spectrum of mutations in the androgen receptor gene that causes androgen resistance. J Clin Endocrinol Metab 1993;76:17–23.
113. LaSpada AR, Wilson EM, Lubahn DB, Harding AE, Fischbeck KH. Androgen receptor gene mutations in X-linked spinal and bulbar muscular atrophy. Nature 1991;352:77–79.
114. Brooks BP, Fischbeck KH. Spinal and bulbar muscular atrophy: a trinucleotide-repeat expansion neurodegenerative disease. Trends Neurosci 1995;18:459–461.
115. Neuschmid-Kaspar F, Gast A, Peterziel H, Schneikert J, Muigg A, Ransmayr G, et al. CAG-repeat expansion in androgen receptor in Kennedy's disease is not a loss of function mutation. Mol Cell Endocrinol 1996;117:149–156.
116. Veldscholte J, Ris-Stalpers C, Kuiper GG, Jenster G, Berrevoets C, Claassen E, et al. A mutation in the ligand binding domain of the androgen receptor of human LNCaP cells affects steroid binding characteristics and response to anti-androgens. Biochem Biophys Res Commun 1990;173:534–540.
117. Newmark JR, Hardy DO, Tonb DC, Carter BS, Epstein JI, Isaacs WB, et al. Androgen receptor gene mutations in human prostate cancer. Proc Natl Acad Sci USA 1992;89:6319–6323.
118. Evans BA, Harper ME, Daniells CE, Watts CE, Matenhelia S, Green J, et al. Low incidence of androgen receptor gene mutations in human prostatic tumors using single strand conformation polymorphism analysis. Prostate 1996;28:162–171.
119. Taplin ME, Bubley GJ, Shuster TD, Frantz ME, Spooner AE, Ogata GK, et al. Mutation of the androgen-receptor gene in metastatic androgen-independent prostate cancer. N Engl J Med 1995;332: 1393-1398.
120. Tilley WD, Buchanan G, Hickey TE, Bentel JM. Mutations in the androgen receptor gene are associated with progression of human prostate cancer to androgen independence Clin Cancer Res 1996;2: 277–285.
121. McPhaul MJ. The androgen receptor and prostate cancer. In: Pasqualini JR, Katzenellenbogen BS, eds. Hormone-Dependent Cancer. Marcel Dekker, New York, 1996, pp. 307–321.
122. Culig Z, Hobisch A, Cronauer MV, Cato AC, Hittmair A, Radmayr C, et al. Mutant androgen receptor detected in an advanced-stage prostatic carcinoma is activated by adrenal androgens and progesterone. Mol Endocrinol 1993;7:1541–1550.
123. Elo JP, Kvist L, Leinonen K, Isomaa V, Henttu P, Lukkarinen O, et al. Mutated human androgen receptor gene detected in a prostatic cancer patient is also activated by estradiol. J Clin Endocrinol Metab 1995;80:3494–3500.
124. Peterziel H, Culig Z, Stober J, Hobisch A, Radmayr C, Bartsch G, et al. Mutant androgen receptors in prostatic tumors distinguish between amino-acid-sequence requirements for transactivation and ligand binding. Int J Cancer 1995;63:544–550.

125. Suzuki H, Sato N, Watabe Y, Masai M, Seino S, Shimazaki J. Androgen receptor gene mutations in human prostate cancer. J Steroid Biochem Mol Biol 1993;46:759–765.
126. Suzuki H, Akakura K, Komiya A, Aida S, Akimoto S, Shimazaki J. Codon 877 mutation in the androgen receptor gene in advanced prostate cancer: relation to antiandrogen withdrawal syndrome. Prostate 1996;29:153–158.
127. Koivisto P, Kononen J, Palmberg C, Tammela T, Hyytinen E, Isola J, et al. Androgen receptor gene amplification: a possible molecular mechanism for androgen deprivation therapy failure in prostate cancer. Cancer Res 1997;57:314–319.
128. Crocitto LE, Henderson BE, Coetzee GA. Identification of two germline point mutations in the 5'UTR of the androgen receptor gene in men with prostate cancer. J Urol 1997;158:1599–1601.
129. Paz A, Lindner A, Zisman A, Siegel Y. A genetic sequence change in the 3'-noncoding region of the androgen receptor gene in prostate carcinoma. Eur Urol 1997;31:209–215.

12 Retained Müllerian Ducts
AMH Resistance Syndrome

Nathalie Josso, MD, Jean-Yves Picard, PhD, and Rodolfo Rey, MD, PhD

CONTENTS

PHENOTYPE OF AMH-RESISTANT INDIVIDUALS
MOLECULAR BASIS OF AMH RESISTANCE
MUTATIONS OF THE AMH TYPE II RESISTANCE
CONCLUSION
REFERENCES

Sexual differentiation is implemented by the action of two distinct hormones, produced by the fetal testis, testosterone, and anti-Müllerian hormone (AMH) (Fig. 1). Testosterone, secreted by Leydig cells, masculinizes the external genitalia and urogenital sinus. AMH, also known as Müllerian inhibiting substance (MIS) or factor (MIF), a Sertoli cell product, is a dimeric glycoprotein belonging to the transforming growth factor-β (TGF-β) family and is responsible for the regression of Müllerian ducts in male fetuses *(1)*. In the absence of AMH, Müllerian ducts give rise to the uterus, fallopian tubes, and upper part of the vagina. Normally, Müllerian duct derivatives develop only in females owing to the fact that in males, fetal Sertoli cells produce AMH very soon after testicular differentiation, i.e., at 8 wk in the human male fetus *(2)*, at a time Müllerian ducts are still responsive to the hormone *(3)*. However, if Sertoli cells are unable to produce AMH or if target organs are unresponsive to its action, Müllerian duct derivatives will be retained in a subject otherwise normally masculinized. This condition is known as the persistent Müllerian duct syndrome (PMDS) or internal pseudohermaphroditism. In the French literature, the term "homme à uterus" (man with an uterus) is also common. A similar syndrome has been generated in mice by targeted deletions of the AMH *(4)* or AMH-receptor *(5)* gene.

PHENOTYPE OF AMH-RESISTANT INDIVIDUALS

In both mice and men, the phenotype of AMH-resistant and AMH-deficient individuals is identical. It is characterized by the retention of Müllerian derivatives in otherwise normally masculinized XY males. By definition, it is never associated with hypospadias.

From: *Contemporary Endocrinology: Hormone Resistance Syndromes*
Edited by: J. L. Jameson © Humana Press Inc., Totowa, NJ

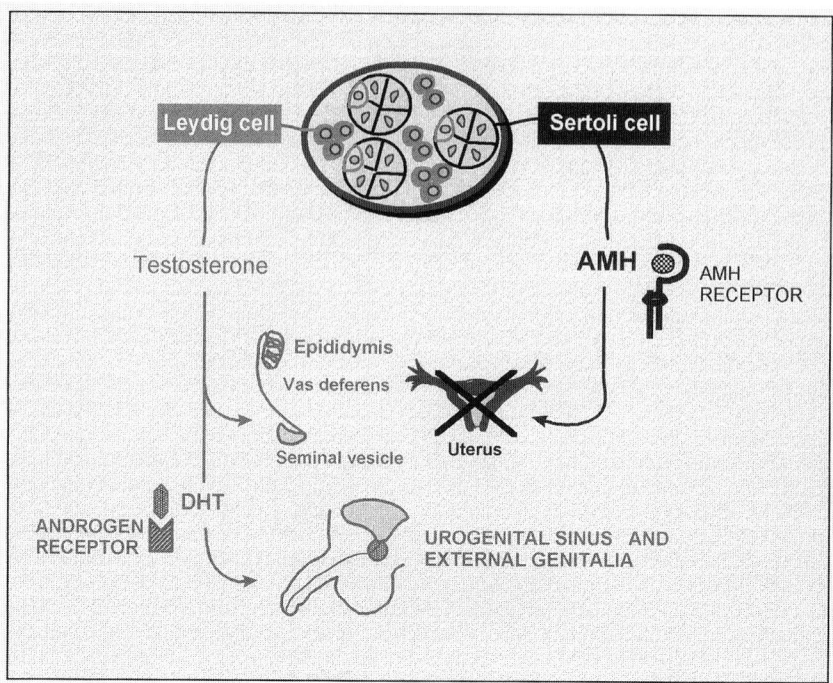

Fig. 1. Hormones involved in sex differentiation: Testosterone, secreted by Leydig cells, virilizes the external genitalia and urogenital sinus, whereas anti-Müllerian hormone, a glycoprotein produced by Sertoli cells, represses the development of Müllerian ducts into uterus and tubes.

Retention of Müllerian ducts associated with incomplete masculinization of the external genitalia is usually the result of testicular dysgenesis and is not genetically transmitted.

Clinical and Anatomical Features of AMH-Resistance in Humans

Patients are externally phenotypic males with either bilateral cryptorchidism or unilateral testicular ectopia associated with inguinal hernia on the contralateral side. The hernia may appear incarcerated, in spite of the lack of symptoms of intestinal obstruction, but this discrepancy is usually not correctly interpreted. Sonography could be useful, but is rarely performed, unless an older sibling has been diagnosed with the condition.

The presence of an uterus and tubes is usually discovered at surgery. Two anatomical forms of PMDS have been described. The most common one, encountered in 80% of the cases, is characterized by the association of unilateral cryptorchidism and contralateral hernia. One testis has descended into the scrotum and the ipsilateral uterus, and fallopian tube have either entered the inguinal canal, a condition known as *hernia uteri inguinalis*, or can be dragged into it by gentle traction, bringing the contralateral testis and fallopian tube in their wake. Often, no traction is necessary, because the contralateral testis is already in the hernial sac. Tranverse testicular ectopia, as this condition is called, is extremely frequent in PMDS *(6)*. More rarely, PMDS presents as bilateral cryptorchidism, the uterus is fixed in the pelvis and both testes are embedded in the broad ligament (Fig. 2). These clinical variants are not genetically determined and may occur within the same sibship *(7)*.

As pointed out by Hutson et al. *(8)*, in PMDS, the testes are abnormally mobile because they are not anchored to the bottom of the processus vaginalis by a normal male guber-

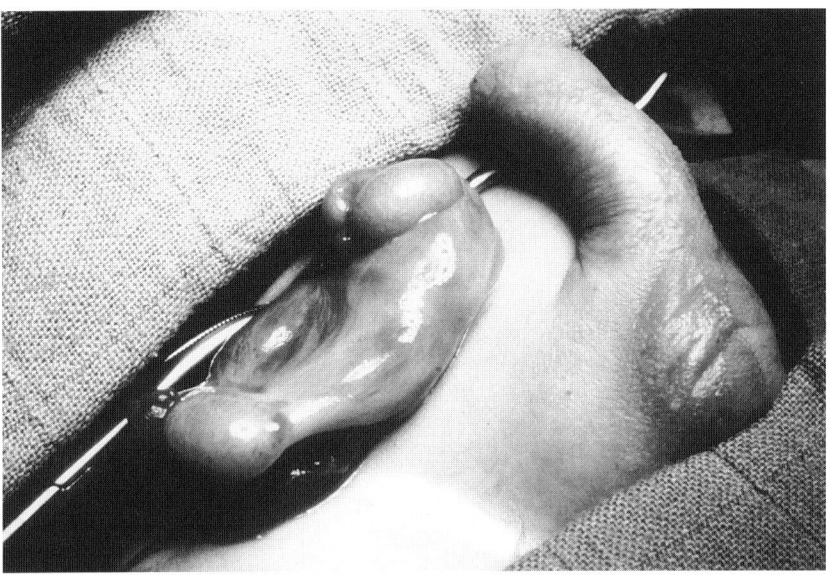

Fig. 2. Internal reproductive tract of a patient with persistent Müllerian duct syndrome owing to a receptor mutation. Note the close attachment of testes to fallopian tubes and the normal appearing external genitalia.

naculum, but are connected instead to elongated thin round ligaments. This hypermobility could favor testicular torsion and subsequent testicular degeneration, a high incidence of which has been reported in PMDS *(9)*.

Testes are normally differentiated and, in the absence of long-standing cryptorchidism, usually contain germ cells. However, the testes are often not properly connected to male excretory ducts; aplasia of the epididymis and upper part of the vas deferens has also been reported *(10)*. Furthermore, it is usually difficult to bring the testes down to a normal position; even after careful dissection, the free segment of the spermatic cord is very short because the vasa deferentia are embedded in the mesosalpynx before entering the lateral uterine wall and cervix. Lack of proper communication between the testis and excretory ducts and difficulties at orchidopexy probably explain why fertility is rare in PMDS patients—11% according to a Kuwait study *(11)*.

No other clinical abnormalities have been described in male patients with PMDS due to receptor mutations, apart from a testicular germ cell tumor in a 24-yr-old patient; another patient developed colon adenocarcinoma and medullary thyroid cancer at age 77. Testicular tumors have also been described in PMDS patients of undetermined molecular etiology *(12,13)*. Sisters of AMH-resistant patients who share their genetic background are phenotypically normal and undergo puberty at the expected age. This is relatively surprising, in view of the fact that AMH is produced by granulosa cells during the period of reproductive activity *(14,15)* and exerts a suppressive effect on follicle maturation *(16)*. Probably, in the absence of AMH action, other growth factors also present in the ovary take over.

Endocrinology of AMH Resistance

Except in patients with testicular degeneration, serum testosterone levels and response to hCG are normal in PMDS patients, whatever the molecular basis of the condition.

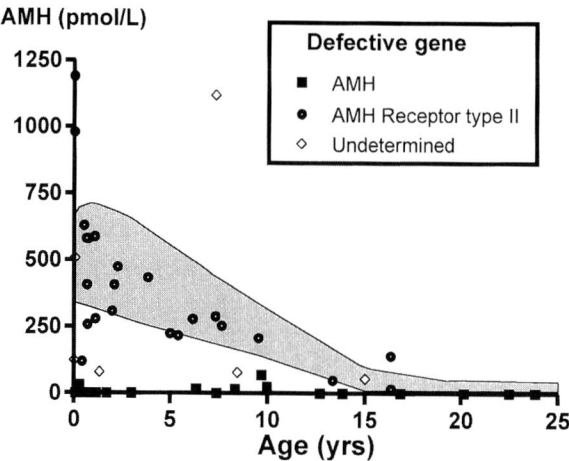

Fig. 3. AMH levels in patients with PMDS, measured using AMH/MIS enzyme immunoassay kit (Immunotech-Coulter). Normal range is shaded.

In contrast, assay of circulating AMH is extremely informative in this regard *(10)*. In normal subjects, mean serum AMH levels, assayed by an enzyme immunoassay kit (Immunotech-Coulter, Marseilles, France) are close to 460 nmol/L (65 ng/mL) in the perinatal period and slowly decrease up to puberty, at which time testosterone downregulates AMH production by Sertoli cells and serum AMH falls to near-undetectable levels *(17)*. Patients in whom mutations of the AMH receptor have been detected usually have AMH serum levels within normal limits. In contrast, patients who fail to produce AMH because of mutations of the AMH gene have abnormally low or undetectable circulating AMH levels. Intermediate levels are found in patients in whom neither AMH nor AMH-receptor mutations have been detected (Fig. 3).

MOLECULAR BASIS OF AMH RESISTANCE

Hormone resistance syndromes are usually the result of mutations in genes responsible for the binding of hormone to their receptor on the cell membrane or nucleus, and for transduction of the signal to the hormone-responsive cell machinery. The mechanism of the response of target cells to AMH is far from being fully elucidated at the present time. A preview of the factors probably involved in AMH signaling can be deduced from studies of other members of the TGF-β family.

Signaling Cascade for the Transforming Growth Factor-β Family

RECEPTORS

AMH is a member of the TGF-β family, which signals through a receptor complex formed by two distantly related serine/threonine kinases (reviewed in *18*). The primary receptor, called type II receptor on the basis of its molecular weight, binds its ligand, but is incapable of signal transduction, unless coexpressed with an appropriate signal transducer, known as the type I receptor. Binding of ligand to the type II receptor induces the formation of a complex, most likely a heterotetramer containing two molecules each of

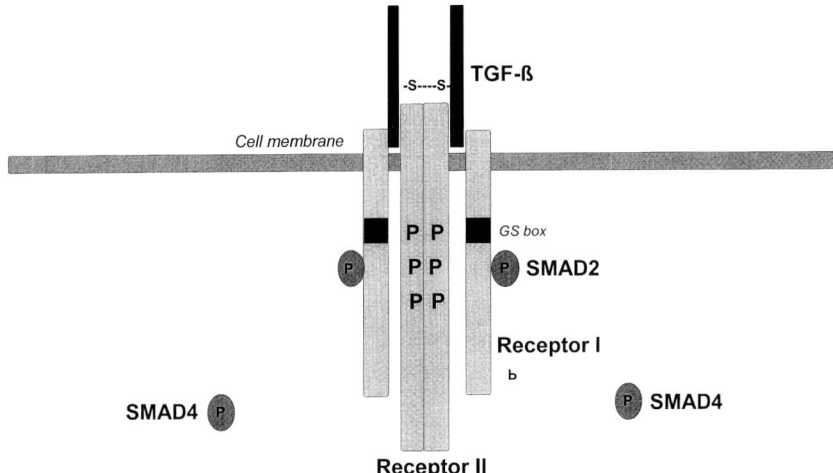

Fig. 4. Possible mechanism of signal transduction by TGF-β. The ligand could bind to two molecules of type II receptor and two molecules of type I receptor. A Smad2 molecule binds temporarily to the activated type I receptor and then interacts with a Smad 4 molecule, before entering the nucleus. Modified from ref. 58 with permission.

TβR-I and TβR-II *(19–21)* (Fig. 4). The central biochemical event in ligand-induced TGF-β receptor activation is the transphosphorylation of the type I receptor (TβR-I) by the type II (TβR-II) at threonines and serines located in a conserved glycine/serine-rich domain located immediately upstream of the kinase consensus domains *(22)*. The model for activation of the TGF-β receptor complex has recently been extended to activin *(23–25)*. The composition of the receptor complex and more particularly the identity of the type I receptor determine the nature of the signal.

Absence of TβR-I or mutations thereof leading either to loss of kinase activity or inability to be transphosphorylated by TβR-II block signaling responses *(19)*. A missense mutation of TβR-II, which blocks the recognition of TβR-I as a substrate, has the same effect *(26)*. Numerous other mutations of TβR-II have been obtained by chemical mutagenesis in TGF-β-responsive cell lines; most impair receptor function only in the homozygous state, but truncation immediately after the transmembrane domain produces dominant negative molecules that bind to ligand and type I receptors, but do not induce biological responses (reviewed in ref. 27).

Downstream Proteins

Yeast two-hybrid screens have allowed the identification of various proteins associating with TGF-β receptors, the WD protein TRIP1 *(28)*; and the α-subunit of farnesyl-protein transferase *(29,30)*. Another protein interacting in a two-hybrid screen with type I receptors is the immunophilin FKBP-12. FKBP-12 binds to ligand-free type I receptors, from which it is released on ligand-induced activation *(31)*.

A new family of proteins, the Smads, appears to play a key role in downstream signaling. Numerous homologs of the prototypic Drosophila gene, mothers against decapentaplegic (MAD), have now been cloned. Some, specific of a TGF-β family member, are said to be "pathway-restricted." For instance, Smad1 is specific for the BMP family, whereas Smad2 propagates signals for TGF-β and activin. In contrast, Smad4, alias

DPC4 for "deleted in pancreatic carcinoma", is a shared mediator of TGF-β family responses, since it associates with pathway-restricted Smads in response to the corresponding agonists *(32,33)*.

The AMH Signaling Cascade

In contrast to other TGF-β family members, little is known concerning the proteins mediating AMH responsiveness; the molecular spectrum of AMH resistance is therefore restricted to the "emerged part of the iceberg" and will hopefully be completed in the near future. Up to now, only the cDNA *(34,35)* and gene *(36)* for the type II receptor has been cloned *(32–34)*. As expected for a primary receptor, it binds iodinated AMH and is expressed in AMH target organs, such as the fetal Müllerian duct and fetal and adult ovary. It is also expressed in testicular Sertoli *(34,35,37)* and Leydig cells *(38)*. AMH represses Leydig cell differentiation and steroidogenesis *(38)* cells, but no function for AMH in Sertoli cells has yet been determined. The AMH receptor gene contains 11 exons: exons 1–3 code for the signal sequence and extracellular domain, exon 4 for most of the transmembrane domain, and exons 5–11 for the intracellular serine/threonine kinase domains. The promoter lacks TATA and CAAT boxes. Exon 2 undergoes alternative splicing in rabbit *(34)* and dog (Picard JY, unpublished), but not in humans *(35)* or rat *(36)*. The wild-type human AMH receptor consists of 573 amino acids, and homology with receptors of other species is approx 80%, but only 30% with receptors for activin and TGF-β. The mature form of the human receptor is expressed at the cell surface and has a mass of 82 kDa *(39)*.

As yet, the type I receptor for AMH has not been identified, nor does one know whether AMH shares one with other family members or whether it has a specific signal transducer of its own. Establishing which type I receptor associates with which ligand/type II receptor complex is tricky, because a given ligand may signal through different receptor complexes and a given receptor can recognize different ligands. Furthermore, given that type I and II receptors, which probably do not associate under normal cellular conditions, may do so when overexpressed in cell lines, binding data alone are not sufficient to assume the existence of a functional signaling complex. Demonstration of coimmunoprecipitation *(40)* or proof of the capacity of the receptor under investigation to restore signaling in mutant cell lines is required *(41)*, but the latter is not yet available for AMH receptors.

Possible candidates include TSR-1 *(42)*, alias Alk1 *(43)*, an orphan receptor *(18)*, but the fact that it is highly expressed only in the endothelium and that mutations produce a hereditary bleeding disorder, but apparently no persistence of Müllerian duct derivatives argues against its involvement in the AMH signaling cascade. ActR-I, alias Alk2 *(43)* or R1 *(44)*, an activin receptor that also transduces signals for bone morphogenetic protein-7, is expressed in brain, lung, the fetal urogenital ridge, and oocytes, but only poorly if at all in granulosa cells *(44)*, whereas the type II AMH receptor is strongly expressed in granulosa cells, but not oocytes *(35,45)*. However, since ActR-I is also a receptor for activin and bone morphogenetic proteins (BMPs), its pattern of expression cannot be expected to coincide with AMH responsiveness.

AMH resistance could also stem from mutations in the cleavage enzyme required to release the bioactive C-terminal fragment from the full-length protein, a processing step required for all members of the TGF-β-family. PC5, a member of the proprotein convertase family, is able to process testicular AMH *(46)*, but its physiological relevance remains to be established.

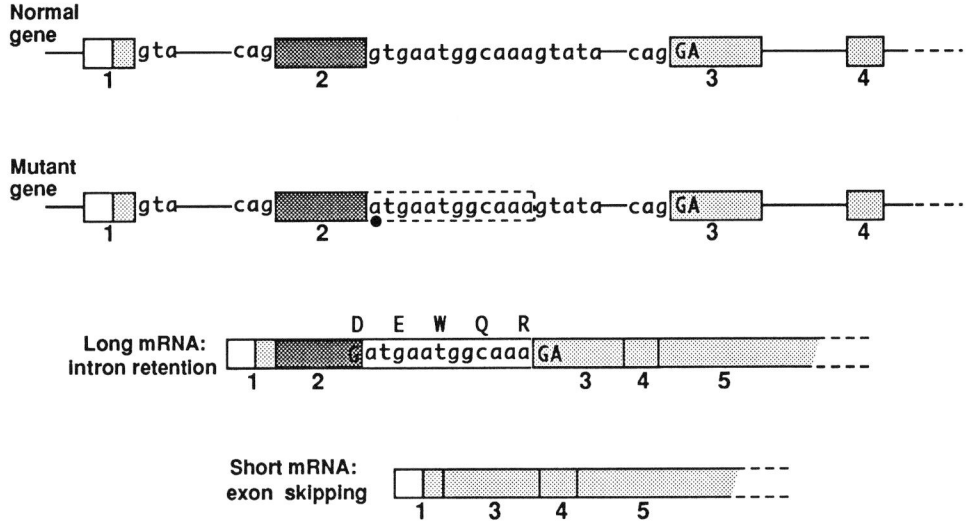

Fig. 5. Splice mutation leading to the formation of two variants of the AMH type II receptor. The mutation changes the g of the donor splice site to an a. In some cases, the splicing mechanism recognizes a cryptic splice site (gtata). This leads to incorporation of 12 additional bases (shown on a white background) into the second exon, and to the incorporation of five amino acids into the protein. Alternatively, skipping of the second exon occurs, yielding short mRNA.

MUTATIONS OF THE AMH TYPE II RECEPTOR

Description

Mutations of the type II AMH receptor have now been described in 16 families. Most, but not all, are accessible to screening with a single-strand conformation polymorphism polymerase chain reaction (SSCP-PCR), but some require extensive sequencing, particularly when present in the heterozygous state. Mutations affect both the extracellular and intracellular domains of the receptor *(10)*.

A splice mutation that destroys the invariant nucleotide at the 5'-end of the second intron generates two abnormal mRNAS, one (short) missing the second exon and the other (long) incorporating the first 12 bases of the second intron (Fig. 5). The short mRNA is similar to the one generated by alternative splicing in the rabbit AMH type II receptor *(35)*. Both the short and long mutant proteins are unable to reach the cell surface when expressed in COS cells, probably because they are retained in the endoplasmic reticulum *(39)*.

Other mutations affecting the extracellular region of the AMH type II receptor include a 4-bp microdeletion in the first exon (Δ84–87), which disrupts the reading frame, three stop mutations, and 2 missense ones (Fig. 6). When appropriate, mutant recombinant proteins were generated in COS cells; none were able to bind AMH owing to nonspecific retention in the endoplasmic reticulum (Messika-Zeitoun, unpublished results). The nonsense mutations are expected to lead to early termination of translation, preventing the insertion of the protein in the membrane.

Among mutations of the intracellular domain, a 27-base deletion in exon 10 is present on one or both alleles in 10 out of 16 families, and is detectable by acrylamide electrophoresis of PCR fragments. The most common cause of small recurrent deletions, DNA

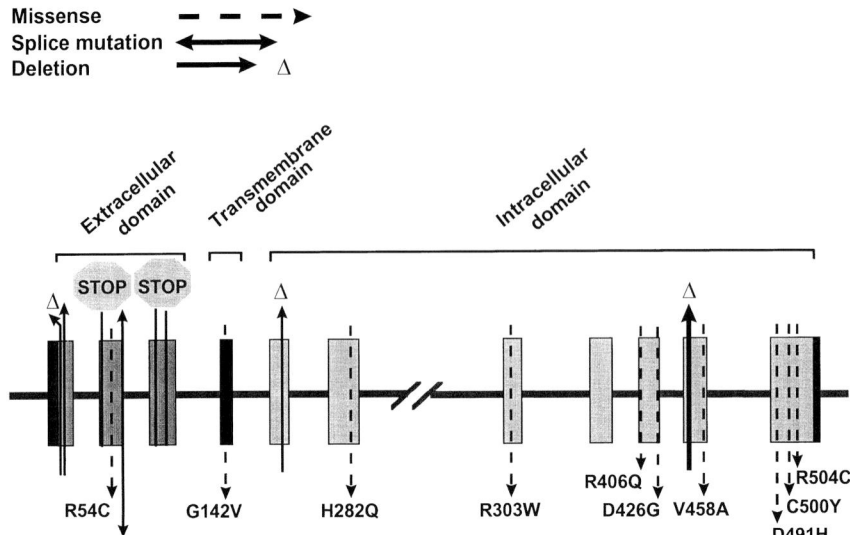

Fig. 6. Mutations of the AMH type II receptor. From ref. *10* with permission.

repeats, such as *Alu* sequences, do not seem to be implicated in this mutation, which could be due to a founder effect, since all alleles carrying this deletion possess the same haplotype, compared to only 50% in the general population. The haplotype was determined from intronic polymorphisms affecting 4–19% of normal subjects *(10)*. Missense mutations were also detected in exons 2–11, with the exception of exons 3, 5, and 8.

Transmission and Incidence

Up to now, all mutations detected in the AMH type II receptor have been compatible with a recessive mode of transmission. Dominant negative mutations described for the TGF-β *(47,48)* and activin receptor *(49)* have not been found. These mutations truncate the protein immediately after the transmembrane domain and are thought to bind ligand and signal transducer, thus preventing the normal allele from gaining access to molecules required for signaling activity. One AMH receptor mutation, a 1-base deletion in exon 5, which leads to a stop mutation immediately downstream, although structurally very similar to these dominant negative mutations, leads to PMDS only when associated with another mutation on the second allele. However, all the dominant negative mutations of TGF-β family members have been produced artificially and the recombinant proteins have been overexpressed; it will be of interest to determine whether natural mutations, when overexpressed relative to the normal allele, do inhibit the function of the wild-type protein. Mutations of the AMH gene itself also exhibit a recessive pattern of inheritance *(50)*. Two pedigrees of PMDS of unknown molecular etiology suggest an X-linked recessive transmission *(51,52)*.

AMH type II receptor mutations account for 42% of cases of the PMDS. A similar proportion are the result of mutations of the AMH gene. Receptor mutations are usually found in patients from Northern Europe and the US, but subjects with mutations of the AMH gene itself originate predominantly from Arab or Mediterranean regions, with a higher rate of consanguinity. This is reflected in a higher rate of homozygosity (81%) in the latter, compared with mutations of the AMH receptor, only 45% of which are homozygous. As discussed above, AMH and AMH receptor mutations can usually be distinguished by the level of serum AMH, which is undetectable in AMH mutations and normal in receptor mutations; there are, however, limitations to the usefulness of this assay: patients with AMH resistance, like normal subjects, undergo repression of AMH production after puberty; AMH levels fall in the few days following a testicular biopsy, which is usually carried out when Müllerian derivatives are discovered at surgery *(53)*.

In 16% of cases of PMDS, screening by SSCP-PCR of the AMH and AMH receptor gene yields normal results. This technique is extremely efficient for detecting mutations in the AMH gene *(50)*, but can miss base changes in the AMH receptor. However, it is unlikely that all patients in this category have undetected mutations of the AMH receptor on both alleles. Since, like other family members, AMH signaling probably requires a type I receptor and Smad proteins, defects in the latter probably account for a significant proportion of unexplained AMH resistance. Testing for mutations of candidate genes in this group will be extremely helpful for the definitive identification of factors required for AMH signaling.

ANIMAL MODELS OF AMH RESISTANCE

Targeted Mutation of the AMH Receptor Locus in Mice

Targeted deletion of the mouse AMH receptor locus, resulting in the replacement of the first six exons by a neomycin resistance vector, produces retention of Müllerian derivatives very similar to that observed in the human patients, except that all male mutant homozygous mice had fully descended testes. By 2 mo of age, however, although spermatogenesis remained normal, proliferation of Leydig cells was observed, together with focal atrophy of the germinal epithelium and vacuolization of Sertoli cells. These abnormalities increased with age, so that at 10 mo of age, spermatogenesis was greatly reduced *(5)*. Fertility was further impeded by the presence of the Müllerian derivatives, which physically blocks sperm transport into the female reproductive tract. Female mutants are normal and fertile.

In double-mutant AMH receptor/α-inhibin mice, Leydig cell tumors appear earlier and are associated with Sertoli cell tumors characteristic of inhibin-deficient mice. Leydig cell hyperplasia in AMH-deficient mice is the mirror image of the Leydig cell hypoplasia observed in mice overexpressing the AMH gene under the control of the metallothionein promoter *(54)*, which is due to a block in the differentiation of the Leydig cell precursors into mature Leydig cells *(38)*. Although testicular tumors have been described in humans (*see above*), most are seminomas due to long-standing cryptorchidism; the incidence of Leydig cell tumors does not appear to be increased.

PMDS in Dogs

PMDS has also been described in a strain of miniature Schnauzer dogs *(55)* and more recently in German Basset hounds *(56)*. In both breeds, it is inherited as an autosomal-recessive trait. Clinical findings are also similar. Affected dogs present primarily with cystitis and pyometra, owing to communication between the vagina and urethra; cryptorchidism is relatively rare. PMDS dogs with scrotal testes are fertile *(55)*. AMH RNA and immunoreactive protein are present in embryonic testes during the critical period of Müllerian regression *(57)*, and after birth, the concentration of serum AMH is similar in affected and normal animals (Picard et al., unpublished). Taken together, these data suggest that AMH itself is normal and that PMDS in dogs is the result of AMH resistance. The dog AMH receptor has now been totally sequenced, and DNA from affected animals is being screened for mutations.

CONCLUSION

In conclusion, in both human and animal models, AMH resistance due to deletion or mutations in the AMH type II receptor gene lead exclusively to the persistence of Müllerian derivatives, arguing against a decisive role of AMH in other events of male sex differentiation. Female sex differentiation and/or maturation is not affected. There is no difference between the phenotype of AMH resistance and AMH deficiency, indicating that no other molecules signal through the AMH receptor. If this were the case, the receptor-defective phenotype would be expected to be more severe than the AMH-defective one. Alternatively, if several receptor complexes could bind AMH, the receptor-defective phenotype would be expected to be milder. Up to now, what we know of the AMH signaling machinery seems to be remarkably simple, consisting of only one ligand and one primary receptor. Downstream of that lies *terra incognita*.

REFERENCES

1. Jost A. Problems of fetal endocrinology: the gonadal and hypophyseal hormones. Rec Progr Horm Res 1953;8:379–418.
2. Josso N, Lamarre I, Picard JY, Berta P, Davies N, Morichon N, et al. Anti-Müllerian hormone in early human development. Early Hum Dev 1993;33:91–99.
3. Josso N, Picard JY, Tran D. The anti-Müllerian hormone. Rec Progr Horm Res 1977;33:117–160.
4. Behringer RR, Finegold MJ, Cate RL. Müllerian-inhibiting substance function during mammalian sexual development. Cell 1994;79:415–425.
5. Mishina Y, Rey R, Finegold MJ, Matzuk MM, Josso N, Cate RL, et al. Genetic analysis of the Müllerian-inhibiting substance signal transduction pathway. Genes Dev 1996;10:2577–2587.
6. Thompson ST, Grillis MA, Wolkoff LH, Katzin WE. Transverse testicular ectopia in a man with persistent mullerian duct syndrome. Arch Pathol Lab Med 1994;118:752–755.
7. Guerrier D, Tran D, van der Winden JM, Hideux S, Van Outryve L, Legeai L, et al. The persistent Müllerian duct syndrome: a molecular approach. J Clin Endocrinol Metab 1989;68:46–52.
8. Hutson JM, Davidson PM, Reece L, Baker ML, Zhou B. Failure of gubernacular development in the persistent Müllerian duct syndrome allows herniation of the testes. Pediatr Surg Int 1994;9:544–546.
9. Imbeaud S, Rey R, Berta P, Chaussain JL, Wit JM, Lustig RH, et al. Progressive testicular degeneration in the persistent Müllerian duct syndrome. Eur J Pediatr 1995;154:187–190.
10. Imbeaud S, Belville C, Messika-Zeitoun L, Rey R, di Clemente N, Josso N, et al. A 27 base-pair deletion of the anti-Müllerian type II receptor gene is the most common cause of the persistent Müllerian duct syndrome. Hum Mol Genet 1996;5:1269–1279.
11. Farag TI. Familial persistent müllerian duct syndrome in Kuwait and neighboring populations. Am J Med Genet 1993;47:432–434.
12. Snow BW, Rowland RG, Seal GM, Williams SD. Testicular tumor in patient with persistent Müllerian duct syndrome. Urology 1985;26:495–497.
13. Nishioka T, Kadowaki T, Miki T, Hanai J. (1992) Persistent Müllerian duct syndrome. Hinyokika Kiyo 1992;38:89–92.
14. Lee MM, Donahoe PK, Hasegawa T, Silverman B, Crist GB, Best S, et al. Mullerian inhibiting substance in humans: normal levels from infancy to adulthood. J Clin Endocrinol Metab 1996;81:571–576.
15. Rey R, Lhommé C, Marcillac I, Lahlou N, Duvillard P, Josso N, et al. Anti-Müllerian hormone as a serum marker of granulosa-cell tumors of the ovary: comparative study with serum alpha-inhibin and estradiol. Am J Obstet Gynecol 1996;174:958–965.
16. di Clemente N, Goxe B, Remy JJ, Cate RL, Josso N, Vigier B, et al. Inhibitory effect of AMH upon the expression of aromatase and LH receptors by cultured granulosa cells of rat and porcine immature ovaries. Endocrine 1994;2:553–558.

17. Al-Attar L, Noël K, Dutertre M, Belville C, Forest MG, Burgoyne PS, Josso N, Rey R. Hormonal and cellular regulation of Sertoli cell anti-Müllerian hormone production in the postnatal mouse. J Clin Invest 1997;100:1335–1343.
18. Massagué J, Weis-García F. Serine/threonine kinase receptors: mediators of transforming growth factor beta family signals. Cancer Surveys. Cell Signaling (Lond) 1996;27:41–64.
19. Weis-García F, Massagué J. Complementation between kinase-defective and activation-defective TGF-β receptors reveals a novel form of receptor cooperativity essential for signaling. Embo J 1996;15:276–289.
20. Luo KX, Lodish HF. Signaling by chimeric erythropoietin-TGF-β receptors: Homodimerization of the cytoplasmic domain of the type I TGF-β receptor and heterodimerization with the type II receptor are both required for intracellular signal transduction. Embo J 1996;15:4485–4496.
21. ten Dijke P, Miyazono K, Heldin CH. Signaling via hetero-oligomeric complexes of type I and type II serine/threonine kinase receptors. Curr Opinion Cell Biol 1996;8:139–145.
22. Franzén P, Heldin CH, Miyazono K. The GS domain of the transforming growth factor-β type I receptor is important in signal transduction. Biochem Biophys Res Commun 1995;207:682–689.
23. Willis SA, Zimmerman CM, Li L, Mathews LS. Formation and activation by phosphorylation of activin receptor complexes. Mol Endocrinol 1996;10:367–379.
24. Attisano L, Wrana JL, Montalvo E, Massagué J. Activation of signalling by the activin receptor complex. Mol Cell Biol 1996;16:1066–1073.
25. De Winter JP, De Vries CJM, van Achterberg TAE, Ameerun RF, Feijen A, Sugino H, et al. Truncated activin type II receptors inhibit activin bioactivity by the formation of heteromeric complexes with activin type I receptors. Exp Cell Res 1996;224:323–334.
26. Cárcamo J, Zentella A, Massagué J. Disruption of transforming growth factor-β signaling by a mutation that prevents transphosphorylation within the receptor complex. Mol Cell Biol 1995;15:1573–1581.
27. Brand T, Schneider MD. Transforming growth factor-β signal transduction. Circ Res 1996;78:173–179.
28. Chen RH, Miettinen PJ, Maruoka EM, Choy L, Derynck R. A WD domain protein that is associated with and phosphorylated by the type II TGF-β receptor. Nature 1995;377:548–552.
29. Kawabata M, Imamura T, Miyazono K, Engel ME, Moses HL. Interaction of the transforming growth factor-β type I receptor with farnesyl-protein transferase-α. J Biol Chem 1995;270:29,628–29,631.
30. Wang T, Danielson PD, Li BY, Shah PC, Kim SD, Donahoe PK. The p21^{RAS} farnesyltransferase alpha unit in TGF-β and activin signaling. Science 1996;271:1120–1122.
31. Wang TW, Li BY, Danielson PD, Shah PC, Rockwell S, Lechleider R.J, et al. The immunophilin FKBP12 functions as a common inhibitor of the TGF β family type I receptors. Cell 1996;86:435–444.
32. Massagué J, Hata A, Liu F. TGF-β signalling through the Smad pathway. Trends Cell Biol 1997;7:187–192.
33. Kretzschmar M, Liu F, Hata A, Doody J, Massagué J. The TGF-β family mediator Smad1 is phosphorylated directly and activated functionally by the BMP receptor kinase. Gene Dev 1997;11:984–995.
34. Baarends WM, van Helmond MJL, Post M, van der Schoot PCJM, Hoogerbrugge JW, de Winter JP, et al. A novel member of the transmembrane serine/threonine kinase receptor family is specifically expressed in the gonads and in mesenchymal cells adjacent to the müllerian duct. Development 1994;120:189–197.
35. di Clemente N, Wilson CA, Faure E, Boussin L, Carmillo P, Tizard R, et al. Cloning, expression and alternative splicing of the receptor for anti-Müllerian hormone. Mol Endocrinol 1994;8:1006–1020.
36. Imbeaud S, Faure E, Lamarre I, Mattei MG, di Clemente N, Tizard R, et al. Insensitivity to anti-Müllerian hormone due to a spontaneous mutation in the human anti-Müllerian hormone receptor. Nature Genet 1995;11:382–388.
37. Baarends WM, Hoogerbrugge JW, Post M, Visser JA, Derooij DG, Parviven M, Themmen APN, Grootegoed JA. Anti-mullerian hormone and anti-mullerian hormone type II receptor messenger ribonucleic acid expression during postnatal testis development and in the adult testis of the rat. Endocrinology 1995;136:5614–5622.
38. Racine C, Rey R, Forest MG, Louis F, Ferré A, Huhtaniemi I, Josso N, di Clemente N. Receptors for anti-Müllerian hormone on Leydig cells are responsible for its effects on steroidogenesis and cell differentiation. Proc Natl Acad Sci USA 1998;95:594–599.
39. Faure E, Gouédard L, Imbeaud S, Cate RL, Picard JY, Josso N, et al. Mutant isoforms of the anti-Müllerian hormone type II receptor are not expressed at the cell membrane. J Biol Chem 1996;271:30,571–30,575.

40. Vivien D, Wrana JL. Ligand-induced recruitment and phosphorylation of reduced TGF-β type I receptor. Exp Cell Res 1995;221:60–65.
41. ten Dijke P, Yamashita H, Ichijo H, Franzen P, Laiho M, Miyazono K, et al. Characterization of type-I receptors for transforming growth factor-β and activin. Science 1994;264:101–104.
42. Attisano L, Cárcamo J, Ventura F, Weis FM, Massagué J, Wrana JL. Identification of human activin and TGFβ type I receptors that form heteromeric kinase complexes with type II receptors. Cell 1993; 75:671–680.
43. ten Dijke P, Ichijo H, Franzen P, Schulz P, Saras J, Toyoshima H, et al. Activin receptor-like kinases —a novel subclass of cell-surface receptors with predicted serine/threonine kinase activity. Oncogene 1993;8:2879–2887.
44. He WW, Gustafson ML, Hirobe S, Donahoe PK. Developmental expression of four nouvel serine-threonine kinase receptors homologous to the activin/transforming growth factor-β type II receptor family. Dev Dynamics 1993;196:133–142.
45. Baarends WM, Uilenbroek JTJ, Kramer P, Hoogerbrugge JW, Vanleeuwen ECM, Themmen APN, et al. Anti-mullerian hormone and anti-mullerian hormone type II receptor messenger ribonucleic acid expression in rat ovaries during postnatal development, the estrous cycle, and gonadotropin-induced follicle growth. Endocrinology 1995;136:4951–4962.
46. Nachtigal MW, Ingraham HA. Bioactivation of Mullerian inhibiting substance during gonadal development by a kex2/subtilisin-like endoprotease. Proc Natl Acad Sci USA 1996;93:7711–7716.
47. Wieser R, Attisano L, Wrana JL, Massagué J. Signaling activity of transforming growth factor β type II receptors lacking specific domains in the cytoplasmic region. Mol Cell Biol 1993;13:7239–7247.
48. Brand T, Maclellan WR, Schneider MD. A dominant-negative receptor for type-β transforming growth factors created by deletion of the kinase domain. J Biol Chem 1993;268:11,500–11,503.
49. Tsuchida K, Vaughan JM, Wiater E, Gaddy-Kurten D, Vale WW. Inactivation of activin-dependent transcription by kinase-deficient activin receptors. Endocrinology 1995;136:5493–5503.
50. Imbeaud S, Carré-Eusèbe D, Rey R, Belville C, Josso N, Picard JY. Molecular genetics of the persistent Müllerian duct syndrome: a study of 19 families. Hum Mol Genet 1994;3:125–131.
51. Sloan WR, Walsh PC. Familial persistent müllerian duct syndrome. J Urol 1976;115:459–461.
52. Naguib KK, Teebi AS, Al-Awadi SA, El-Khalifa MY, Mahfouz ES. Familial uterine hernia syndrome: report of an Arab family with four affected males. Am J Hum Genet 1989;33:180,181.
53. Korsch E, Rey R, Imbeaud S, Picard JY, Josso N. Persistent Müllerian duct syndrome: decrease of serum levels of antiMüllerian hormone shortly after orchidopexy procedure [Abstract]. Horm Res 1997; 48(Suppl. 2):117.
54. Behringer RR, Cate RL, Froelick GJ, Palmiter RD, Brinster RL. Abnormal sexual development in transgenic mice chronically expressing Müllerian inhibiting substance. Nature 1990; 345:167–170.
55. Meyers-Wallen VN, Donahoe PK, Ueno S, Manganaro TF, Patterson DF. Mullerian inhibiting substance is present in testes of dogs with persistent mullerian duct syndrome. Biol Reprod 1989;41: 881–888.
56. Jones H, Nickel RF. Das Syndrom des persistierenden Müllerschen Gänge: eine erbliche Form von männlichem Pseudohermaphroditismus beim Bassethound. Kleintierpraxis 1996;41:911–918.
57. Meyers-Wallen VN, Lee MM, Manganaro TF, Kuroda T, MacLaughlin D, Donahoe PK. Mullerian inhibiting substance is present in embryonic testes of dogs with persistent mullerian duct syndrome. Biol Reprod 1993;48:1410–1418.
58. Josso N, di Clemente N. Serine/threonine kinase receptors and ligands. Curr Opinion Genet Dev 1997;7:371–377.

13 Adrenal Insufficiency
ACTH Resistance

Adrian J. L. Clark, DSc, FRCP

CONTENTS

> INTRODUCTION
> INHERITED ACTH-RESISTANT SYNDROMES
> FAMILIAL GLUCOCORTICOID DEFICIENCY
> TRIPLE A SYNDROME
> SUMMARY
> REFERENCES

INTRODUCTION

This chapter shall briefly review the recognized causes of primary adrenal failure, and then focus on those disorders that appear to result primarily from failure of the adrenal to respond to adrenocorticotropin (ACTH). Hypopituitarism may present clinically as apparent adrenal failure, but will not be discussed here. True ACTH resistance, as will be seen, is essentially the result of two distinct disease processes. The first of these was known originally as familial glucocorticoid deficiency, but subsequently was referred to by several terms, including isolated glucocorticoid deficiency and hereditary unresponsiveness to ACTH. This chapter shall only use the first of these terms, which is abbreviated as FGD. The second of the diseases to be discussed is the triple A syndrome or Allgrove's syndrome in which ACTH resistance is accompanied by alacrima (absence of tears) and achalasia of the cardia.

Primary Adrenal Failure

Primary adrenal failure is an uncommon disease. Estimates suggest it has an annual incidence of 4.5/million/yr in New Zealand *(1)* and 5.6 million/million/yr in Britain *(2)*. In Westernized communities, the principal cause is autoimmune destruction of the gland, although in circumstances of high prevalence of tuberculosis, this becomes an important cause. The frequency of causes in various surveys conducted over the last 25 years is listed in Table 1, which highlights this trend away from tuberculosis, and the relative rarity of other causes. In Brazil, paracoccidiomycosis is a frequent cause of adrenal failure *(3)*. Table 2 lists the recognized causes of this disorder, including those that occur so infrequently that they make little or no impact on the prevalence studies summarized in Table 1.

From: *Contemporary Endocrinology: Hormone Resistance Syndromes*
Edited by: J. L. Jameson © Humana Press Inc., Totowa, NJ

Table 1
Causes of Primary Adrenal Failure in Various Surveys

	Mason et al. (51)	Nerup, (52)	Eason et al. (1) Cauc.	Eason et al. (1) Poly.	Kong and Jeffcoate (2)	Mantero and Betterle (53)	Nomura et al. (54)
Number surveyed	82	108	53		86	117	74
Autoimmune destruction	69%	66%	92%	25%	93%	76%	49%
Tuberculosis	31%	17%	4%	63%	0%	16%	38%
Adrenoleukodystrophy	—	—	—	—	3.5%	2%	—
Neoplastic destruction	—	—	—	—	2.5%	2%	—

Table 2
Causes of Adrenal Failure

Autoimmune destruction of adrenals
  Infective
    Tuberculosis
    Coccidiomycosis
    Histoplasmosis
    Blastomycosis
    HIV
    Meningococcal septicemia
  Infiltrations
    Neoplastic infiltration
    Amyloidosis
    Sarcoidosis
  Congenital
    Adrenoleukodystrophy
    Congenital adrenal hypoplasia
    Familial glucocorticoid deficiency
    Triple A syndrome
  Others
    Adrenal infarction
    Drug induced
    Iatrogenic (e.g., therapeutic surgical removal of adrenals)

DIAGNOSIS

Adrenal failure is characterized by a primary deficiency of cortisol and aldosterone, with consequent elevation in ACTH and renin, indicating an intact drive to the gland. Establishment of true steroid deficiency may be difficult without measurement of ACTH or renin, and if the diagnosis is suspected, some sort of dynamic testing is warranted, such as an ACTH stimulation test or measurement of lying and standing renin and aldosterone. The application and interpretation of these and other tests are discussed in detail in the major texts on clinical endocrinology (e.g., 4,5).

Having established the functional diagnosis, i.e., primary adrenal failure, investigation of the etiology is important. Factors of particular significance include the age of the patient, the presence of other infective or neoplastic disease, and adrenal imaging. Investigations that will help in the diagnosis include measurement of adrenal antibodies, although these will be negative in around 50% of patients with autoimmune Addison's disease, and in cases with an appropriate history, measurement of the long-chain fatty acids in the circulation will indicate patients with adrenoleucodystrophy. Again, a detailed description of these aspects is contained in the major textbooks of clinical endocrinology.

ACTH resistance differs diagnostically from these other conditions in that the renin–angiotensin–aldosterone axis is essentially unaffected. Thus, patients with one of these rare disorders will usually have normal electrolytes and near-normal aldosterone and renin levels combined with complete or partial cortisol deficiency.

Acquired ACTH Resistance

ACTH resistance is almost always an inherited disorder, the only possible exceptions being patients with autoimmune Addison's disease in which antibodies block the ACTH receptor. Kendall-Taylor et al. reported a study of IgG isolated from a patient with Addison's disease associated with the type 1 polyglandular syndrome, and demonstrated that this inhibited ACTH-stimulated cortisol secretion by guinea pig adrenal cells *(6)*. Dibutyryl cAMP (dbcAMP) was however able to bypass this block, and this was seen as evidence for the existence of anti-ACTH receptor antibodies. In a more extensive study, Wardle et al. examined the effect of 29 Addison's disease patients' IgGs in the same type of bioassay *(7)*. Although 12 samples inhibited ACTH-stimulated cortisol secretion, only 1 of these could be bypassed by dbcAMP. It was concluded that ACTH receptor blocking antibodies were not a common cause of Addison's disease. It is now well recognized that antibodies to 21-hydroxylase are one of the primary pathogenic mechanisms in this disease *(8)*.

INHERITED ACTH-RESISTANT SYNDROMES

Inherited ACTH-resistant syndromes are an uncommon group of disorders most likely to be encountered by the pediatric endocrinologist. In recent years, at least three separate genetic causes of these diseases, which are inherited in an autosomal-recessive manner, have been defined. They include familial glucocorticoid deficiency (FGD), also known as isolated glucocorticoid deficiency, or hereditary unresponsiveness to ACTH, and a separate entity known as the triple A syndrome or Allgrove's syndrome. As will be discussed, FGD has at least two distinct genetic causes, one of which is mutation of the ACTH receptor. The triple A syndrome gene locus has now been identified, although the culprit gene remains unknown.

FAMILIAL GLUCOCORTICOID DEFICIENCY

Clinical Presentation

The typical presentation of FGD is the result of glucocorticoid deficiency. Thus, in the neonatal period, many patients will exhibit hypoglycemia. This may not be profound and often responds to more frequent feeding (e.g., *9–11*). Less commonly in the neonatal period a picture of hepatitis with mild jaundice can be found, which reverses after glucocorticoid replacement *(12)*. Excessive pigmentation of the skin resulting from elevated ACTH levels takes longer to be manifest and is usually first noted after 4 or 5 mo of life *(9)*. Children in whom the diagnosis is not made by this time tend to be especially prone to infection and take longer to recover from relatively minor infective episodes. In some cases, this may result in profound and sometimes fatal septic events at any time in the childhood years (e.g., *13–15*).

Affected patients do not suffer from major defects in mineralocorticoid or sex steroid pathways, and therefore, there is no hypotension and hyperkalemia as in Addison's disease, no salt craving, hypertension, or androgenization as in congenital adrenal hyperplasia, and no defect in gonadal differentiation as in congenital adrenal hypoplasia. The absence of such symptoms means that children who escape diagnosis in the neonatal period may wait many years before a formal diagnosis is made. Perhaps predictably, children with FGD fail to develop an adrenarche, the prepubertal secretion of adrenal androgens *(16)*.

In FGD the diagnostic biochemical defects are present from birth. In the triple A syndrome, the adrenal component of the disorder may take many years to develop, and these patients may have normal ACTH and cortisol indices in early childhood. When the adrenal disorder does develop in the triple A syndrome, it is often indistinguishable from FGD, but a small proportion of patients (5%) also exhibit mineralocorticoid deficiency *(17,18)*. Probably the earliest and most consistent diagnostic feature of the triple A patients is the absence of tears or alacrima. Features of dysphagia and delayed gastric emptying may also be a prominent early feature. In many affected patients, however, it is the varied neurological features that take prominence. These have been reviewed in a series of 20 patients by Grant et al. *(18)*, and include hyperreflexia, increased muscle tone, muscle weakness and wasting, dysarthria/nasal speech, ataxia, sensory impairment, optic atrophy, nerve deafness, seizures, mental retardation, and dementia. Occasionally these patients show abnormalities of palmar and plantar keratinization, which may have some pathological significance (*see below*).

DIAGNOSIS AND DIFFERENTIAL DIAGNOSIS

The salient feature of these ACTH-resistant syndromes is the finding of subnormal or undetectable plasma cortisol in combination with an elevated plasma ACTH. Frequently, the cortisol values at 9 AM are between 100 and 300 nmol/L, which may not in itself be very remarkable, but these values respond poorly or not at all to injection of 250 µg of synthetic $ACTH_{1-24}$. In other cases, 9 AM cortisol values are clearly undetectable and fail to respond to stimulation. Plasma ACTH is usually markedly elevated with values usually over 1000 pg/mL.

These features could be found in other adrenal disorders, such as autoimmune Addison's disease, but can clearly be distinguished by measurement of the renin and aldosterone concentrations. Although ACTH can stimulate aldosterone production, its principal regulator is angiotensin. Since there is no disorder of the renin–angiotensin system in FGD, this axis is not significantly disturbed, and renin and aldosterone values are usually normal or close to normal, clearly distinguishing this disease from other adrenal pathologies that will be characterized by aldosterone deficiency and a compensatory elevation of renin. Aldosterone need not necessarily be entirely normal in these patients probably for two reasons: (1) ACTH stimulates aldosterone production by glomerulosa cells (less significantly than angiotensin II). Thus patients with ACTH resistance will lack this stimulus *(19,20)*. (2) At presentation, patients with ACTH resistance may be unwell as a result of infection or frequent hypoglycemia. As in any other patient, these disturbances may stimulate the renin–aldosterone axis. These influences may therefore result in a broad range of renin and aldosterone levels in patients with ACTH resistance, which may occasionally confuse the diagnosis.

Table 3
Summary of Published ACTHR Mutations and Their Probable Functional Consequences

Mutation	Functional consequence	Reference
Pro 27 Arg	Benign polymorphism	(39)
Ile 44 Met	Signal transduction defect	(26)
Ser 74 Ile	Intramolecular bond/signal transduction defect	(24)
Asp 107 Asn	Loss of ligand binding	(28)
Ile 118 frameshift	Truncated receptor	(55)
Ser 120 Arg	Structural disruption of transmembrane domain 3	(25)
Arg 128 Cys	Loss of signal transduction	(26)
Arg 146 His	Loss of signal transduction	(26)
Leu 192 frameshift	Truncated receptor	(26)
Arg 201 X	Truncated receptor	(25)
Gly 217 frameshift	Truncated receptor	(28)
Cys 251 Phe	Loss of disulfide loop	(28)
Tyr 254 Cys	Interference with disulfide bonds	(27)
Pro 273 His	Disruption of conserved region of transmembrane domain 7	(56)

TREATMENT

After any initial resuscitation attempts, the treatment of ACTH insensitivity is very simple in that only glucocorticoid replacement with oral hydrocortisone is required. Occasionally oral dexamethasone has advantages in view of its longer action. As with any patient taking glucocorticoids, the dose needs to be increased in times of stress or injury. No mineralocorticoid replacement is needed.

Pathogenesis

A number of hypotheses have been put forward over the years to explain the origin of FGD. These mostly include proposals of a defect in the receptor for ACTH, (e.g., ref. 9), although a defect in the ACTH signal transduction system, or a defect in adrenal gland development has also been postulated. Evidence favoring the first of these came from Smith et al., who demonstrated defective ACTH binding to peripheral blood mononuclear cells in a patient with FGD, in contrast to normal binding characteristics in cells from a control subject (21). However, Yamaoka et al. demonstrated a normal mononuclear cell ACTH binding and cAMP generation, and a failure to generate cortisol when cAMP was infused into the patient. They concluded that the disease resulted from a postreceptor defect (22).

ACTH Receptor Mutations

Following the cloning of the human ACTH receptor by Mountjoy et al. in 1992 (23), we were able to demonstrate a homozygous missense point mutation in two affected siblings (24). This mutation converted Ser74, which lies in the second transmembrane domain to Ile (S74I), and segregated with the disease in the family. Subsequently, we and others have reported a number of different missense and nonsense mutations in this gene, which occur in homozygous or compound heterozygous form in patients with the disorder (25–28). In all cases, these mutations cosegregate with the disease in the family. The current status of these published mutations is summarized in Table 3.

EXPRESSION STUDIES

Confirmation that these mutations cause the disease depends on expression studies in which plasmid DNA incorporating the mutant receptor gene is introduced into cells that lack endogenous ACTH receptors. For reasons that remain unclear, it has been exceptionally difficult to do this with the ACTH receptor. We had some limited success in expressing the normal and S74I mutant receptor using a very high-efficiency system in Cos 7 cells *(29)*. This system provided only poor levels of expression, which was partly confounded by an endogenous melanocortin receptor in these cells. We were able to show that the mutant receptor functioned with a significantly right-shifted dose–response curve. This finding is consistent with the clinical observation in some patients with this mutation that by elevating the plasma ACTH markedly, an adequate amount of cortisol can be produced by the adrenal to render the disease less severe than that with some other mutations.

Naville et al. used a different system in which the mutant ACTH receptors were expressed in the M3 melanoma cell line, which has endogenous MC-1 (α-MSH) receptors *(28)*. They showed that the D107N, C251 F, and G217 frameshift mutations lacked all cAMP-generating function in this system in contrast to the normal sequence receptor. Again, however, these results are confounded by the endogenous response derived from the MC-1 receptor. Currently our own strategy is to use the mouse Y6 cell line, a mutant derivative of the mouse Y1 adrenocortical cell line that fails to express the endogenous ACTH receptor *(30)*. This line seems to be capable of expressing transfected ACTH receptors in the absence of any background and is providing promising results (e.g., *31,32* and Elias et al., submitted).

GENOTYPE-PHENOTYPE CORRELATIONS

Since these expression studies suggest that different mutations are likely to disable the receptor to different degrees, making phenotype–genotype comparisons is difficult. However, the S74I mutation that we originally described has proven to be the most prevalent of these mutations, and we have now identified it in 10 individuals from 6 families in homozygous form and as a compound heterozygote with a more severe mutation in two cases. Many of these cases have a Scottish family background, and it seems highly likely that the founder mutation occurred in this region probably in the past few centuries.

ACTH receptor mutations provide some interesting physiological points. Since a point mutation in both alleles of this gene can result in complete failure to secrete cortisol, there can be little doubt that the gene originally cloned by Mountjoy et al. *(23)* encodes the ACTH receptor, and that it is the only ACTH receptor. There is good evidence that adrenal cortical cells also express the MC-5 (MSH) receptor, which has been shown to respond in vitro to ACTH at high concentrations *(33,34)*. However, it appears that the high concentrations of ACTH found in untreated FGD are not sufficiently high to recruit this receptor for stimulation of cortisol production.

Another interesting insight provided by these mutations is into the role of ACTH in adrenal growth. There has long been a debate on the relative contributions of ACTH, other pro-opiomelanocortin-derived peptides, and other factors in adrenal growth. Since the ACTH receptor is highly specific for ACTH *(35)*, it would seem that patients with FGD would provide an excellent model with which the role of ACTH could be defined. These patients consistently have small adrenal glands, suggesting an important role for

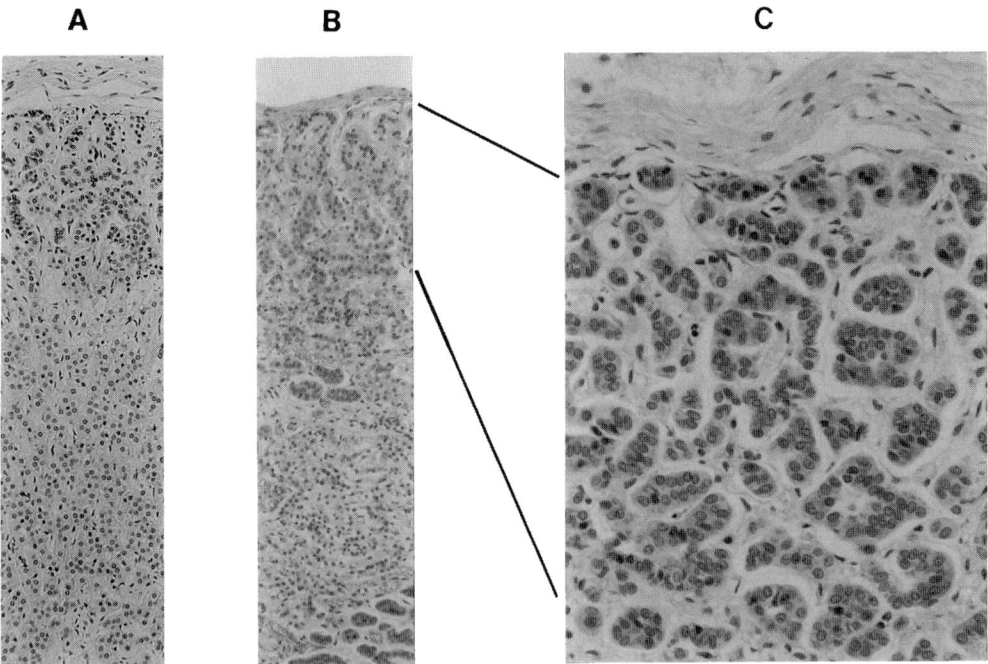

Fig. 1. Appearance of the adrenal cortex in FGD. The photomicrographs show low-power sections of (**A**) a normal adrenal gland obtained from a 3-yr-old child dying of an acute nonadrenal cause, (**B**) the adrenal of a 3-yr-old child homozygous for an ACTH receptor missense mutation who died of acute sepsis, and (**C**) a higher-power section of the glomerulosa-like cell layer. Note the absense of any organized fasciculata/reticularis cell layer in (B) in contrast to (A).

ACTH. Few of these glands have become available for histological examination. However, adrenal sections obtained from a 3-yr-old girl who died from FGD caused by an S74I mutation are shown in Fig. 1 and compared with those from a child of the same age dying of nonadrenal disease. The histological appearance suggests the presence of slightly disorganized glomerulosa cells surrounding a normal adrenal medulla with no evidence of fasciculata or reticularis cells.

It appears that the heterozygous state for these ACTH receptor mutations is not associated with any disorder of the pituitary–adrenal axis. No obvious abnormalities of circulating ACTH or cortisol concentrations have been reported in heterozygotes. However, it may be that a more subtle test is required to reveal subclinical abnormalities, and for this reason, the standard CRH test has been used. Tsigos et al. reported that the heterozygote parents of a patient with the S120R and R201X mutations had slightly exuberant responses to CRH in a conventional CRH stimulation test *(25)*. However, this effect may be explained by ethnic differences in the CRH response *(36)*. We reported the responses to CRH in four subjects who were each heterozygous for a different ACTH receptor mutation *(26)*. All had a normal response.

Normal Receptor FGD

Not all cases of FGD are associated with mutations within the coding region of the ACTH receptor. Of 37 families that we have studied to date, in only 14 families are affected cases associated with homozygous or compound heterozygous mutations. One possibility is that there are mutations in regions of the gene apart from the coding exon (such as the promoter). However, it is generally true that mutations in promoters are not a common cause of genetic disease

As a result of the human genome mapping project, it is now relatively straightforward to identify highly polymorphic microsatellite repeat sequences at given locations in the genome. The ACTH receptor was mapped to the short arm of chromosome 18 (18p11.2) *(37,38)*, and so we investigated the proximity of a number of repeats in this region by performing linkage analysis in the families with ACTH receptor mutations. This approach revealed that the markers *D18S40* and *D18S44* were positioned on either side of the ACTH receptor gene at distances of 3 and 4 centimorgans, respectively *(39)*. Such a distance, although large in physical terms, is satisfactory for the segregation studies proposed.

The results of this analysis indicate that in the case of several of the families without ACTH receptor mutations, the segregation analysis was not compatible with an etiological role for the ACTH receptor gene. This result is important, since it makes it clear that the clinical phenotype of FGD can be caused by a second genetic defect. For ease of reference, we have adopted the term FGD type 2 for this syndrome, which is not linked to the ACTH receptor locus. It is hoped that ultimately the identity of the causative gene for this syndrome will be identified, which may allow a more descriptive distinction between the etiologies for this disease.

Phenotypic Distinctions

We have examined the clinical features of the two types of FGD in the hope of identifying distinctions between them, which may help elucidate the nature of the defect in FGD type 2. No differences in ACTH, cortisol, ACTH stimulation tests, or other biochemical aspects are apparent. Likewise, the clinical presentation did not distinguish these disorders *(40)*. However, an unexpected difference was found in the patients' heights. Thus, although those patients with FGD type 2 have heights that fall within the normal distribution (± 2 SD), patients with ACTH receptor mutations have a mean height that is about 2 SDs greater than normal. This difference is statistically significant *(40)*. These findings are summarized in Fig. 2.

Although it has not been studied systematically, no abnormalities in growth hormone, growth hormone dynamics, or IGF-1 have been identified in any of these tall patients to date. There is a suggestion in some patients that the rate of growth slows after glucocorticoid replacement is commenced, but it is difficult to rule out a negative effect on growth of excessive replacement. It seems probable that this observation is telling us something about the etiology of FGD type 2, as well as revealing unexpected aspects of ACTH physiology. One hypothesis is that ACTH acts on a melanocortin receptor in the cartilaginous growth plate to influence linear growth, and this question is under investigation.

TRIPLE A SYNDROME

Clinical Features

The triple A syndrome was first described in 1978 as the triad of alacrima (absence of tears), achalasia of the cardia of the esophagus, and adrenal failure *(17)*. The adrenal disorder is usually in the form of ACTH resistance with preservation of the renin–aldosterone axis, although about 5% of cases also manifest aldosterone deficiency *(18)*. The adrenal disorder is not usually apparent in the neonatal period, and there are clear

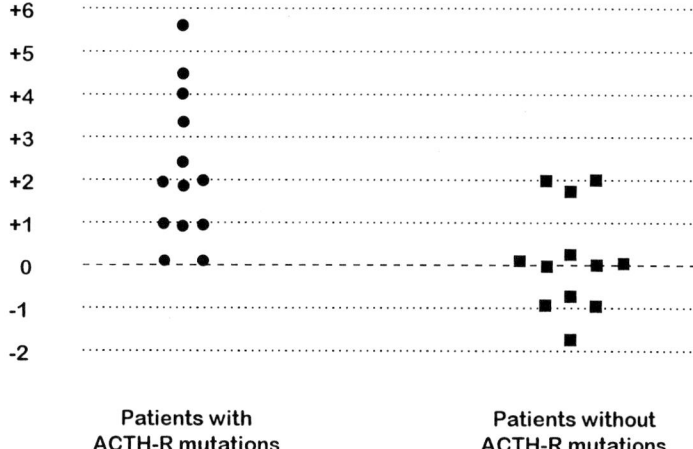

Fig. 2. Height of patients with two forms of FGD. The height of patients without ACTH receptor mutations is normal, i.e., falls within ±2 SD of the mean. Height of those patients with ACTH receptor mutations is significantly greater ($p < 0.05$).

instances in which normal adrenal function has progressed to later adrenal failure *(41)*. Histological examination of the adrenals has been reported in two cases and reveals an absent fasciculata and reticularis cell layers with a normal glomerulosa cell layer—similar to the findings in FGD *(17, 42)*.

The earliest symptom to be manifest in most cases is the absence of tears, although this may not be detected unless it is looked for. A standard Schirmer's test is a simple and adequate diagnostic test. The precise cause of the alacrima is not clear, although CT scan of the orbits in one case revealed absence of the lacrimal glands themselves, rather than any obstruction to tear flow *(41)*. The gastro-esophageal motility problems also often appear in childhood and vary from asymptomatic abnormalities on barium swallow examination to severe immobility of the upper gut, including the stomach and duodenum. A variety of surgical procedures may be required, most notably Hellers cardiomyotomy *(43)*. Severe cases may require long-term gastrostomy feeding.

A fourth component of the syndrome includes a variety of neurological defects, which range from peripheral nerve defects, autonomic nerve abnormalities to a wide range of central defects, including cerebellar ataxia, movement disorders, and dementia *(44)*. The range of neurological defects was reviewed in a series of 20 affected patients by Grant et al. *(18)* and resembles that of no other disease pattern.

The severity of each of these disease components is highly variable, even between affected siblings (e.g., *17,45*), although the disorder seems to be progressive. At one extreme, a patient has presented with a neurological disorder in middle age, only to be found to have the other components of the disease when tested for them. Additionally, there are case reports in the literature of patients with apparently only two components of the syndrome, who may have a variant of the disorder, but most probably have not developed the complete syndrome at the time of investigation (e.g., *46,47*). That this is a genetic disease was implied by the first report, which described, among others, two affected brothers *(17)*. Subsequently, other family clusters have come to light, including the large Canadian Metis kindred reported initially by Moore et al. *(41)*. Genetic analysis suggests this is an autosomal-recessive disorder resulting from a defect in an unidentified gene. The variable penetrance, even between siblings, may suggest that modification of the progression of the disease is possible, which perhaps opens the way for some beneficial therapeutic options when we have better understanding of its pathogenesis.

Candidate Genes

As discussed above, the triple A syndrome exhibits several manifestations outside the adrenal gland, and it would seem unlikely to result from ACTH receptor defects, unless this were part of a contiguous gene defects in which genes adjoining the receptor gene were deleted. That this was not the case has been shown by ourselves and by Tsigos et al., who demonstrated a normal ACTH receptor sequence in patients with this syndrome *(27;* unpublished observations). In addition, we were able to demonstrate significant absence of linkage between the triple A syndrome and the ACTH receptor locus *(48)*. Other good candidate genes are not especially forthcoming. Moore et al. proposed that the vasoactive intestinal peptide (VIP) gene could be a candidate in view of its role in esophageal relaxation and lacrimation *(41)*. However, in a large collection of affected families from several continents, we also demonstrated absence of linkage to the VIP locus (chromosome 6q24) and the VIP-1 receptor locus (chromosome 3p22) *(48)*.

Identification of Chromosomal Locus

We have used this family collection to conduct a genome search for the triple A locus using 345 anonymous highly polymorphic DNA markers scattered widely throughout the genome such that they were spaced on average every 10 centimorgans (cM). Using this strategy, we have been able to identify a single locus on chromosome 12q13 that is very strongly linked to the disease in every family we have studied. We have been able to narrow this locus down to about 6 cM, a region that contains several mapped genes that are potential candidates *(48)*.

This locus is very close to the type II keratin gene cluster, although probably distinct from it. It is therefore conceivable that a contiguous gene deletion in some patients could disrupt not only the triple A gene, but also some of the keratin genes resulting in the keratitis seen in some patients. Other interesting genes localized in this area include the retinoic acid receptor γ, which when knocked out in mice produces a syndrome that includes defects in development of the eye, but not specifically alacrima *(49)*.

Arguably the most interesting candidate gene in this region is the SCN8A gene encoding a sodium channel. This was discovered to be deleted in the mouse motor end-plate disease *(med)*, a syndrome characterized by muscular atrophy, ataxia, and juvenile lethality *(50)*. This gene is under investigation. The eventual identification of the triple A gene will be of considerable interest not only in understanding more clearly the pathogenesis of a very distressing disease, but also in terms of identifying a gene with an important role in a variety of tissues.

SUMMARY

The syndromes of ACTH resistance provide a fascinating insight into several aspects of adrenal disease and the function of the ACTH receptor. Potentially novel aspects of physiology are highlighted, such as the unexplained tall stature of patients with ACTH receptor mutations. The unusual combination of defects in the triple A syndrome result from an unknown gene on chromosome 12. The eventual identification of this gene should reveal a factor involved in a diverse array of biological functions.

REFERENCES

1. Eason RJ, Croxson MS, Perry MC, Somerfield SD Addison's disease, adrenal autoantibodies and computerised adrenal tomography. N Z Med J 1982;95:569–573.
2. Kong MF, Jeffcoate W. Eighty-six cases of Addison's disease. Clin Endocrinol 1994;41:757–761.
3. Moreira AC, Martinez R, Castro M, Elias LLK. Adrenocortical dysfunction in paracoccidiomycosis: comparison between plasma beta-lipoprotein/adrenocorticotropin levels and adrenocortical tests. Clin Endocrinol 1992;36:545–551.
4. DeGroot LJ. Endocrinology, 2nd ed. W.B. Saunders, Harcourt Brace Jovanovich, Philadelphia, 1989.
5. Grossman AB. Clinical Endocrinology, 2nd ed. Blackwell Scientific, Oxford, 1977.
6. Kendall-Taylor P, Lambert A, Mitchell R, Robertson WR. Antibody that blocks stimulation of cortisol secretion by adrenocorticotropic hormone in Addison's disease. Br Med J 1988;296:1489–1491.
7. Wardle CA, Weetman AP, Mitchell R, Peers N, Robertson WR. Adrenocorticotropic hormone receptor-blocking immunoglobulins in serum from patients with Addison's disease: a reexamination. J Clin Endocrinol Metab 1993;77:750–753.
8. Winqvist O, Karlsson FA, Kampe O. 21-hydroxylase, a major autoantigen in idiopathic Addison's disease. Lancet 1992;339:1559–1562.
9. Migeon CJ, Kenny FM, Kowarski A, Snipes CA, Spaulding JS, Finkelstein JW, et al. The syndrome of congenital adrenocortical unresponsiveness to ACTH. Report of six cases. Pediatr Res 1968;2:501–513.
10. Monteleone JA, Monteleone PL. Heredirtary adrenal unresponsiveness to ACTH–another case. Pediatrics 1970;45:321–322.
11. Kershnar AK, Roe TF, Kogut MD. Adrenocorticotropic hormone unresponsiveness: report of a girl with excessive growth and review of 16 reported cases. J Pediatr 1972;80:610–619.
12. Lacy DE, Nathavitharana KA, Tarlow MJ. Neonatal hepatitis and congenital insensitivity to adrenocorticotropin (ACTH). J Pediatr Gastroenterol Nutr 1993;17:438–440.
13. Shepard TH, Landing BH, Mason DG. Familial Addison's disease. Am J Dis Child 1959;97:154–162.
14. Franks RC, Nance WE. Hereditary adrenocortical unresponsiveness to ACTH. Pediatrics 1970;45:43–48.
15. Moshang T, Rosenfield RL, Bongiovanni AM, Parks JS, Amrhein JA. Familial glucocorticoid insufficiency. J Pediatr 1973;82:821–826.
16. Weber A. Clark AJL, Perry LA, Honour JW, Savage MO. Diminished adrenal androgen secretion in familial glucocorticoid deficiency implicates a significant role for ACTH in the induction of adrenarche. Clin Endocrinol 1997;46:431–437.
17. Allgrove J, Clayden GS, Grant DB, Macaulay JC. Familial glucocorticoid deficiency with achalasia of the cardia and deficient tear production. Lancet 1978;I:1284–1286.
18. Grant DB, Barnes ND, Dumic M, Ginalska-Malinowska M, Milla PJ, v Petrykowski W, et al. Neurological and adrenal dysfunction in the adrenal insufficiency/alacrima/achalasia (3A) syndrome. Arch Dis Child 1993;68:779–782.
19. Spark RF, Etzkorn JR. Absent aldosterone response to ACTH in familial glucocorticoid deficiency. N Engl J Med 1977;297:917–921.
20. Davidai G, Kahana L, Hochberg Z. Glomerulosa failure in congenital adrenocortical unresponsiveness to ACTH. Clin Endocrinol 1984;20:515–520.
21. Smith EM, Brosnan P, Meyer WJ, Blalock JE. An ACTH receptor on human mononuclear leukocytes: relation to adrenal ACTH-receptor activity. N Engl J Med 1987;317:1266–1269.
22. Yamaoka T, Kudo T, Takuwa Y, Kawakami Y, Itakura M, Yamashita K. Hereditary adrenocortical unresponsiveness to adrenocorticotropin with a postreceptor defect. J Clin Endocrinol Metab 1992;75:270–274.
23. Mountjoy KG, Robbins LS, Mortrud MT, Cone RD. The cloning of a family of genes that encode melanocortin receptors. Science 1992;257:1248–1251.
24. Clark AJL, McLoughlin L, Grossman A. Familial glucocorticoid deficiency caused by a point mutation in the ACTH receptor. Lancet 1993;341:461–462.
25. Tsigos C, Arai K, Hung W, Chrousos GP. Hereditary isolated glucocorticoid deficiency is associated with abnormalities of the adrenocorticotropin receptor gene. J Clin Invest 1993;92:2458–2461.
26. Weber A, Toppari J, Harvey RD, Klann RC, Shaw NJ, Ricker AT, et al. Adrenocorticotropin receptor gene mutations in familial glucocorticoid deficiency: relationships with clinical features in four families. J Clin Endocrinol Metab 1995;80:65–71.
27. Tsigos C, Arai K, Iatronico AC, DiGeorge AM, Rapaport R, Chrousos GP. A novel mutation of the adrenocorticotropin receptor (ACTH-R) gene in a family with the syndrome of isolated glucocorticoid

deficiency, but no ACTH-R abnormalities in two families with the triple A syndrome. J Clin Endocrinol Metab 1995;80:2186-2189.
28. Naville D, Barjhoux L, Jaillard C, Faury D, Despert F, Esteva B, et al. Demonstration by transfection studies that mutations in the adrenocorticotropin receptor gene are one cause of the hereditary syndrome of glucocorticoid deficiency. J Clin Endocrinol Metab 1996;81:1442–1448.
29. Weber A, Kapas S, Hinson J, Grant DB, Grossman A, Clark AJL. Functional characterization of the cloned human ACTH receptor: impaired responsiveness of a mutant receptor in familial glucocorticoid deficiency. Biochem Biophys Res Commun 1993;197:172–178.
30. Schimmer BP, Kwan WK, Tsao J, Qiu R. Adrenocorticotropin-resistant mutants of the Y1 adrenal cell line fail to express the adrenocorticotropin receptor. J Cell Physiol 1995;163:164–171.
31. Yang Y-K, Ollmann MM, Wilson BD, Dickinson C, Yamada T, Barsh GS, et al. Effects of recombinant agouti-signalling protein on melanocortin action. Mol Endocrinol 1997;11:274–280.
32. Clark AJL, Elias LLK, Pullinger G. Expression of the transfected human ACTH receptor (MC-2R) in mouse Y6 cells. Abstract presented at FEBS Cell Signalling Meeting, Amsterdam, P5-029, 1997.
33. Griffon N, Mignon V, Facchinetti P, Diaz J, Schwartz JC, Sokoloff P. Molecular cloning and characterization of the rat fifth melanocortin receptor. Biochem Biophys Res Commun 1994;200:1007–1014.
34. Labbe O, Desarnaud F, Eggerickx D, Vassart G, Parmentier M. Molecular cloning of a mouse melanocortin 5 receptor gene widely expressed in peripheral tissues. Biochemistry 1994;33:4543–4549.
35. Kapas S, Cammas FM, Hinson JP, Clark AJL. Agonist and receptor binding properties of adrenocorticotropin peptides using the cloned mouse ACTH receptor expressed in a stably transfected HeLa cell line. Endocrinology 1996;137:3291–3294.
36. Yanovski JA, Yanovski SZ, Gold PW, Chrousos GP. Differences in the hypothalamic-pituitary-adrenal axis of black and white women. J Clin Endocrinol Metab 1993;77:536–541.
37. Gantz I, Tashiro T, Barcroft C, Konda Y, Shimoto Y, Miwa H, et al. Localization of the genes encoding the melanocortin-2 (adrenocorticotropic hormone) and melanocortin-3 receptors to chromosomes 18p11.2 and 20q13.2–q13.3 by fluorescent in situ hybridization. Genomics 1993;18:166,167.
38. Magenis RE, Smith L, Nadeau JH, Johnson KR, Mountjoy KG, Cone RD. Mapping of the ACTH, MSH, and neural (MC3 & MC4) melanocortin receptors in the mouse and human. Mammalian Genome 1994;5:503–508.
39. Weber A, Clark AJL. Mutations of the ACTH receptor gene are only one cause of familial glucocorticoid deficiency. Hum Mol Genet 1994;3:585–588.
40. Clark AJL, Cammas FM, Watt A, Kapas S, Weber A. Familial glucocorticoid deficiency: one syndrome, but more than one gene. J Mol Med 1997;75:394–399.
41. Moore PSJ, Couch RM, Perry YS, Shuckett EP, Winter JSD. Allgrove syndrome: an autosomal recessive syndrome of ACTH insensitivity, achalasia and alacrima. Clin Endocrinol 1991;34:107–114.
42. Werder EA, Haller R, Vetter W, Zachmann M, Siebenmann R. Isolated glucocorticoid deficiency. Helv Paediatr Acta 1975;30:175–183.
43. Nihoul-Fekete C, Bawab F, Lorat-Jacob S, Arhan P, Pellerin D. Achalasia of the esophagus in childhood: surgical treatment in 35 cases with special reference to familial cases and glucocorticoid deficiency association. J Pediatr Surg 1989;10:1060–1063.
44. Grant DB, Dunger DB, Smith I, Hyland K. Familial glucocorticoid deficiency with achalasia of the cardia associated with mixed neuropathy, long-tract degeneration and mild dementia. Eur J Pediatr 1992;151:85–89.
45. Heinrichs C, Tsigos C, Deschepper J, Drews R, Collu R, Dugardeyn C, et al. Familial adrenocorticotropin unresponsiveness associated with alacrima and achalasia: biochemical and molecular studies in two siblings with clinical heterogeneity. Eur J Pediatr 1995;154:191–196.
46. Efrati Y, Mares AJ. Infantile achalasia with deficient tear production. J Clin Gastroenterol 1985;7:413–415.
47. Haverkamp F, Zerres K, Rosskamp R. Three sibs with achalasia and alacrima: a separate entity different from the triple-A syndrome. Am J Med Genet 1989;34:289–291.
48. Weber A, Wienker TF, Jung M, Easton D, Dean HJ, Heinrichs C, et al. Linkage of the gene for the triple A syndrome to chromosome 12q13 near the type II keratin gene cluster. Hum Mol Genet 1996;5:2061–2066.
49. Kastner P, Grondona JM, Mark M, Gansmuller A, LeMeur M, Decimo D, et al. Genetic analysis of RXR alpha developmental function: convergence of RXR and RAR signalling pathways in heart and eye morphogenesis. Cell 1994;78:987–1003.

50. Burgess DL, Johrman DC, Galt J, Plummer NW, Jones JM, Spear B, et al. Mutation of a new sodium channel gene, Scn8a in the mouse mutant motor endplate disease. Nature Genet 1995;10:461–465.
51. Mason AS, Meade TW, Lee JA, Morris JN. Epidemiological and clinical picture of Addison's disease. Lancet 1968;ii:744–747.
52. Nerup J. Addison's disease–clinical studies: a report of 108 cases. Acta Endocrinol 1974;76:127–141.
53. Mantero F, Betterle C. Surveillance of Addison's disease in Padua from 1967–1993. In: Bhatt HR, James VHT, Besser GM, Bottazzo GF, Keen H, eds. Advances in Thomas Addison's Diseases. J Endocrinol, Bristol, 1994, pp. 131–135.
54. Nomura K, Demura H, Saruta T. Addison's disease in Japan: characteristics and changes revealed in a nationwide survey. Intern Med 1994;33:602–606.
55. Elias LLK, Klann R, Prarasam G, Pullinger G, Canas JA, Clark AJL. Severe ACTH insensitivity resulting from a novel compound heterozygote mutation of the ACTH receptor. Abstract P2-506 presented to the Endocrine Society Annual Meeting, Minneapolis, 1997.
56. Stratakis CA, Wu SM, Bourdony CJ, Cohen D, Rennert OM, Chan W-Y. Hereditary isolated glucocorticoid deficiency: description of a kindred with adrenocorticotropin (ACTH) unresponsiveness and identification of a novel mutation (P273H) of the ACTH receptor (MC2R). Abstract P3-483 presented to the Endocrine Society Annual Meeting, Minneapolis, 1997.

14
Glucocorticoid Resistance and Hypersensitivity

*Denis P. Franchimont, MD
and George P. Chrousos, MD*

CONTENTS

ROLES AND ACTIONS OF GLUCOCORTICOIDS-GENERALIZED
   GLUCOCORTICOID RESISTANCE
THE GLUCOCORTICOID RECEPTOR TRANSDUCTION SYSTEM
   AND TARGET TISSUE SENSITIVITY TO GLUCOCORTICOIDS
DISEASE STATES POTENTIALLY ASSOCIATED WITH ALTERATIONS
   IN TARGET TISSUE SENSITIVITIES TO GLUCOCORTICOIDS
"FAMILIAL" OR "SPORADIC" GENERALIZED GLUCOCORTICOID
   RESISTANCE
"TISSUE-SPECIFIC" GLUCOCORTICOID RESISTANCE
REFERENCES

ROLES AND ACTIONS OF GLUCOCORTICOID-GENERALIZED GLUCOCORTICOID RESISTANCE

Target tissue resistance to steroid hormones implies inability or decreased sensitivity of the tissues to respond to these hormones *(1,2)*. This resistance can be transient or permanent, incomplete (partial) or complete, and compensated or noncompensated. Two patients, a father and a son, with long-term "hypercortisolism" not associated with clinical manifestations of Cushing's syndrome were described by Vingerhoeds et al. in 1976 (reviewed in ref. *3*). We studied this family extensively as well as several other families with the syndrome of familial glucocorticoid resistance reported since. Two New World primate species, the squirrel monkey (*Saimiri sciureus*) and the marmosets *Callithrix argentatus* and *jacchus*, and *Sanguinus oedipus*, also have elevated plasma cortisol values, when compared to Old World primates, without any evidence of glucocorticoid hormone excess *(2)*. In addition, these New World primates have elevated plasma levels of aldosterone, progesterone, testosterone, and 1,25-$(OH)_2$-vitamin D, suggesting that these species have generalized or "pan-steroid" hormone resistance. Two other New World animal species also have glucocorticoid resistance: the guinea pig, for which there is considerable amount of information available *(4)*, and the little-studied prairie vole

From: *Contemporary Endocrinology: Hormone Resistance Syndromes*
Edited by: J. L. Jameson © Humana Press Inc., Totowa, NJ

Microtus ochrogaster, an interesting monogamous rodent with many features reminiscent of the social primate *Callithrix jacchus jacchus (5)*.

A complex feedback system exists that regulates glucocorticoid homeostasis *(6)*. Of principal importance is the ability of glucocorticoids to exert negative feedback effects on the hypothalamic secretion of corticotropin-releasing hormone (CRH) and arginine vasopressin (AVP) and on the pituitary secretion of ACTH, as well as on suprahypothalamic centers controlling the activity of the hypothalamic–pituitary–adrenal (HPA) axis, such as the hippocampus and amygdala. This complex system is activated in states of glucocorticoid resistance, resulting in compensatory increases of ACTH and cortisol secretion *(3)*. Although adequate compensation for the inability of target tissues to respond to glucocorticoids appears to be achieved by the elevated cortisol concentrations in the majority of cases reported, excess ACTH secretion results in increased production of adrenal steroid intermediates with salt-retaining (mineralocorticoid) activity, such as deoxycorticosterone (DOC) and corticosterone (B), and enhanced production of adrenal androgens, such as 4-androstenedione, dehydroepiandrosterone (DHEA), and DHEA-sulfate. Although clinical manifestations of glucocorticoid deficiency were not present in the majority of patients with glucocorticoid resistance reported and several affected subjects evaluated in the context of family studies were asymptomatic, members of two particular families presented with the chief complaint of chronic fatigue, which might indicate incomplete compensation by the elevated glucocorticoids in certain resistant target tissues, perhaps parts of the central nervous system (CNS) or the muscles. The elevated concentrations of cortisol, DOC, and B, on the other hand, caused symptoms and signs of mineralocorticoid excess, such as hypertension and/or hypokalemic alkalosis, in several subjects. Finally, the increased levels of adrenal androgens caused symptoms and signs of androgen excess, including masculinization in women, with manifestations from the skin, such as acne, hirsutism, and male pattern baldness, and from the reproductive system, such as menstrual irregularities (oligo-amenorrhea), oligo-anovulation, and infertility. Also, early and excessive "adrenarche" was associated with precocious puberty, and interference of adrenal androgens with feedback regulation of FSH was associated with abnormal spermatogenesis in men with this syndrome.

Not all the above clinical manifestations were reported in all patients with symptomatic glucocorticoid resistance, and there has been variability in the clinical presentation, even between members of the same family *(3)*. Two main explanations can be given for this. First, the degree of resistance—as indicated by the compensatory elevations of cortisol production indices or by the degree of resistance to dexamethasone—has generally varied between subjects, with associated variability in the production of mineralocorticoid and androgen compounds and, consequently, mineralocorticoid and androgen excess. Second, the degree of target tissue sensitivity to mineralocorticoids and androgens can also be quite variable among individuals, resulting in unequal responsiveness to similar elevations of circulating steroids. The factors responsible for the different sensitivity of target tissues to these hormones could be individual differences in the activity of key hormone inactivating or activating enzymes, such as 11β-hydroxysteroid dehydrogenase—which inactivates cortisol, DOC, or B- or 5α-reductase and 17-ketosteroid reductase—which convert adrenal androgens to more potent metabolites. Additionally and/or alternatively, the mineralocorticoid and androgen receptor transduction systems could differ among subjects, producing further variability of target tissue responsiveness to hormones.

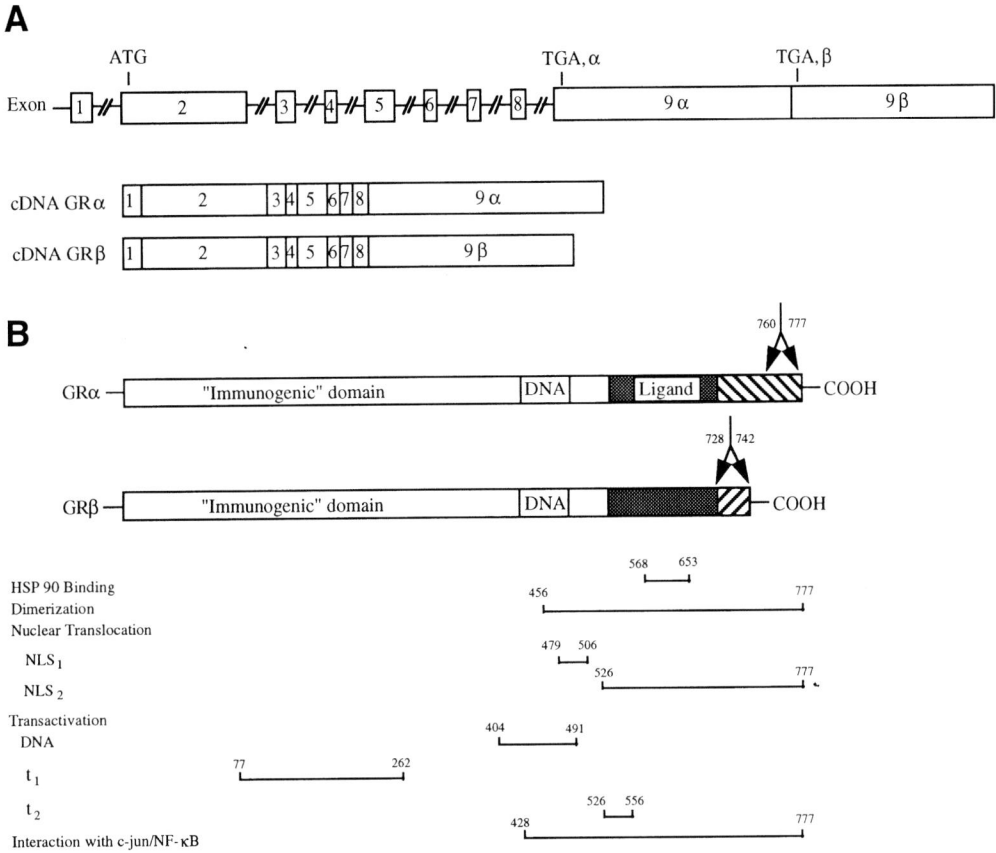

Fig. 1. (**A**) The human glucocorticoid receptor gene and the cDNAs of its alternatively spliced products GR-α and GR-β. (**B**) The linearized structures of the GR-α and GR-β proteins and their three main functional domains. Arrows indicate epitopes employed for generation of specific anti-isoform antibodies. Underneath are the additional functional subdomains of the GR.

THE GLUCOCORTICOID RECEPTOR TRANSDUCTION SYSTEM AND TARGET TISSUE SENSITIVITY TO GLUCOCORTICOIDS

The cloning of the human glucocorticoid receptor cDNA in 1985 allowed deduction of its primary structure and definition of its functional domains, and permitted further studies on the molecular mechanisms of glucocorticoid resistance *(7)*. The 777 amino acid long human glucocorticoid receptor protein has three main functional domains (Fig. 1): the ligand binding domain, which lies in the carboxy-terminal portion of the protein, the DNA binding domain, which is present in the middle, and a third, amino-terminal domain, which was initially characterized by its property of providing antigenic epitopes for generation of antireceptor antibodies and, therefore, called the "immunogenic domain." Further characterization of the functional domains of the receptor allowed not only better characterization of the sequences responsible for steroid binding and DNA recognition, but also revealed a number of domains representing other functions involved in the cascade of events, from ligand binding to modulation of gene transcription. These are described below.

There is a gross similarity between the structures of the members of the steroid/sterol family of receptors, with most of the homologies occurring primarily in the DNA and, secondarily, in the ligand binding domains. This explains both the known existing crossreactivities of ligand binding between the natural steroids and their receptors, and the sharing of common DNA sequences in the promoter regions or enhancers of steroid-regulated genes, with which these receptors interact. Relevant to glucocorticoid resistance is the virtually 100% crossreactivity of the mineralocorticoid receptor for cortisol. Normally, the kidney mineralocorticoid receptor is partially protected from the mineralocorticoid effects of circulating cortisol by the intracellular enzyme 11-hydroxysteroid dehydrogenase in the distal convoluted tubule, which converts 11β-hydroxylated compounds, such as cortisol, into inactive 11-ketosteroids, such as cortisone. The presence of increased amounts of circulating cortisol, as well as deoxycorticosterone and corticosterone, in patients with glucocorticoid resistance appears to overwhelm the protective enzyme and produce mineralocorticoid excess effects in some of the patients via the mineralocorticoid receptor.

A wealth of new information on the molecular mechanisms of action of glucocorticoids has added tremendously to the initial model, which continues to be rapidly modified and to offer new explanations on the pathophysiology of glucocorticoid resistance *(3)*. Currently, the cascade of events leading to gene transcription could be summarized as follows: The inactivated ligand-free glucocorticoid receptor in the cytoplasm is complexed with other proteins, including heat-shock protein 90 (HSP 90), HSP 70, HSP 56 or "immunophilin," and HSP 22, in the form of a hetero-oligomer. It appears that the anchoring of the receptor to HSP 90 facilitates binding of the steroid to the receptor, improving the affinity of the protein for the ligand. Binding of the glucocorticoid to its receptor results in release of the steroid–receptor complex from HSP 90 and unmasking of domains responsible for dimerization, nuclear localization, DNA binding, and transactivation. The dimerization domains are within the DNA and steroid binding domains.

In theory, receptor homodimers can be present both in the cytoplasm and in the nucleus. Ligand-bound receptor monomers or dimers appear to cross into the nucleus via recognition of their "nuclear localization domains" (or "nuclear localization sequences" [NLS]) by specialized proteins, such as importin α, which recognizes the NLS, and importin β, which interacts with the nuclear pore *(8)*. One of these sequences is homologous to the NLS of SV40 T-antigen and other proteins known to translocate into the nucleus. Intracellular fibrils might participate in the alignment of receptors and their translocation.

Inside the nucleus, the ligand-bound glucocorticoid receptor exerts its genomic effects on transcription in at least three potential ways. First, in the form of a dimer, it interacts with specific DNA sequences—"glucocorticoid responsive elements" (GREs)—in the promoter regions or enhancers of glucocorticoid-responsive genes. The DNA binding domain of the receptor consists of two protein folds, the "zinc fingers," which selectively interact with the GREs. The ligand-bound receptor dimers with, perhaps, the assistance of other proteins called coactivators, such as CBP and P300 *(9,10)*, appear to stabilize the polymerase II initiation complex of positively regulated genes, possibly by binding to both the glucocorticoid receptor dimer and to ancillary factors and general transcription factors (GTFs) of RNA polymerase II, such as TFIIB and/or TFIID, whose interaction with the partial initiation complex appears to represent the rate-limiting step of transcription. The ability of coactivators of the steroid receptor superfamily, such as SRC-1 *(11)*, GRIP-1 (also named TIF2) *(12)*, and p/CIP (also named ACTR) *(13)* to form a link between the GRE-bound receptor and the RNA polymerase II complex results in local chromatin remodeling through their histone acetyltransferase activity (HAT) and enhancement of target gene expression. Recently, a partial pan-steroid resistance was observed in mice with disruption of SRC-1 gene. The homozygous mutants were viable and fertile, but several steroid target tissues, such as the uterus, prostate, testis, and mammary glands, exhibited decreased growth and development in response to the appropriate steroid hormones. This partial pan-steroid resistance might be explained by the presence of the highly homologous coactivator TIF2, compensating for the loss of the coactivator function of SRC-1 *(14)*. These SRC-1 knock-out animals have a polysteroid resistance similar to that of New World primates.

The above interaction has been described with "positive" GREs. However, the presence of "negative" GREs has been described as well. Second, mostly negative regulation by glucocorticoids on transcription can be effected with the glucocorticoid receptor dimer interacting with part of the responsive element of another positive transcription factor, by displacing this factor, or by hindering it from properly exerting its transcription-promoting effect. Third, another recently described major molecular path, via which glucocorticoids appear to exert part of their major antigrowth and anti-inflammatory effects *(15)*, appears to be the formation of intranuclear complexes of the glucocorticoid-bound receptor with other transcription factors. The spectrum of transcription factors present determines whether the gene may be transcribed or not. The pattern of gene expression is quite complex and the ratio of the concentrations of different transcription factors is very important, particularly if competing positive and negative factors are involved. Glucocorticoid-bound receptor interacts with several transcription factors, such as c-jun (AP-1) *(16)*, P65 or Rel A (NF-κB) *(17–19)*, CREB *(20)*, NF-IL-6 *(21)*, and STAT5 *(22)*, and alters the ability of these factors to exert their ubiquitous activating, growth-promoting, and pro-inflammatory effects.

One should note that the above model of glucocorticoid actions at the cellular level is still evolving and that many crucial questions remain *(3)*. Thus, phosphorylation and dephosphorylation of the receptor may participate in the activation–inactivation, recycling, and turnover of the receptor, and a number of early/rapid effects of glucocorticoids, such as changes in membrane potential and intracellular cGMP, take place in as yet undefined mechanisms. Finally, the mechanism(s) by which glucocorticoids alter the stability of specific mRNAs, an unequivocal and important function of these hormones, which has been observed with several genes, is relatively unclear.

Given the large variety of functions carried by a single glucocorticoid receptor protein, it is not surprising that the genomic structure of its gene, located on chromosome 5, is quite complex, consisting of 10 different exons *(23,24)* (Fig. 1). One exon codes for the entire immunogenic domain, two for the DNA binding domain—one for each zinc finger—and five for the steroid binding domain—which carries most of the identified functions. Alternative splicing appears to result in synthesis of another "receptor" isoform, glucocorticoid receptor (GR)-β, which does not bind glucocorticoid and whose function we are currently studying. We pursued the nonligand binding isoform of the human GR-β, in order to define whether it plays a role in determining the sensitivity of human tissues to glucocorticoids *(25)*. As a first step, we coexpressed it with the glucocorticoid receptor α isoform and a reporter gene, luciferase, coupled to the MMTV-LTR, and examined its ability to express dominant negative activity. Indeed, it did have such activity, which was specific and dose-dependent. As a second step, we examined the ability of this isoform to alter the affinity of GR for dexamethasone, which it did not, and its ability to bind GREs, which it did. As a third step, we demonstrated the expression of its mRNA in a large panel of human tissues and we developed specific antibodies against GR-α and GR-β, using oligopeptides corresponding to antigenic epitopes in the C-terminal domain of these proteins (Fig. 1). This isoform protein could play a significant role in physiologic and pathophysiologic conditions, since we have observed a ratio hGR-β/hG-α from 1/1 to 5/1 in human tissues (26). However, there is no actual evidence that it could be regulated in a generalized or tissue-specific manner. No interaction was found with c-jun/c-fos or NF-κB (27). It has been shown that hGR-β weakly binds HSp90, but whether hGR-β is nuclear, cytoplasmic, or both is still controversial (28). Currently, we are studying the potential role and the possible regulation of hGR-β in vivo in physiologic and pathophysiologic conditions. Recently, Leung et al. observed that peripheral blood mononuclear cells (PBMCs) and cells of broncho-alveolar lavage (BAL) from patients with steroid-resistant asthma showed a stronger immunostaining for hGR-β than the PBMCs and the BAL cells from patients with steroid-sensitive asthma (29). Whether this is a primary state or secondary, is induced by inflammation, is not known.

The apparent importance of HSP 90 to allow better ligand binding of the glucocorticoid receptor at lower ligand concentrations and consequently, to alter the effect of glucocorticoids on gene transcription stimulated us to examine potential changes of HSP 90 mRNAs or protein concentrations in several tissues, basally, at the zenith and nadir of circadian corticosterone secretion, and following acute or chronically repeated immobilization stress (30,31). Generally, the mRNA and protein concentration changes observed were concordant. There was tissue specificity in the basal level of HSP 90 or HSP 90β and in the change of one or the other with the time of the day or during immobilization stress. In the periphery, there was a two- to threefold decrease in the level of HSP 90 in the liver and spleen at the time of highest stress system activity (night or postimmobilization), suggesting that the glucocorticoid sensitivity of these organs to glucocorticoids is decreased at these times. In contrast, there was an increased expression of HSP 90 mRNA in the subregions of the hippocampus known to contain the highest levels of glucocorticoid receptor mRNA, suggesting that there is coordinate regulation of HSP 90 and glucocorticoid receptor mRNA in this major regulatory organ (32).

NF-κB is a key regulator of the transcriptional activity of many cytokines. Conversely, many cytokines activate NF-κB, leading to a positive reverberating cycle in an inflammatory site. Glucocorticoids inhibit NF-κB in different ways, by direct interaction of GR with P65 or Rel A (17–19) or by increasing I-κB, which inactivates NF-κB (32). The overexpression of P65 has been shown to inhibit the ability of glucocorticoids to stimulate the transcription rate of the MMTV GRE-containing promoter. Many cytokines exert their cellular effects through the Janus kinases that phosphorylate STAT proteins on tyrosine residues. Recently, Stocklin et al. showed that STAT 5 directly interacts with the glucocorticoid receptor and that the STAT 5/glucocorticoid receptor complex strongly suppresses the response of a GRE-containing promoter to glucocorticoids (22). These two examples help to explain the "acquired" glucocorticoid resistance of several inflammatory diseases.

DISEASE STATES POTENTIALLY ASSOCIATED WITH ALTERATIONS IN TARGET TISSUE SENSITIVITIES TO GLUCOCORTICOID

The current state of knowledge on the mechanism of action of glucocorticoids at the cellular/molecular level, and the profound effects of glucocorticoids on CNS function

Table 1
How Changes in the Glucocorticoid Sensitivity
of Target Tissues Could Lead to Disease Manifestations

	Resistance	Hypersensitivity
Generalized	Glucocorticoid deficiency, mineralocorticoid excess, androgen excess	Cushing's syndrome
Tissue-specific		
CNS[a]	Asthenia/irritability Susceptibility to depression and/or addiction	Susceptibility to depression and/or addiction
Immune system	Susceptibility to autoimunity and/or allergy	Susceptibility to infections and/or neoplasms
Cardiovascular	Hypotension	Hypertension
Adipose tissue	Low visceral fat content	High visceral fat-insulin resistance, metabolic syndrome X, polycystic ovaries, diabetes type 2
Bone	—	Osteoporosis

[a]CNS, central nervous system.

and behavior as well as on the metabolic and immune responses, offer new potential avenues of research in pathophysiologic states of major emotional and socioeconomic impact (Table 1).

Glucocorticoid Resistance

Given the complexity by which glucocorticoid receptors exert their effects, and the varying impact expected by different potential mutations of the receptor and/or the cellular factors that interact with it, one can envision not only generalized glucocorticoid resistance states, but also glucocorticoid resistance states that are limited to certain tissue(s) or cellular functions *(3)*. It is theoretically possible that such mutations would not affect the regulation of the HPA axis—in which case the patients could not be classified as having familial generalized glucocorticoid resistance, but would influence the action of glucocorticoids in some other glucocorticoid-dependent system. As an example, the dopaminergic mesocortical/mesolimbic system, which is responsible for pleasure and reward phenomena and/or the locus ceruleus/norepinephrine system, which sustains proper arousal, could be resistant to the "normal" circulating levels of glucocorticoids (both of these systems are replete with glucocorticoid receptors). In both instances, glucocorticoid understimulation could lead to emotional disorders characterized by stimulus- or drug-seeking behaviors and/or chronic fatigue. Similarly, if only portions of the immunosuppressive/anti-inflammatory effects of glucocorticoids were exerted because of a mutation impacting primarily on this set of functions, one would expect phenomena resulting from an unrestrained immune system, such as autoimmune, allergic, or inflammatory disease.

Glucocorticoid Hypersensitivity

In the same line of reasoning, mutations of the glucocorticoid receptor causing hypersensitivity of a target tissue or function to glucocorticoids, without altering the feedback regulation of the HPA axis, could also lead to metabolic, cardiovascular, behavioral, or immune function alterations, depending on the tissue(s) or function(s) influenced by the mutation. Thus, hypersensitivity of the adipose tissue would lead to obesity, insulin resistance, hyperlipidemia, and hypertension (metabolic syndrome X), hypersensitivity of the blood vessels to hypertension, hypersensitivity of the mesocorticolimbic system to depression, hypersensitivity of the immune system to immunosuppression, and hypersensitivity of the bone to osteoporosis. Extricating potential mutations of the glucocorticoid receptors associated with normal HPA axis activity and proving their pathogenetic potential in causing glucocorticoid resistance- or hypersensitivity-mediated disease will be an interesting and quite provocative challenge. Two recent reports, in which linkage between a polymorphism of the glucocorticoid receptor gene, defined by Southern blotting, and essential hypertension or severe visceral obesity were found, further encouraged us to pursue this concept *(33,34)*.

"Familial" or "Sporadic" Generalized Glucocorticoid Resistance

We had the opportunity to study the first, as well as several additional reported families and/or sporadic cases with glucocorticoid resistance, that had a glucocorticoid receptor with decreased affinity for the hormone or decreased concentrations of the receptor, as well as several New World primate species with generalized "pan-steroid" hormone resistance. We obtained permanent cell lines from these individuals and species, and have been studying them in an attempt to understand the molecular mechanisms of hormone resistance. The animal model is of special interest, because its pan-steroid hormone resistance suggests a common link among the transduction systems of the six classes of steroid/sterol hormones, possibly a common coactivator or corepressor molecule. We completed our molecular studies with the first kindred and defined the pathological mutation, as well as the mode of genetic transmission in this family *(35)*. In affected members there was a T for A base substitution at nucleotide 2054, leading to a nonconservative amino acid mutation from aspartate, an acidic amino acid, to valine, a hydrophobic one, at position 641. This mutation in the steroid binding domain is close to a cysteine that plays a major role in ligand binding. The propositus, who presented with severe mineralocorticoid excess, was a homozygote for the mutation. His biochemically only affected asymptomatic brother and son were heterozygote. This mutation was confirmed to be in exon 7 of the genomic DNA, and no normal individual examined in parallel had such an alteration. Introduction of the Val 641 mutation in a GR expression vector resulted in a low-affinity, dysfunctional glucocorticoid receptor, confirming the pathogenetic importance of the mutation detected.

Differential glucocorticoid-dependent transcriptional activation and repression was observed among human glucocorticoid receptor variants associated with glucocorticoid resistance. The dexamethasone-induced repression of transcription from elements in the promoter of the intercellular adhesion molecule-1 via NF-κB through the D641V variant was in fact more efficient than the the wild-type GR; however the patients with this mutation did not have clinical immunosuppression *(27)*.

Koper et al. have defined five glucocorticoid receptor cDNA polymorphisms in the normal population, which are at variance with the original sequence reported by Hollenberg et al. *(36)*. In a population of 216 healthy subjects, a reduced response in a short dexamethasone suppression test as a marker of relative glucocorticoid insensitivity was observed in 20 otherwise healthy persons. However, this relative glucocorticoid insensitivity was not associated with any of these polymorphisms *(36)*. Nevertheless, this selected sample of the general population could represent a part of a normal distibution of glucocorticoid sensitivity.

We studied in great detail the molecular pathophysiology of glucocorticoid resistance in four additional propositi and/or families with proven glucocorticoid resistance. In the first of these, the propositus presented with hyperandrogenism and infertility *(37)*. Her father and two brothers had hypercortisolism of the same degree, but no pathological symptomatology. In this family, all affected members had a decrease in their glucocorticoid receptors by 50%. This receptor had normal affinity for dexamethasone. We found a 4-base microdeletion in a splice junction between exon 6 and the ensuing intron in all affected members, which precluded the expression of one of the two alleles. In the propositi of the third and fourth families we studied, we found no abnormalities in the coding sequence of the DNA of the glucocorticoid receptor or in the exon-intron junctions (unpublished information). This suggests that in these families, the molecular defect may either be in the regulatory region of the GR gene or in one of the molecules that influence its function, specifically HSP 90 *(38,39)*. In the fifth propositus, a hypercortisolemic, non-Cushingoid male who presented with hypertension and infertility in his early 30s and who developed pituitary Cushing's disease in his mid-30s, we found a heterozygotic nonconservative amino acid substitution in the "hinge" region between the DNA binding domain and the ligand binding domain, which abolished the binding affinity and functional activity of this mutant receptor *(40)*. Family studies of this patient failed to show glucocorticoid resistance in his parents or siblings. This suggested that this patient was a sporadic case, with a *de novo* mutation. Using "single-cell PCR," we demonstrated that 50% of this patient's sperm and all of his cultured fibroblasts or B-lymphoblasts contained this mutation. Further studies revealed that this patient's mutation had marked negative inhibitory activity on the normal receptor, which could explain the increased severity of his resistance compared to heterozygotes from our first kindred, who were biochemically only affected. Subsequent study of this patient's pituitary corticotropinoma revealed the excess presence of p53, which was not seen in normal cells, suggesting that the latter might represent a somatic mutation and a "second hit" in the process of corticotroph tumorigenesis in this subject *(40)*.

"TISSUE-SPECIFIC" GLUCOCORTICOID RESISTANCE

Tissue-Specific Glucocorticoid Resistance in Neoplasms

The early beneficial effect of glucocorticoid therapy, through induction of apoptosis, disappears with the occurence of glucocorticoid-resistant clones. Glucocorticoid receptor and postreceptor abnormalities, such as mutations of the hGR gene or hGR variants from alternative splicing, have been reported in malignant cells from patients suffering from leukemia and myeloma *(41–43)*.

It has been known that pituitary tumors secreting ACTH (corticotropinomas) are glucocorticoid-resistant. In an attempt to find potential defects of the glucocorticoid receptor in such tumors, we sequenced each of the nine exons of the glucocorticoid receptor in large corticotropinomas from patients with Nelson's syndrome *(44)*. One of four tumors had a somatic base insertion within the coding region of the receptor, resulting in a frame shift and early termination. This mutation was not present in circulating leukocytes from the same patient. Thus, this tumor expressed only one of the glucocorticoid receptor alleles, effectively having resistance to glucocorticoids of a degree similar to that in our second family. None of these tumors studied had abnormalities of the p53 gene. Recently, Huizenza et al. demonstrated loss of GR gene heterozygosity in the glucocorticoid receptor gene locus in corticotropinomas in 6 of 22 patients with Cushing's disease. The loss of heterozygosity in 25% of the tumors seems to be a relatively frequent phenomenon in these patients and might explain the relative resistance of these tumors to the negative feedback action of cortisol on ACTH secretion *(45)*.

Tissue-Specific Glucocorticoid Resistance in Autoimmune and Inflammatory Diseases

The glucocorticoid resistance observed in patients suffering from autoimmune and inflammatory diseases could be induced by inflammation or genetically determined. Leung et al. described two different populations of steroid-resistant asthma patients, Type 1, defined by decreased GR affinity binding, and Type 2, defined by decreased GR number per cell. Type 1 SR-asthma was shown to be reversible and secondary to inflammation, whereas Type 2 SR-asthma appeared to be genetically determined *(46)*. Schlaghecke et al. reported a 50% decrease of GR number /cell in patients with rheumatoid arthritis *(47)*; however, surprisingly, this did not appear to influence the corticosensitivity of these patients *(48)*.

Corticosensitivity was shown to be modulated by cytokines in various ligand binding studies *(49)*; some cytokines, such as IL-1β, IL-6, TNF-α *(49,50)*, and IFN-γ *(51)*, increased sensitivity to glucocorticoids, whereas others, such as IL-2, IL-4, and IL-13, decreased sensitivity to glucocorticoids *(52,53)*. NF-κB can be viewed as intracellular amplification mechanism that exacerbates chronic inflammatory processes *(54)*. Excessive NF-κB induction in an inflammatory site might inhibit glucocorticoid activity contributing, thus, to a decrease in corticosensitivity *(55)*.

Resistance to glucocorticoid therapy could be owing to a primary defect in the GR signal transduction pathway. No consistent polymorphism was observed in the reverse-transcribed hGR cDNA from corticosteroid-sensititive and corticosteroid-resistant asthma patients *(56)*. However, Adcock et al. have reported that the ability of the glucocorticoid receptor to bind to GREs is impaired in steroid-resistant asthma patients because of a reduced number of receptors available for binding to DNA *(57)*. Leung et al. have suggested that this decreased binding to DNA could be related to the increased hGRβ / hGRα ratio in these patients *(29)*. We cloned and sequenced the cDNA and gene of the glucocorticoid receptor of two patients with severe glucocorticoid-resistant asthma who had low concentrations of receptor in their leukocytes (unpublished information). We found no abnormalities in the 5'-regulatory region, open-reading frame, and splice junctions of the receptor gene in these patients. We are continuing their study by examining the alternative splicing rate of their glucocorticoid receptor gene and the cellular concentration of HSP 90 in an attempt to explain the low receptor number and profound resistance of their cells to dexamethasone.

REFERENCES

1. Chrousos GP, Loriaux DL, Gold PW. (eds.) Mechanisms of physical and emotional stress. In: Advances in Experimental Medicine and Biology, vol. 245, Plenum, New York, 1988.
2. Chrousos GP, Loriaux DL, Lipsett MB. (eds.) Steroid hormone resistance: mechanisms and clinical aspects. In: Advances in Experimental Medicine and Biology, vol. 196, Plenum, New York, 1986.

Chapter 14 / Glucocorticoid Resistance

3. Chrousos GP, Detera-Wadleigh S, Karl M. Syndromes of glucocorticoid resistance. Ann Int Med 1993; 119:113–1124.
4. Keightley MC, Fuller PJ. Cortisol resistance and the guinea pig glucocorticoid receptor. Steroids 1995; 60:87–92.
5. Taymans SE, De Vries AC, De Vries MB, Nelson RJ, Friedman TC, Castro M, et al. The hypothalamic–pituitary–adrenal axis of Prairie Voles (*Microtus Ochrogaster*): Evidence for target tissue glucocorticoid resistance. Comp Endocrinol 1997;106:48–61.
6. DeKloet ER. Brain corticosteroid receptor balance and homeostatic control. Frontiers Neuroendocrinol 1991;12:95–164.
7. Evans RM. The steroid and thyroid hormone receptor superfamily. Science 1988;240:889–895.
8. Ulman KS, Powers MA, Forbes DJ. Nuclear export receptors: from importin to exportin. Cell 1997;90: 967–970.
9. Chakravarti D, LaMorte VJ, Nelson MC, Nakajima T, Shulman IG, Juguilon H, et al. Role of CBP/P300 in nuclear receptor signaling. Nature 1996;383:99–103.
10. Kamei Y, Xu L, Heinzel T, Torchia J, Kurokawa R, Gloss B, et al. A CBP integrator complex mediates transcriptional activation and AP-1 inhibition by nuclear receptors. Cell 1986;85:403–414.
11. Onate SA, Tsai SY, Tsai MJ, O'Mallay BW. Sequence and characterization of a co-activator for the steroid hormone receptor superfamily. Science 1995;270:1354–1357.
12. Hong H, Kohli K, Trivedi A, Johnson DL, Stallcup MR. GRIP1, a novel mouse protein that serves as a transcriptional coactivator in yeast for the hormone binding domains of steroid receptors. Proc Natl Acad Sci USA 1996;93:4948–4952.
13. Torchia J, Rose DW, Inostroza J, Kamei Y, Westin S, Glass CK, et al. The transcriptional co-activator p/CIP binds CBP and mediates nuclear-receptor function. Nature 1997;387:677–684.
14. Xu J, Qiu Y, De Mayo FJ, Tsai SY, Tsai MJ, O'Malley BW. Partial hormone resistance in mice with disruption of the steroid receptor co-activator-1 (SRC-1) gene. Science 1998;279:1922–1925.
15. Boumpas DT, Chrousos GP, Wilder RL, Cupps TR, Balow JE. Glucocorticoid Therapy of Immune-Related Diseases: Basic and Clinical Correlates. Ann Intern Med 1993;119:1198–1208.
16. Schule R, Rangarajan P, Kliewer S, Ransone LJ, Bolado J, Yang N, et al. Functional antagonism between oncoprotein c-Jun and the glucocorticoid receptor. Cell 1990;62:1217–1226.
17. Ray A, Prefontaine KE. Physical association and functional antagonism between the p65 subunit of transcription factor NF-κB and the glucocorticoid receptor. Proc Natl Acad Sci USA 1994;91:752–756.
18. Caldenhoven E, Liden J, Wissnik S, Van de Stolpe A, Raaijmakers J, Koenderman L, et al. Negative cross talk between rel A and glucocorticoid receptor. a possible mechanism for the antiinflammatory action of glucocorticoids. Mol Endocrinol 1995;9:401–412.
19. Scheinman RI, Gualberto A, Jewell CM, Aldowski JA, Baldwin AS Jr. Characterization of mechanisms involved in transrepression of NF-κB by activated glucocorticoid receptors. Mol Cell Biol 1995;15: 943–953.
20. Imai E, Miner JN, Mitchell JA, Yamamoto KR, Granner DK. Glucocorticoid receptor-cAMP response element-binding protein interaction and the response of the phosphoenolpyruvate carboxykinase gene to glucocorticoids. J Biol Chem 1993;268:5353–5356.
21. Nishio Y, Isshiki H, Kishimoto T, Akira S. A nuclear factor for interleukin-6 expression (NF-IL6) and the glucocorticoid receptor synergically activate transcription of the rat a1-acid glycoprotein gene via direct protein-protein interaction. Mol Cell Biol 1993;13:1854–1862.
22. Stocklin E, Wissler M, Gouilleux F, Groner B. Functional interactions between Stat 5 and the glucocorticoid receptor. Nature 1996;383:726–728.
23. Encio IJ, Detera-Wadleigh SD. The genomic structure of the human glucocorticoid receptor. J Biol Chem 1991;266:7182–7188.
24. Nobukumi Y, Smith CL, Hager GL, Detera-Wadleigh SD. Characterization of the human glucocorticoid receptor promoter. Biochemistry 1995;34:8207–8214.
25. Bamberger CM, Bamberger AM, De Castro M, Chrousos GP. Glucocorticoid receptor-beta, a potential endogenous inhibitor of glucocorticoid action in humans. J Clin Invest 1995;95:2435–2441.
26. Castro M, Elliot S, Kino T, Bamberger C, Karl M, Webster E, et al. The non-ligand binding B-isoform of the human glucocorticoid receptor (hGRb): tissue levels, mechanism of action, and potential physiologic role. Mol Med 1996;5:1076–1551.
27. De Lange P, Koper JW, Huizenga NATM, Brinkman AO, De Jong FH, Karl M, et al. Differential hormone-dependent transcriptional activation and repression by naturally occurring human glucocorticoid receptor variants. Mol Endocrinol 1997;11:1156–1164.

28. Oackley RH, Webster JC, Sar M, Parker CR, Cidlowski JA. Expression and subcellular distribution of the β isoform of the human glucocorticoid receptor. Endocrinology 1997;138:5028–5038.
29. Leung DYM, Hamid Q, Vottero A, Szefler SJ, Surs W, Minshall E, et al. Association of glucocorticoid insensitivity with increased expression of glucocorticoid receptor β. J Exp Med 1997;186:1567–1574.
30. Vamvakopoulos NC, Fukuhara K, Patchev VK, Chrousos GP. Effect of single and repeated immobilization stress on the heat shock protein (HSP) 70/90 system in the rat: glucocorticoid-independent, reversible reduction of HSP90 in the liver and spleen. Neuroendocrinology 1993;57:1057–1065.
31. Patchev VK, Brady LS., Karl M, Chrousos GP. Gene expression of HSP90 and glucocorticoid receptors in the brain: evidence for coordinate regulation by adrenal steroid levels in vivo. Cell Mol Endocrinol 1994;103:57–64.
32. Auphan N, Didonato JA, Rosette C, Helmberg A, Karin M. Immunosuppression by glucocorticoids: inhibition of NF-kappaB activity through induction of I kappa B synthesis. Science 1995;270:283–286.
33. Watt GCM, Harrap SB, Foy CJW, Holton DW, Edwards HV, Davidson HR, et al. Abnormalities of glucocorticoid metabolism and the renin-angiotensin system: a four corners approach to the identification of genetic determinants of blood pressure. J Hypertens 1992;10:473–482.
34. Weaver JV, Hitman GA, Kopelman PG. An association between a bcp1 restriction fragment length polymorphism of the glucocorticoid receptor locus and hyperinsulinaemia in obese women. J Mol Endocrinol 1992;9:295–300.
35. Hurley D, Accilli D, Stratakis C, Karl M, Vamvakopoulos N, Rorer E, et al. Mutation of the glucocorticoid receptor gene in familial glucocorticoid resistance. J Clin Invest 1991;87:680–686.
36. Koper JW, Stolk RP, de Lange P, Huizenga NATM, Molijn GJ, Pols HAP, et al. Lack of association between five polymorphisms in the human glucocorticoid receptor gene and glucocorticoid resistance. Hum Genet 1997;99:663–668.
37. Karl M, Lamberts SW, Detera-Wadleigh S, Encio IJ, Stratakis CA, Hurley DM, et al. Familial glucocorticoid resistance caused by a splice site deletion in the human glucocorticoid receptor gene. J Clin Endocrinol Metab 1993;76:683–689.
38. Nathan DF, Lindquist S. Mutational analysis of HSP 90 function: interactions with a steroid receptor and a protein kinase. Mol Cell Biol 1995;15:3917–3925.
39. Kang KI, Devin J, Cadepond F, Jibard N, Guiochon-Mantel A, Baulieu EE, et al. In vivo functional protein-protein interaction: nuclear targeted HSP 90 shifts cytoplasmic steroid receptors into the nucleus. Proc Natl Acad Sci USA 1994;91:340–344.
40. Karl M, Lamberts SW, Koper JW, Katz DA, Huizenga NE, Kino T, et al. Cushing's disease preceded by generalized glucocorticoid resistance: clinical consequences of a novel, dominant-negative glucocorticoid receptor mutation. Proc Assoc Am Phys 1996;108:296–307.
41. Strasser-Wozak EMC, Hattmannstoffer R, Hala M, HartmanBL, Fiegl SG, Kofler R. Splice site mutation in the glucocorticoid receptor gene causes resistance to glucocorticoid-induced apoptosis in a human acute leukemic cell line. Cancer Res 1995;55:348–353.
42. Powers JH, Hillman AG, Tang DC, Harmon JM. Cloning and expression of mutant glucocorticoid receptors from glucocorticoid-sensitive and resistant human leukemic cells. Cancer Res 1993;53:3877–3879.
43. Moalli PA, Pillay S, Krett NL, Rosen ST. Alternative spliced glucocorticoid receptor messenger RNAs in glucocorticoid-resistant human myeloma cells. Cancer Res 1993;53:3877–3879.
44. Karl M, Von Wichert G, Kempter E, Katz DA, Reincke M, Monig H, et al. Nelson's syndrome associated with a somatic frame shift mutation in the glucocorticoid receptor gene. J Clin Endocrinol Metab 1996;81:124–129.
45. Huizenza NATM, de Lange P, Koper JW, Clayton RN, Farrel WE, van der Lely AJ, et al. Human adrenocorticotropin-secreting pituitary adenomas show frequent loss of heterozygosity at the glucocorticoid receptor gene locus. J Clin Endocrinol Metab 1998;83:917–921.
46. Sher ER, Leung DVM, Surs W, Kam JC, Zieg G, Kamada AK, Szefler SJ. Steroid-resistant asthma. cellular mechanisms contributing to inadequate response to glucocorticoid therapy. J Clin Invest 1994;93:33–39.
47. Schlaghecke R, Kornely E., Wollenhaupt J, Specker C. Glucocorticoid receptors in rheumatoïd arthritis. Arthritis Rheum 1994;35:740–744.
48. Schlaghecke R, Beuscher D, Kornely E, Specker C. Effects of glucocorticoids in rheumatoid arthritis. Arthritis Rheum 1994;37:1127–1131.
49. Costas M, Trapp T, Pereda MP, Sauer J, Rupprecht R, Nahmod VE, et al. Molecular and functional evidence for in vitro cytokine enhancement of human and murine target cell sensitivity to glucocorticoïds. J Clin Invest 1996;98:1409–1416.

50. Rakasz E, Gal A, Biro J, Balas G, Falus A. Modulation of glucocorticosteroid binding in human lymphoid, monocytoid and hepatoma cell lines by inflammatory cytokines IL-1β, IL-6 and TNFα. Scand J Immunol 1993;37:684–689.
51. Salkowski CA, Vogel SN. IFNγ mediates increased glucocorticoid receptor expression in murine macrophages. J Immunol 1992;148:2770–2777.
52. Kam JC, Szefler SJ, Surs W, Sher ER, Leung DY. Combination of IL-2 and IL-4 reduces glucocorticoïd receptor-binding affinity and T cell response to glucocorticoïds. J Immunol 1993;7:3460–3466.
53. Spahn DS, Szefler J, Surs W, Doherty DE, Leung DY. A novel action of interleukin-13; induction of diminished monocyte glucocorticoïd receptor binding affinity. J Immunol 1996;157:2654–2659.
54. Barnes PJ, Karin M. Nuclear Factor-κB. A pivotal transcription factor in chronic inflammatory diseases. N Engl J Med 1997;336:1066–1071.
55. Van der Burg B, Liden J, Okret S, Delaunay F, Wissink S, Van der Saag PT, et al. Nuclear factor-κB repression in anti-inflammation and immunosuppression by glucocorticoids. Trends Endocrinol Metab 1997;8:152–157.
56. Lane SJ, Arm PM, Staynov DZ, Lee TH. Chemical mutational analysis of the human glucocorticoïd receptor cDNA in glucocorticoïd-resistant bronchial asthma. Am J Respir Mol Biol 1994;11:42–48.
57. Adcock IM, Lane SJ, Brown CR, Peters MJ, Lee MJ, Barnes PJ. Differences in binding of glucocorticoïd receptor to dna in steroïd-resistant asthma. J Immunol 1995;154:3500–3505.

INDEX

A

Acquired adrenocorticotropin (ACTH) resistance, 247
ACTH; *see* Adrenocorticotropin
Adenylyl cyclase, 47, 48
ADHD; *see* Attention-deficit hyperactivity disorder
Adrenal failure,
  causes, 246
Adrenal insufficiency,
  245–254
Adrenocorticotropin (ACTH), 260
  receptor mutations, 249–251
    expression studies, 250
    genotype-phenotype correlations, 250, 251
  resistance, 245–254
Adult,
  luteinizing hormone (LH) insensitivity syndrome, 186, 187
AHO; see Albright hereditary osteodystrophy
Albright hereditary osteodystrophy (AHO), 39, 44, 45
  neurosensory abnormalities, 42, 43
  obesity, 42
  skeletal deformities, 40–42
  treatment, 53
Allgrove's syndrome, 247
Amenorrhea,
  luteinizing hormone (LH) receptor gene mutation, 190
Ames mouse, 6–9
AMH; *see* Anti-Mullerian hormone
Amino acid substitutions,
  androgen receptor (AR) mutations, 216, 217
Androgen receptor (AR),
  defects,
    analyzing methods, 212, 213
    phenotypes, 211, 212
  DNA binding domain (DBD), 219
  function,
    genetic alterations, 209–226
    functional domains, 213
    homopolymeric repeats and heterogeneity, 222–223
    mutations, 214–222
    mutations and phenotype relationships, 221–223
    phenotype variations, 222
    schematic structure, 214
Androgen receptor (AR) mutations,
  amino acid substitutions, 216, 217
  ligand binding,
    absence, 214, 215
    decreased levels, 220, 221
    normal levels, 219, 220
    qualitative abnormalities, 217–219
  mRNA splicing, 215, 216
  premature termination codons, 216
  prostate cancer, 224
  virilization defects, 221
Androgen resistance,
  and mutation, 211, 212
Anti-Mullerian hormone (AMH), 186
  deficiency, 241
  resistance,
    clinical and anatomical features, 234, 235
    dog models, 241, 42
    endocrinology, 235, 236
    mouse models, 241
    molecular basis, 236–238
    mutations, 239–241
    phenotype, 233–236
    signaling cascade, 238
Anti-Müllerian hormone (AMH) resistance syndrome, 233–242
Anti-Müllerian hormone (AMH) type II receptors,
  mutations,
    descriptions, 239, 240
    transmission and incidence, 240, 241
AR; *see* Androgen receptor
Archibald's sign, 41

273

Attention-deficit hyperactivity disorder (ADHD),
  thyroid hormone resistance (RTH), 149, 150, 158
Autosomal-dominant familial hyperthyroidism,
  germline mutations, 135

B
Brachydactyly,
  Albright hereditary osteodystrophy (AHO), 40, 41
Bulbar muscular atrophy, 224

C
Calcitonin (CT),
  calcium (Ca^{2+}) ions, 88
Calcitriol resistant rickets (CRR), 60
Calcium (Ca^{2+}) ions, 87–105
  acquired tissue specific resistance, 104, 105
  homeostasis,
    schematic representation, 88
  parathyroid hormone (PTH), 88
  resistance syndromes, 91–104
Calcium ion sensing receptor (CaR), 88–90
  cloning, 89
  familial hypocalciuric hypercalcemia (FHH) mutations,
    HEK293 cells, 97, 98
  mutations,
    characteristics, 94–99
  physiological role, 89, 90
  schematic representation, 95
  tissue distribution, 89
Calcium therapy,
  hereditary vitamin D-resistant rickets (HVDRR), 75, 76
CAMP, 6, 40
CaR; *see* Calcium ion sensing receptor
Congenital hyperthyroidism,
  sporadic germline mutations, 135–139
Congenital hypothyroidism,
  thyroid stimulating hormone (TSH) receptor mutations, 129–131
Cortical collecting ducts,
  calcium ion sensing receptor (CaR), 90
CRR; *see* Calcitriol resistant rickets
CT; *see* Calcitonin

D
DBD; *see* DNA binding domain
Diabetes mellitus, 165–180
  insulin deficiency, 165–167
1,25-dihydroxyvitamin D [1,25(OH)2D], 59
  action mechanism, 66, 67
  calcium (Ca2+) ions, 88
  catabolism, 62
    cellular evidence, 67–69
  familial hypocalciuric hypercalcemia (FHH), 92
  hereditary vitamin D-resistant rickets (HVDRR), 67
  metabolic pathway, 61
  neonatal severe hyperparathyroidism (NSHPT), 101
  schematic representation, 64
DNA binding domain (DBD), 62, 63, 115
  affecting mutations, 69–71
  androgen receptor (AR), 219
  glucocorticoid resistance, 263
  mutations, 65
Dog models
  anti-Mullerian hormone (AMH) resistance, 241, 242
Dwarfism, 1–12
Dwarfism of Sindh, 9–11
Dwarf phenotype mouse; *see* Little mouse

E
Endocrine changes,
  luteinizing hormone (LH) insensitivity syndrome, 186, 187
Epstein-Barr virus (EBV) immortalized B-lymphoblasts,
  hereditary vitamin D-resistant rickets (HVDRR), 69
Euthyroid hyperthyrotropinemia,
  thyroid stimulating hormone (TSH) receptor mutations, 125–129
Exon deletion,
  insulin-like growth factor-1 (IGF-1), 32, 33

F
Familial glucocorticoid deficiency (FGD), 247, 251, 252
  clinical presentation, 247, 248

diagnosis and differential diagnosis, 248
normal, 251, 252
pathogenesis, 249
treatment, 249
Familial hypocalciuric hypercalcemia (FHH),
biochemical features, 91–94
clinical features, 91
genetics, 94
mouse models, 103, 104
Familial hypocalciuric hypercalcemia (FHH) mutations,
calcium ion sensing receptor (CaR), HEK293 cells, 97
Familial hypocalciuric hypercalcemia (FHH) 3
calcium ion sensing receptor (CaR), 94–99
Familial male-limited precocious puberty (FMPP),
luteinizing hormone (LH) receptors gene mutation, 191, 193
FGD; *see* Familial glucocorticoid deficiency
FHH; *see* Familial hypocalciuric hypercalcemia
FMPP; *see* Familial male-limited precocious puberty
Follicle-stimulating hormone (FSH),
glucocorticoid resistance, 260
luteinizing hormone (LH) insensitivity syndrome, 186, 187, 190
ovarian and testicular function, 198
Follicle-stimulating hormone (FSH) alpha gene, 199
Follicle-stimulating hormone (FSH) beta gene,
mutations, 199–202
Follicle-stimulating hormone (FSH) resistance, 197–205
clinical features,
animals, 203, 204
female, 202
male, 202
Follicle-stimulating hormone receptor (FSHR),
mutations, 200–202
signaling pathways, 198, 199
structure, 200

FSH; see Follicle-stimulating hormone
FSHR; *see* Follicle-stimulating hormone receptor

G

Generalized glucocorticoid resistance, 266, 267
Generalized resistance thyroid hormone (GRTH), 148, 149
Genes encoding transcription factors, diabetes mellitus, 167
Genetic alterations,
androgen receptor function, 209–226
Germline mutations,
autosomal-dominant familial hyperthyroidism, 135
GH; *see* Growth hormone
GHBP; *see* Growth hormone binding protein
GH-R; *see* Growth hormone receptors
GHRH; *see* Growth hormone releasing hormone
GHRH-R; *see* Growth hormone releasing hormone receptor
Glomerular filtration rate,
hypercalcemia, 93
Glucocorticoid,
target tissue sensitivity, 261–264
disease states, 264–267
Glucocorticoid hypersensitivity, 265, 266
Glucocorticoid receptor genes, 261
Glucocorticoid receptor transduction system, 261–264
Glucocorticoid resistance, 259–268
actions, 259–261
tissue specific, 267, 268
Glucokinase gene,
diabetes mellitus, 166, 167
GNAS1 gene, 44, 45
guanine nucleotide binding proteins (G-proteins), 47
GPCR; *see* G-protein coupled receptors
G-protein coupled receptors (GPCR), 89
mutations, 119, 120
G-protein of adenylyl cyclase (Gs), 40, 47, 48
alpha subunits,
mutations, 48–51

luteinizing hormone (LH) receptors gene
mutation, 188, 191
pseudohypoparathyroidism (PHP), 115
thyroid stimulating hormone (TSH)
receptors, 116
G-proteins; *see* Guanine nucleotide binding
proteins
Growth hormone binding protein (GHBP)
characteristics,
19–22
Growth hormone (GH),
little mouse, 6
Growth hormone (GH) secretion,
regulation, 1–2
Growth hormone receptors (GH-R),
genetic abnormalities, 19
mutations, 20, 21
protein and gene structure, 18
schematic illustration, 22
Growth hormone releasing hormone
(GHRH),
pituitary growth and development, 5
resistance, 6–12
animals, 6–9
humans, 9–11
resistance syndromes, 12
potential implications, 12
structure and activity, 3
Growth hormone releasing hormone
(GHRH) system, 1, 2
Growth hormone releasing hormone receptor
(GHRH-R),
cloning, 6, 7
expression, 3, 5
signaling, 5
structure, 3, 4
Growth hormone releasing peptide (GHRP)
system, 1, 2
GRTH; *see* Generalized resistance thyroid
hormone
Gs; *see* G-protein of adenylyl cyclase
Guanine nucleotide binding proteins
(G-proteins), 46
GNASI, 47
stimulation, 47
structure, 47

H

HCG; *see* Human chorion gonadotropin
HEK293 cells,
calcium ion sensing receptor (CaR), 97, 98

luteinizing hormone (LH) receptors gene
mutation, 191
Hereditary 1 alpha-hydroxylase deficiency
(VDDRI), 62
hereditary vitamin D-resistant rickets
(HVDRR), 67
Hereditary resistance to vitamin D,
59–78
Hereditary vitamin D-resistant rickets
(HVDRR), 59–78
clinical features, 67
hereditary 1 alpha-hydroxylase deficiency
(VDDRI), 67
HTLV-1 virus immortalized
T-lymphoblasts, 69
treatment, 74–76
Heterotopic ossification, 41, 53
Heterotrimeric G-proteins,
46–48
Hill coefficient, 98
Homeostasis,
calcium ion sensing receptor (CaR),
89, 90
Hormone resistance, 111
HTLV-1 virus immortalized
T-lymphoblasts,
hereditary vitamin D-resistant rickets
(HVDRR), 69
Human chorion gonadotropin (hCG),
physiological functions, 185–187
Humans,
growth hormone releasing hormone
(GHRH) resistance,
biochemical phenotype, 11
clinical phenotype, 10, 11
discovery and genetics, 9
molecular lesion, 9, 10
HVDRR; *see* Hereditary vitamin D-resistant
rickets
1 alpha-hydroxylase, 60, 62
24-hydroxylase, 62
Hypercalcemia, 93
neonatal severe hyperparathyroidism
(NSHPT), 100
Hyperparathyroidism, 67
calcium (Ca^{2+}) ions resistance,
104, 105
Hypersensitivity, 259–268
Hypocalcemia, 39, 43, 44, 51
severe or symptomatic, 52
Hypogonadism,
43, 44

Hypoparathyroidism, 39
  biochemical, 51
Hypothalamic pituitary level,
  growth hormone (GH) secretion, 2
Hypothyroidism, 43, 44
HYT/HYT mouse, 124, 125

I

IGF-1; *see* Insulin-like growth factor-1
IGFBP-3; *see* Insulin-like growth factor binding protein-3
Inherited adrenocorticotropin (ACTH) resistance, 247
Insulin action,
  downstream effectors, 177–180
  tyrosine phosphorylation, 174, 175
Insulin deficiency,
  diabetes mellitus, 165–167
Insulin gene,
  diabetes mellitus, 166
Insulin-like growth factor binding protein-3 (IGFBP-3), 19
  Laron Syndrome (LS) treatment, 28, 29
Insulin-like growth factor-1 (IGF-1), 19
  gene deletion, 32, 33
  Laron Syndrome (LS) treatment, 28, 29
Insulin receptors,
  insulin resistance, 167–177
    action, 167, 168
    mutations, 168
      biosynthesis impairment, 170
      clinical syndromes, 169, 170
      decreasing binding affinity, 171, 172
      degradation acceleration, 173, 174
      transportation impairment, 170, 171
      tyrosine kinase activity impairment, 172, 173
    oligomeric structure, 168
    phosphorylated,
      integral membrane proteins, 176
    phosphorylation substrates, 174–180
Insulin receptor substrate (IRS),
  mutations, 176, 177
  phosphotyrosine binding (PTB) domains, 175, 176
  Insulin resistance, 167–180
  associated clinical syndromes, 179

Insulin signaling pathway,
  proteins, 178, 179
Integral membrane proteins,
  phosphorylated,
    insulin receptor, 176
IRS; *see* Insulin receptor substrate

J

Jackson mouse, 6–9
Janus kinase 2 (JAK2), 18

L

Laron dwarfism, 18
Laron Syndrome (LS), 17–33
  biochemical features, 28
  body proportions, 25
  carbohydrate metabolism, 28
  congenital malformations, 23
  features, 23, 24
  geographical distribution, 22
  hormonal features, 28
  linear growth, 25, 30
  lipid metabolism, 28
  neonatal period, 23
  neuromuscular development, 26, 27
  nutritional state, 24, 25
  psychological development, 26, 27
  sexual maturation, 26
  skeletal development, 26
  treatment, 28–32
LBD; *see* Ligand binding domain
LCH; *see* Leydig cell hypoplasia
Leprechaunism, 169
Leydig cell hypoplasia (LCH), 187, 188, 192, 193
Leydig cells, 187, 233
  testosterone, 186
LH; *see* Luteinizing hormone
Ligand binding domain (LBD), 62, 63
  affecting mutations, 72–74
  mutations, 65
Lipoatrophy,
  associated clinical syndromes, 179
Little mouse,
  biochemical phenotype, 7, 8
  central nervous system, 7
  discovery and genetics, 6
  fertility, 7, 8
  molecular defect, 6, 7
  morphology, 8, 9

physical phenotype, 7, 8
pituitary development, 8, 9
LS; *see* Laron Syndrome,
Luteinizing hormone (LH), 197
  physiological functions,
    185–187
  receptors gene mutation,
    188–190
Luteinizing hormone (LH) insensitivity
    syndrome, 185–194
  LH resistance, 187–191
  molecular aspects, 191–193

M

MAPK; *see* Mitogen activated protein kinase
MAS; *see* McCune-Albright syndrome
Maturity-onset-type diabetes on the young
    (MODY),
  forms, 167
McCune-Albright syndrome (MAS), 50
Medullary collecting ducts,
  calcium ion sensing receptor (CaR), 90
Mice models,
  Ames, Jackson, and Snell, 6–9
  anti-Mullerian hormone (AMH)
    resistance, 241
  FHH and NSHPT, 103, 104
  little, 6–9
MIF; *see* Mullerian inhibiting factor
MIS; *see* Mullerian inhibiting substances
Mitogen activated protein kinase (MAPK),
  18
Mitogen-activated protein (MAP) kinase,
  activation, 178
MODY; *see* Maturity-onset-type diabetes on
    the young
Müllerian inhibiting factor (MIF),
  233–242
Müllerian inhibiting substances (MIS),
  233–242
Myeloid progenitor cells,
  hereditary vitamin D-resistant rickets
    (HVDRR), 69

N

Neonatal hyperparathyroidism (NHPT),
  neonatal severe hyperparathyroidism
    (NSHPT), 100
Neonatal Severe Hyperparathyroidism
    (NSHPT), 90
  biochemical features, 100, 101

clinical features, 99, 100
familial hypocalciuric hypercalcemia
    (FHH), 94
genetics, 101
heterozygous calcium ion sensing
    receptor (CaR) mutations,
    101, 102
mouse models, 103, 104
Neurosensory abnormalities,
  Albright hereditary osteodystrophy
    (AHO), 42, 43
NHPT; *see* Neonatal hyperparathyroidism
Normal familial glucocorticoid deficiency
    (FGD), 251, 252
NSHPT; *see* Neonatal Severe
    Hyperparathyroidism

O

Obesity,
  Albright hereditary osteodystrophy
    (AHO), 42
Osteosis cutis, 42
Ovarian function,
  follicle-stimulating hormone (FSH), 198

P

Parathyroid hormone (PTH),
  calcium (Ca^{2+}) ions, 88
  deficiency, 39
  familial hypocalciuric hypercalcemia
    (FHH), 92
  1 alpha-hydroxylase, 60, 62
  vs. pseudohypoparathyroidism (PHP), 40
  receptors, 45, 46
  related peptides, 45
PDDR; *see* Hereditary 1 alpha-hydroxylase
    deficiency
Peripheral mononuclear cells,
  hereditary vitamin D-resistant rickets
    (HVDRR), 69
Persistent Mullerian duct syndrome
    (PMDS),
  233, 234, 240–242
Phospholipases, 89
Phosphorylated,
  insulin receptor,
    integral membrane proteins, 176
Phosphotyrosine binding (PTB),
  domains,
    insulin receptor substrate (IRS),
      175, 176

PHP; *see* Pseudohypoparathyroidism
PHPT; *see* Primary hyperparathyroidism
Phytohemagglutinin (PHA)-stimulated
  lymphocytes,
    hereditary vitamin D-resistant rickets
      (HVDRR), 69
PI 3-kinase pathway,
  177, 178
Pit-1, 115
Pit-1 gene, 7
Pituitary,
  growth hormone releasing hormone
    receptor (GHRH-R), 5
Pituitary dwarfism with high serum GH, 18
Pituitary resistance to thyroid hormone
  action (PRTH), 148, 149
Primary adrenal failure, 245
  causes, 246
  diagnosis, 246, 247
Primary amenorrhea,
  luteinizing hormone (LH) receptors gene
    mutation, 190
Primary growth hormone resistance,
  17–33
Primary hyperparathyroidism,
  calcium (Ca^{2+}) ions resistance,
    104, 105
Primary hyperparathyroidism (PHPT),
  familial hypocalciuric hypercalcemia
    (FHH), 91–93
Prolactin,
  little mouse, 6
Prop-1 gene, 7
Prophet of Pit-1 (Prop-1) gene, 7
Prostate cancer,
  androgen receptor (AR) mutations, 224
Pseudohypoparathyroidism (PHP),
  39–53
  clinical features, 40–44
  diagnosis, 51, 52
  genetics, 44, 45
  G-protein of adenylyl cyclase (Gs), 115
  history, 39, 40
  molecular pathophysiology,
    45–51
  vs. parathyroid hormone (PTH), 40
  treatment, 52, 53
  type Ia, 43, 44
  type Ib, 44
  type Ic, 44
  type II, 44
  various forms, 41

PTB; *see* Phosphotyrosine binding
PTH; *see* Parathyroid hormone
Puberty,
  luteinizing hormone (LH) insensitivity
    syndrome, 186, 187

R

Rabson-Mendenhall Syndrome, 170
RAR; *see* Retinoic acid receptor
Ras,
  activation, 178
Renal blood flow,
  hypercalcemia, 93
Resistance to thyroid hormone (RTH),
  145–158
Retained Mullerian ducts,
  233–242
Retinoic acid receptor (RAR), 64
Reverse transcriptase polymerase chain
  reaction (RT-PCR), 118
RTH; *see* Resistance to thyroid hormone;
  Thyroid hormone resistance
RT-PCR; *see* Reverse transcriptase
  polymerase chain reaction

S

SCP; *see* Start codon polymorphism
Sertoli cells, 201, 233
  luteinizing hormone (LH),
    185–187
Sex differentiation, 185, 186
Sex hormone binding globulin (SHBG), 145
SHBG; *see* Sex hormone binding globulin
Signal transducers and activators of
  transcription (STATs), 18, 19
Skeletal deformities,
  Albright hereditary osteodystrophy
    (AHO), 40–42
Snell mouse, 6–9
Somatic thyroid stimulating hormone (TSH)
    receptor mutations,
  toxic adenomas,
    134, 135
Somatostatin system, 1, 2
Spinal muscular atrophy, 224
Spontaneous healing,
  hereditary vitamin D-resistant rickets
    (HVDRR), 76
Sporadic congenital hypothyroidism,
  112–115

causes, 113
etiologies, 114
Sporadic germline mutations,
    congenital hyperthyroidism,
    135–139
Start codon polymorphism (SCP), 63
STATs; see Signal transducers and
    activators of transcription

T

Testicular function,
    follicle-stimulating hormone (FSH), 198
Testosterone,
    Leydig cell, 186
Thyroglobulin levels,
    thyroid stimulating hormone (TSH)
        receptors, 131
Thyroid carcinomas,
    thyroid stimulating hormone (TSH)
        receptor mutations, 136
Thyroid hormone action indices,
    thyroid hormone resistance (RTH), 149
Thyroid hormone resistance (RTH),
    145–158
    animal models, 156, 157
    clinical features,
        146–150
    features, 150
    future directions, 158
    management, 157, 158
    molecular genetics, 150–152
    mutant receptors,
        properties, 152, 153
    mutations,
        schematic representation, 151
    thyroid hormone action indices, 149
tissue resistance pathogenesis,
    153–156
Thyroid hormones, 145, 146
Thyroid stimulating hormone (TSH), 44
    receptor mutations,
        congenital hypothyroidism,
            129–131
        euthyroid hyperthyrotropinemia,
            125–129
        thyroid carcinomas, 136
    receptors,
        activating mutations,
            134–136
        G-protein of adenylyl cyclase (Gs),
            116

intracellular signaling, 119
molecular characteristics,
    115–119
mutations, 118
resistance, 111–127, 120–134
    clinical studies,
        120–124
    significance, 136, 137
resistance receptors,
    with mutations, 129–131
    without mutations,
        131–134
thyroid stimulating hormone (TSH)
    receptor gene mutations,
    124, 125
thyrotropin releasing hormone (TRH),
    112
Thyrotropin releasing hormone (TRH), 44
    receptors, 2
    thyroid stimulating hormone (TSH), 112
Thyroxine T4; see Thyroid hormones
Tissue specific glucocorticoid resistance,
    autoimmune and inflammatory diseases,
    268
    neoplasms, 267, 268
Toxic adenomas,
    somatic thyroid stimulating hormone
        (TSH) receptor mutations,
        134, 135
Transforming growth factor-beta family,
    anti-Mullerian hormone (AMH),
        236–238
    signaling cascade,
        236–238
TRH; see Thyrotropin releasing hormone
Triiodothyronine T3; see Thyroid hormones
Triple A syndrome, 247
    candidate genes, 254
    chromosomal locus identification, 254
    clinical features, 252–254
TSH; see Thyroid stimulating hormone
Type A insulin resistance,
    169, 170
Tyrosine phosphorylation,
    insulin action, 174, 175

U

Uremic hyperparathyroidism,
    calcium (Ca^{2+}) ions resistance,
    104, 105
Urinary cAMP, 41, 52

V

VDDRI; *see* Hereditary 1 alpha-hydroxylase deficiency
VDR; *see* Vitamin D receptors
Virilization defects,
   androgen receptor (AR) mutations, 221
   Vitamin D,
   active form, 59
   hereditary resistance, 59–78
   physiology, 60–62

Vitamin D receptors (VDR), 62–66
   1,25-dihydroxyvitamin D [1,25(OH)2D], 66, 67
   DNA binding domain (DBD), 64, 65
   gene, 62, 63
   ligand binding domain (LBD), 65
   protein, 63–66
Vitamin D therapy, 59, 60
   hereditary vitamin D-resistant rickets (HVDRR), 75

ABOUT THE EDITOR

Dr. J. Larry Jameson is C.F. Kettering Professor of Medicine and serves as Chief of the Division of Endocrinology, Metabolism, and Molecular Medicine at Northwestern University Medical School. He is recognized for his research in molecular endocrinology and has been the recipient of several awards including the Oppenheimer Award from the Endocrine Society and the Van Meter Award from the American Thyroid Association.

Dr. Jameson has published more than 125 original research reports along with numerous reviews and chapters. He has served as an Editor of the journal *Endocrinology* and he is Co-Editor of DeGroot's textbook, *Endocrinology.* He was recently appointed as an Editor for Harrison's *Principles of Internal Medicine,* a book used internationally by most medical students and internists.

His research interests focus on the genetic mechanisms that control the transcription of endocrine genes. In particular, he has investigated pathways that control expression of the gonadotropins, which are genes that control reproduction. He also has a long-standing interest in the genetic basis of endocrine disorders. His research has led to new insights into genetic causes of gonadotropin deficiency disorders and the pathophysiology of thyroid hormone action.